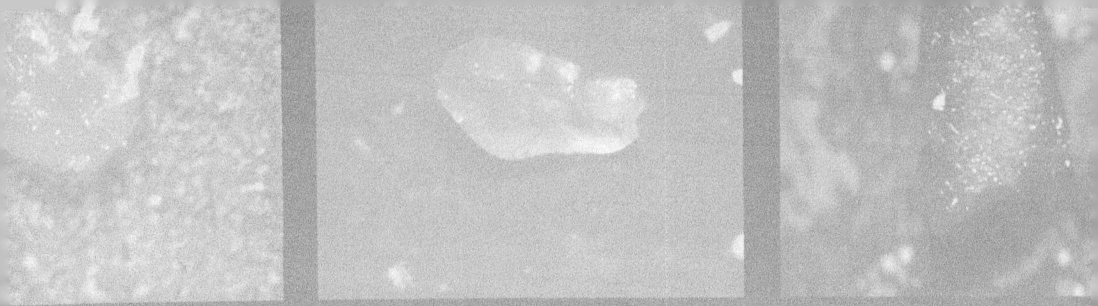

Microplastic Pollution

Microplastic Pollution
Environmental Impact and Solutions

edited by
Julia Martín | Juan Luis Santos
Irene Aparicio | Esteban Alonso

JENNY STANFORD
PUBLISHING

Published by

Jenny Stanford Publishing Pte. Ltd.
101 Thomson Road
#06-01, United Square
Singapore 307591

Email: editorial@jennystanford.com
Web: www.jennystanford.com

British Library Cataloguing-in-Publication Data
A catalogue record for this book is available from the British Library.

Microplastic Pollution: Environmental Impact and Solutions

ISBN 978-981-5129-77-9 (Hardcover)
ISBN 978-1-003-66006-4 (eBook)
DOI 10.1201/9781003660064

Contents

Preface

While working on this book, we were often struck by the rapid pace and sheer volume of new scientific publications on microplastics. At times, it felt like we are aiming at a constantly shifting target—a reflection of the growing global attention this pervasive environmental issue has attracted in recent years. This book contributes meaningfully to the expanding body of literature by seeking unifying patterns and processes to help us better understand, manage, and ultimately resolve the problem of plastic pollution.

The book presents a collection of case studies from diverse geographic regions and scientific disciplines, including materials science, analytical chemistry, environmental toxicology, and public health. The opening chapter addresses the complex and widespread issue of microplastics in the environment, introducing a unified framework for the plastic pollution cycle by integrating knowledge on the transport, fate, and effects of plastics across global environmental systems. Subsequent chapters offer detailed methodologies and protocols for detecting, separating, and identifying microplastics in complex environmental matrices. Chapter 2 reviews sampling, preparation, and pyrolysis-GC/MS analysis techniques for detecting MPs in environmental waters, along with quality assurance and quality control (QA/QC) measures for accurate polymer identification. Chapter 3 provides an example of MP analysis in natural waters using micro-FTIR techniques. Chapters 4 through 6 examine the fate and distribution of microplastics in wastewater treatment plants and several natural environments. Chapter 4 summarizes the occurrence, characteristics, and removal of MPs in sewage treatment systems, emphasizing their role in transporting microplastics to aquatic and terrestrial ecosystems and evaluating removal technologies. Chapter 5 focuses on the Setiu Wetlands in Malaysia, a semi-enclosed lagoon system with rich mangrove cover, which is vulnerable to plastic accumulation. Chapter 6 delves into the current understanding of microplastic sources, transport mechanisms, and fate in mountain environments. The unique topography, climate, and hydrological processes of these regions create complex pathways for microplastic transport and fate, posing potential risks to fragile flora and fauna. Sequentially, in Chapter 7, the presence of microplastics in Antarctic biota and the Southern Ocean is explored, addressing their potential ecological impacts, toxicity, and future research needs. The next chapters delve into the role of microplastics as vectors for hazardous chemicals. Chapter 8 discusses the growing body of research linking microplastics to chemical pollution, reinforcing the call for chemical simplification to safeguard both environmental and human health. Digestive physiology is highlighted as a key site for chemical desorption and release within organisms. Chapter 9 reviews the co-occurrence of MPs and antibiotics in aquatic environments, examining sorption mechanisms, influencing factors, and their interactive toxicity to aquatic organisms. Finally, Chapter 10 broadens the focus to include both aquatic and terrestrial organisms, analyzing the ecotoxicological effects of fossil- and biobased micro(nano)plastics and the consequences of exposure and co-exposure to microplastic-associated pollutants.

As an excellent reference, this book presents in detail the environmental behavior and potential impacts, risks, detection, and removal of microplastics, and provides an outlook for future research directions and developments. We hope that this 10-chapter book will not only provide water scientists and engineers with professional and valuable information about microplastics in urban water management, but more importantly, make the public, graduate and undergraduate students, educators, resource managers, and especially policy makers and technical professionals fully aware of the severity of microplastic pollution and its risks in urban water systems.

Finally, we convey our heartfelt gratitude to all the authors who contributed to this book. They are from diverse backgrounds and have provided holistic information on the topic. We also sincerely thank the team at Jenny Stanford Publishing who helped to produce an incredible and meaningful edited volume on a very relevant theme.

Julia Martín
Juan Luis Santos
Irene Aparicio
Esteban Alonso
Summer 2025

Chapter 1

Microplastic Contamination: An Introduction to an Emerging Issue

Laura Martín-Pozo, Julia Martín, Juan Luis Santos, Irene Aparicio, and Esteban Alonso
Departamento de Química Analítica, Escuela Politécnica Superior, Universidad de Sevilla. C/ Virgen de África, 7, E–41011 Seville, Spain
jbueno@us.es

Microplastics (MPs) are environmental pollutants of emerging concern because they do not degrade easily in the environment, persist for long periods, and can accumulate in aquatic organisms. Given these environmental threats, there is a need to gain a basic understanding and accurate and clear information on this hot topic.

Seawater and freshwater bodies are the main scenarios where MPs have been studied and identified, as rivers are the main routes for their transport from the ocean to terrestrial and remote areas. MPs can be also transported between environmental compartments via atmospheric circulation. And the ingestion of MPs by various types of marine organisms has drawn more and more attention.

Microplastic Pollution: Environmental Impact and Solutions
Edited by Julia Martín, Juan Luis Santos, Irene Aparicio, and Esteban Alonso

ISBN 978-981-5129-77-9 (Hardcover), 978-1-003-66006-4 (eBook)
www.jennystanford.com

DOI: 10.1201/9781003660064-1

In this introductory chapter, we present a general viewpoint on the state of knowledge on MP research efforts, including their pathways, classifications, occurrence, and movement in aquatic, terrestrial, and atmosphere ecosystems, and highlight the future research prospects for addressing this ubiquitous environmental problem.

1.1 Introduction

Plastics are ubiquitous in our lives because of their numerous properties (stability, versatility, flexibility, strength, durability to degradation, and, most notably, economic benefit). In 2018, plastic production reached almost 360 million tons and is projected to rise to 1.1 billion tons by 2050 [1, 2]. Microplastics (MPs) (plastic particles < 5 mm) have been identified as one of the most significant environmental threats because they do not degrade easily in the environment, remain for long periods, and can accumulate in aquatic organisms [3]. Understanding the origins of MPs and their fate in diverse environmental media is basic to enhance our perception of this issue. MPs can enter the environment through three different routes: (a) microbeads used daily by humans (textile, personal care products, washing industry) that are directly discharged into water bodies); (b) plastic beads and fibres discharged by sewage treatment plants; and (c) fibre films and debris generated by water operation industries, navigation, or fisheries [4].

Figure 1.1 schematizes the global cycling pathways of MPs [5]. Anthropogenic causes, human activities (fishing, tourism, etc.), and environmental variables (wind, cyclones, wave currents, tides, etc.) have been associated with the transportation of MPs in the atmosphere, terrestrial, and aquatic ecosystems. Some studies have confirmed MP transport from terrestrial to the marine environment via the freshwater system [5–7]. Figure 1.2 shows the main movements that MPs undergo in their transport through the different compartments of the environment. Apart from urban areas, MPs have been detected in rivers and lakes [5], deepest parts of ocean layers [8], faraway mountain tops [9], and the snows of poles [10].

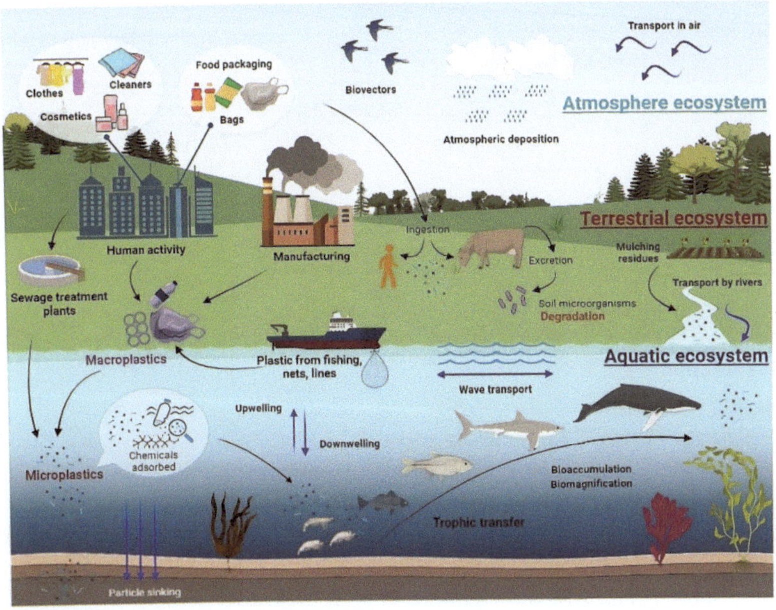

Figure 1.1 MP cycles through different compartments of the ecosystem.

Furthermore, MPs are bioavailable to aquatic species that can take them by ingestion and transfer within food webs [11–13]. MP ingestion has been described in a variety of wild and aquaculture species of commercial interest, including fish, bivalves, and crustaceans consumed by humans [3, 11, 14, 15]. For example, among the 25 species that contribute the most to global sea fishing, 11 were found to contain MPs [16]. Bioavailability increases when the particle size decreases, and because of that, nanoplastics (NPs, plastic particles < 1 um) have a higher ability to penetrate and be absorbed by aquatic species [17, 18].

Since the presence of MPs in the environment was first described in 2004 by Thompson [19], thousands of publications have updated our understanding of the situation worldwide and the global scenario of MP pollution regarding its occurrence, detection, and environmental effects (Fig. 1.3). To date, 5437 documents have been published about MPs. More than 50% of the works were from the last two years (2022, 2023). A quantification of this information is shown in Fig. 1.4.

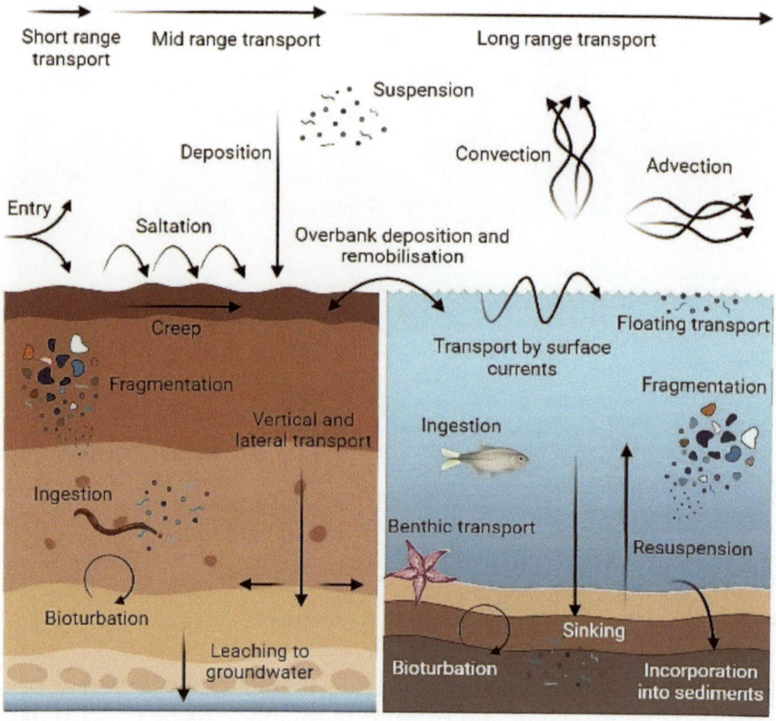

Figure 1.2 Movements of MPs in the environment.

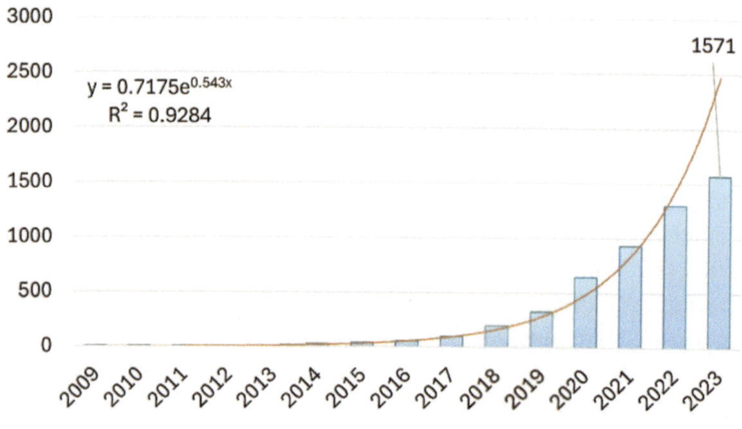

Figure 1.3 Graphical data showing the exponential growth of the numbers of microplastic research from 2009 to 2023. Scopus searches using keywords of [(microplastic) AND (pollution) AND (environment)]. *Search on 25/01/2023.*

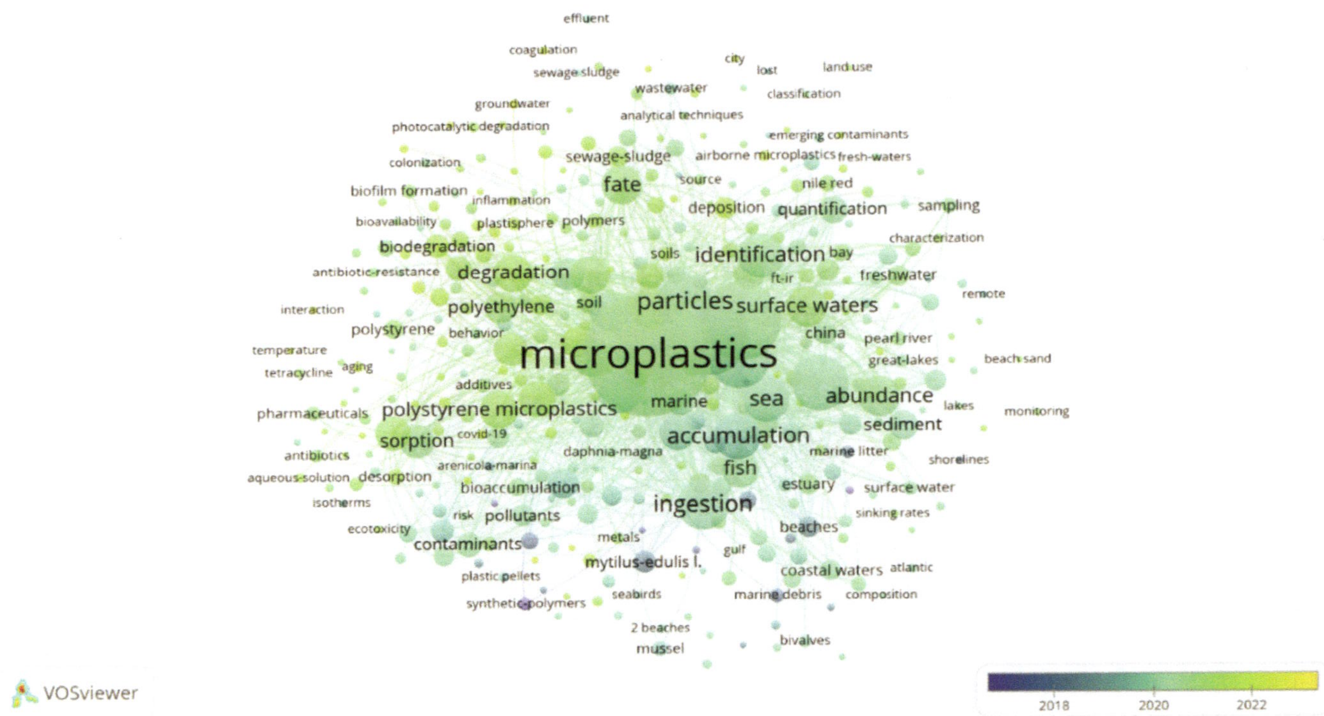

Figure 1.4 Co-occurrence map of the keywords used in articles derived from query Q1 (Q = quarter) with a minimum occurrence of 5.

The primary focus was placed on the identification, abundance, and accumulation of MPs, mainly polyethylene (PE) and polystyrene (PS), in the marine environment, where MPs occur mostly. The development of analytical methods for sampling, extraction, purification, and qualification and quantification of MPs in a multi-compartmental manner is highly relevant to MP-related research, as seen in the co-occurrence map. A lot of methods have been proposed in freshwater, sediment, soil, and biota matrices due to the lack of a standard technique [20, 21]. The growing focus on this research area is explained since MPs are in a wide range of sizes, shapes, colours, and types, together with the complexity of environmental samples [20, 21]. This fact has produced incomparable results, making it challenging to decide what to investigate further. In addition, MPs undergo alterations because of photodegradation, mechanical erosion, or long-time exposure, which can influence their composition and morphology, and consequently hinder their analysis [21].

On the other hand, compared to these general research topics, a fewer were related to the terms wastewater treatment and sludge treatment. Until 2020, wastewater and MP-related research comprised less than 10% of the total published research. However, at present investigations of MPs in wastewater treatment plants (WWTPs), including analytical methods and removal rates of MPs in different types of treatments or adsorption in the sludge, are gaining more attention from researchers, and to date, the percentage of related publications has increased to 13% (Scopus).

Multiple and complex factors are involved in the MP cycle, influencing their fate, transport, and impact on ecosystems [22], such as the sorption of chemicals added during the manufacturing process or adsorbed in MP particles during their permanence in the environment, the degradation or fragmentation of plastic debris, or the influence of environmental conditions [2, 23, 24]. However, at present, the available studies were carried out at a laboratory scale, and MPs utilized in laboratory testing and those found in the environment have different particle types, concentrations, shapes, and size ranges. There is an urgent need for more environmentally realistic studies that address this mismatch [25].

In this introductory chapter, we provide a general overview of the state of knowledge on MP research efforts, including their pathways, classifications, occurrence, and movement in aquatic, terrestrial, and atmosphere ecosystems, and highlight the future research prospects for addressing this ubiquitous environmental problem.

1.2 Microplastic Classification: Sizes, Origin, Polymer Types, Shapes, and Colours

Microplastics are classified according to various criteria as they differ in their polymeric compositions, sizes, shapes, colours, and sources. The most used classification system is based on MP size—mega (> 1 m), macro (2.5 cm to 1 m), meso (5 mm to 2.5 cm), micro (1 μm to 5 mm), and nano (<1 μm). Then MPs are defined as particles smaller than 5 mm, with a lower bound of 1 μm to differentiate MPs from nanoplastics (NPs) (Fig. 1.5) [26, 27].

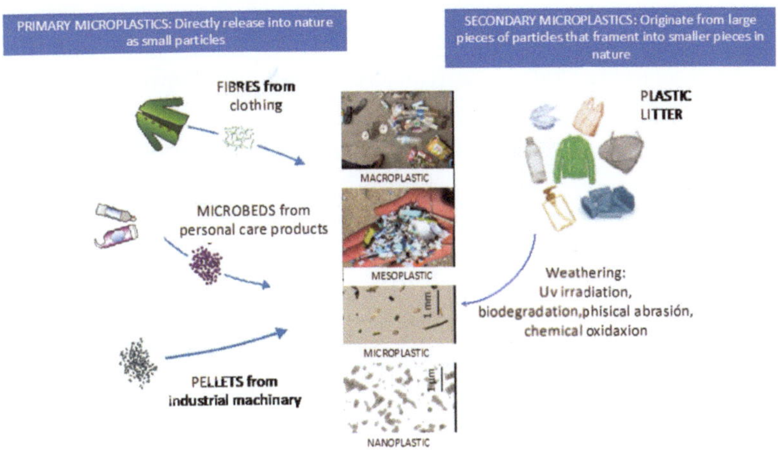

Figure 1.5 Primary and secondary microplastics.

Based on their origin, MPs are divided into two groups: primary and secondary. Primary MPs enter the environment directly as MPs and are produced for specific applications (such as fibres from clothing, microbeads used in cosmetics, or pellets from industrial machinery). Secondary MPs are produced by the fragmentation of MPs (e.g. fibres) due to a weathering process such as biodegradation, UV irradiation, physical abrasion, or chemical oxidation (Fig. 1.5).

Currently, more than 20 types of MPs have been detected in the natural environment, predominantly PE, polypropylene (PP), PS, polyvinyl chloride (PVC), and polyethylene terephthalate (PET) in accordance with the total global plastic production [28, 29]. Table 1.1 summarizes the most broadly found polymer types in the ecosystem and some details such as acronym, structure, density, and their main uses.

Table 1.1 Types of plastics and their uses (adapted from Prasad, 2023) [5].

Scientific Name	Acronym	Chemical Formula	Density	Products and Typical Origin
Polyethylene terephthalate	PET	$(C_{10}H_8O_4)_n$	1.55	Carpet, furniture, drink bottles, etc.
High-density polyethylene	HDPE	$(C_2H_4)_n$	0.96	Carry bags, bottles, recycling bins, agricultural pipe, playground equipment, base cups, etc.
Polyvinylchloride	PVC	$(C_2H_3Cl)_n$	1.40	Shower curtains, window profiles, fencing, pipes, children's toys, flooring, lawn chairs, etc.
Low-density polyethylene	LDPE	$(C_2H_4)_n$	0.92	Plastic bags, dispensing bottles, tubing, wash bottles, etc.
Polypropylene	PP	$(C_3H_6)_n$	0.90	Auto parts, food containers, industrial fibres, dishware, etc.
Polystyrene	PS	$(C_8H_8)_n$	1.02–1.05	Trays, plastic utensils, toys, clamshells, insulation board, etc.
Polyamide	PA	CO-NH$(CH_2)_n$	1.02–1.14	Clothing, other textiles, rope, fishing line

The most common types of MP shapes are microbeads, including spherical or irregular shapes; fibres; pellets; films; fragments; and foams [30]. The dominant shapes are the fibres and fragments coming from domestic or industrial laundry. Pellets with spherical and granular shapes come from personal care products. The films, foams, and fragments (broken small pieces) are derived from the erosion process of plastic products [30]. Long et al. (2019) reported that fragments (91.3%) and granules (91.4%) are more effectively removed from WWTPs than fibres (82.8%) and pellets (78.9%) [31]. In another attempt, Choi et al. (2018) described that irregularly shaped MPs caused higher toxicity to *Cyprinodon variegatus* organs than spherical MPs [32].

The MP colour is also an important feature that can be used to trace the MP source. The most frequently detected colours in WWTPs are those used in human daily life, such as black, blue, white, transparent, and red. The colour of MPs could change during their pass through WWTPs. Jia et al. (2019) reported that the initial colour of MPs changed to transparent (>75%) in WWTPs from Shanghai, which is explained by the effect of UV irradiation and erosion in the environment [33]. However, the transparent colour was normally not the dominant colour in aquatic systems, which requires further investigation [34, 35]. The influence of the types and the colour of MPs can also play an important role in the adsorption of contaminants and toxicity [22, 36, 37]. In nature, the colour of MPs can confuse living organisms [30], which can consume them as food. Frias et al. (2020) observed that black MPs present a higher adsorption capacity of contaminants than white MPs [38]. Weathering degradation leads to increased surface area and changes in polarity by the formation of new carbonyl and ketone groups and, therefore, influences the sorption capacity of MP particles [22, 39]. Weathered MPs tend to absorb greater quantities of pollutants than virgin particles due to a suitable charge surface [40, 41]. Bhagwat et al. (2021) determined that the adsorption capacities of perfluoroalkyl sulphonate (PFOS) were influenced by the type of MPs and the degree of weathering. The concentrations of adsorbed PFOS were higher in weathered MPs, which in addition varied (20–85%) between different types of MPs (PP > PE > PA > PS) [42].

1.3 Evidence of MP Presence in Natural Environment

An increasing number of studies have provided evidence for the ubiquitous presence of MPs across all environmental compartments—air, water, and soil—encompassing the biota within each domain. The type, shape, colour, size, and occurrence of MPs exhibit significant variability among studies, contingent upon the location, source of origin, and the factors influencing these characteristics over time. Table 1.2 presents the distribution of identified MPs in various ecosystem compartments from recent studies, detailing the most prevalent types of MPs, particle sizes, and the abundances found.

Table 1.2 Summary of polymer type, size range, and abundance from the different ecosystem compartments (seawater, fresh water, soil, atmosphere)

Location	Polymer Type	Size Range	Average Abundance (Particles/Unit)	Ref.
Seawater				
Atlantic Ocean	No data	Pump: <2500 μm	~60 p/m^3 (net) ~1829 p/m^3 (pump)	[43]
Baltic Sea, Finland	PE, PP, PET, PA, PS, ABS	20–724 μm (net)/8386 μm (water sampler)	0.41–1.6 p/m^3 (net) 0.02–1.7 p/L (water sampler)	[44]
Thermaic Gulf, North Aegean Sea, Greece	PP, PE, PS, EVA, PA, ABS, BCPP	1000–1500 μm (most abundant)	1.9 ± 2.0 p/m^3	[45]
Groundwater				
Dawashan Island, China	PE, PP, PET	500–5000 μm	38–64 p/L	[46]
Piedmont, Italy	PE, PVA, PS, EVOH, PVC	<1000 μm	12–54 p/L	[47]
Northwest Mexico	PP, HEC, PVC, LDPE	63–1002 μm	10–34 p/L	[48]

Location	Polymer Type	Size Range	Average Abundance (Particles/Unit)	Ref.
Lakes				
Saimaa Lake, Finland	PP, PES, PE, POM, PMMA, PS, PVC, PUR	1100–1300 μm (fibres) 410–430 μm (fragments)	0.58–0.86 p/L (water) 2100 p/kg (sediments)	[49]
Phewa Lake, Nepal	PP, PE, PS, ABS	200–1000 μm (>65%)	55–122.5 p/kg (sediments)	[50]
Rivers				
Neuse River, North Carolina, USA	PE, PP, PS, PET, PVC, ABS, EVA, PA	65–3587 μm <335 μm (90.9%)	0.02–221 p/m^3 (MPs > 335 μm) 20–130 p/m^3 (MPs > 64 μm)	[51]
Plankenburg River, South Africa	PE, cotton, PP, PET	63–200 μm	6.87–9.25 p/L (water) 616–1233 p/kg (sediments)	[52]
WTPPs				
Denmark	PE, PET, acrylate, PP, PU	100 μm (influent) 91 μm (effluent)	2223–18285 p/L (influent) 19–447 p/L (effluent)	[53]
Cádiz, Spain	PVC, PE, HDPE, PMMA, PET, PA, EAA, EVA, PP, PS	<355 μm (53–71%)	645–1567 p/L (influent) 16–297 p/L (effluent)	[54]
Soils				
Mediterranean, Argean, and Marmara regions, Türkiye	PE, PP, PS	<500 μm (49.2% in urban areas, 42.4% in agricultural areas) <1000 μm (66.8%)	Agriculture: 192.7 ± 14.2 p/kg Urban: 127.3 ± 21.6 p/kg	[55]
Coimbra, Portugal	PP, PE, PVC, PS, acrylates/PU, PA	9–2949 μm (116 μm mean)	5000–571000 p/kg	[56]

(Continued)

Table 1.2 (*Continued*)

Location	Polymer Type	Size Range	Average Abundance (Particles/Unit)	Ref.
Amsterdam, the Netherlands	PE, PA, PP, PS, PU, PVC	<250 μm (47%)	4825.3 p/ kg	[57]
Atmosphere				
Nanjing, China	PA, PE, PES, EP, PP	Deposited MPs: 796 μm Suspended MPs: 541 μm	Deposited MPs: 134–500 p/m²/day Suspended MPs: 0.29–4.88 p/m³	[58]
Tibetan Plateau	PET, PE, PA, PS, PP, PVC	<100 μm (~89.0%)	2.5–58.8 p/m³	[59]
Northern Germany	PE, PET, PA, EVA, PVC, PS, PP, PU, PMMA, PHBV, PVA, ABS	<30 μm (53%)	53–166 p/m²/day	[60]
Santa Barbara and Ventura Counties, California	PS, PET, PE, PVC, PC, PA, ABS	25–2061 μm (outdoor) 20–8961 μm (indoor)	0.6 ± 0.6 p/m³ (fibre, outdoor) 5.6 ± 3.2 p/m³ (fragment, outdoor) 3.3 ± 2.9 p/m³ (fibre, indoor) 12.6 ± 8.0 p/m³ (fragment, indoor)	[61]

ABS: acrylonitrile butadiene styrene; BCPP: block copolymer polypropylene; EAA: ethylene acrylic acid; EP: epoxy resin; EVA: ethylene vinyl acetate; EVOH: ethylene-vinyl alcohol; HEC: hydroxyethyl cellulose; LDPE: low-density polyethylene; PA: polyamide; PC: polycarbonate; PE: polyethylene; PES: polyester; PET: polyethylene terephthalate; PHBV: poly(3-hydroxybutyrate-co-3-hydroxyvalerate; PMMA: polymethylmethacrylate; POM: polyoxymethylene; PP: polypropylene; PS: polystyrene; PU: polyurethane; PVA: polyvinylalcohol; PVC: poly(vinyl chloride)

1.3.1 Microplastic in the Aquatic Environment

Contamination by MPs in marine and freshwater ecosystems has garnered substantial research interest in recent decades. The initial evidence dates back to the early 1970s, with the discovery of polystyrene spherules in waters and plankton off the New England coast [62, 63]. Since then, MPs have been identified in various water sources, including oceans, seas, lakes, rivers, streams, and estuaries, making it one of the most extensively studied topics. Given the diverse characteristics of each compartment in terms of hydrology, chemistry, flora, and fauna, the contamination of aquatic ecosystems with MPs is complex as they can function as receptors, reservoirs, and transporters of plastics [64]. Human activities in the coastal zone, such as fishing, aquaculture, and tourism, contribute significantly to the MP pollution of the aquatic environment [65].

Typically, MP research predominantly focuses on marine and surface waters. While research on freshwater sources is in its early stages, investigations within these environments are experiencing exponential growth [47]. Available data suggests that the abundance of MPs in freshwater environments is comparable to that in marine environments with highly heterogeneous distribution in different areas [66].

The presence of MPs in seawater has been widely examined in seas worldwide, including the Atlantic Ocean, the Midwest Pacific, and the Mediterranean. Generally, concentrations of MPs exhibit higher levels in coastal areas and semi-closed bay seas compared to open seas. Furthermore, these concentrations tend to show an increase over time [64]. In this context, rivers play a crucial role as the primary contributors to the transport pathways of MPs from terrestrial ecosystems to the oceans [52]. The proximity of urbanization to rivers has been identified as a significant source of MPs. Previous studies have quantified varying levels of MPs along rivers, with concentrations influenced by factors such as land use, population density, and the construction of dams and reservoirs [51]. On the other hand, lakes play a pivotal role in biogeochemical cycles and as focal points for the accumulation of MPs. The widespread presence of MPs has been documented in numerous lakes across the globe, with abundances reaching nearly 2,090,000 particles per square kilometre [67]. Moreover, sediments within these lakes appear to be potential long-term reservoirs for MPs [67]. MPs can also enter groundwater through various mechanisms, through soil filtration, intrusion from seawater, and the exchange of surface water with groundwater. Their occurrence is primarily determined by their transport dynamics during the recharge process and their ability to persist for extended periods, due to the slow flow velocity and long cycling time of groundwater [68].

A large part of MPs entering the aquatic ecosystem is discharged from the plastic industry and domestic wastewater. In this context, WWTPs play a fundamental role in the elimination of MPs as a final barrier between their entry into the environment, but can also be a potential entry point, due to insufficient containment [64]. Most MPs are eliminated during the primary processes of wastewater treatment, although secondary and tertiary mechanisms, including advanced treatment methods, have been shown to be highly efficient in removal, such as membrane bioreactors and coagulation [69]. Removal efficiency varies depending on size, shape, and type, with ratios differing across studies, although, in general, most MPs tend to be removed [64]. The percentage of removal can vary from 60% to more than 99%, according to previous studies [54]. For instance, in Denmark, 2% of MPs were not removed and were discharged from WWTPs annually, amounting to 3 tons of MPs. Nevertheless, when compared to annual MP emissions (600–3100 tons), this quantity is considered insubstantial [53].

1.3.2 Microplastic in the Terrestrial Environment

The terrestrial ecosystem has been less explored in terms of MP distribution. However, recent studies confirm the ubiquity and significance of MPs in terrestrial ecosystems. It is estimated that soil receives up to 40 times more plastics than water [70], and the accumulated load can be up to 23 times higher [71]. European farmlands are estimated to generate between 63,000 and 430,000 tons of MPs annually, which is alarming [71].

The terrestrial environment becomes contaminated with MPs through various pathways as a result of anthropogenic activities. In general, the main entry routes of MPs into the soil come from the industrial, domestic, and agricultural sectors. Contamination can occur through primary MPs produced for various purposes, such as in the cosmetic industry (microbeads used in cosmetics, pellets, detergents, and similar products) and through MPs degraded by UV light, mechanical abrasion, or biological processes [72]. In the soil, agriculture stands out as a major source of MPs. Agricultural practices play a key role in MP emissions, surpassing emissions from the marine environment. The presence of MPs in soils and sediments is increasing dramatically due to plastic mulching, greenhouses, the use of compost and fertilizers, the application of sewage sludge, and wastewater irrigation [73, 74].

Once in the soil, MPs can be ingested and transferred to soil organisms and can be transported to other areas, even to other ecosystems, through wind and water currents. Within the soil, they can migrate and distribute both vertically and horizontally, with migration influenced by biota action, erosion, soil aggregates, soil cracking, and agricultural practices [75]. Migration mainly occurs through bioturbation, such as plant root growth and ingestion and excretion by soil organisms such as earthworms [76]. Additionally, the characteristics of MPs also influence their movement, such as type and shape. Among different types of MPs, microfibres exhibit higher mobility [77]. In addition to the harmful effects MPs may have on biota, some studies have indicated that their deposition in the soil can result in uptake by plants and entry into the food chain [72]. Other studies have confirmed that MPs can affect phosphorous and nitrogen cycles and microbial communities [78].

It is essential to continue researching the ecological impact of MPs on soil, and their repercussions on crops, and to promote sustainable agricultural and waste management practices that minimize MP pollution.

1.3.3 Microplastic in the Atmosphere

The assessment and quantification of MPs in the atmosphere remain less explored compared to other ecosystem compartments. Attention to atmospheric MPs has emerged recently, with most research published in the last 5 years [58, 79]. MPs are ubiquitous and prevalent in the atmospheric environment, especially in large cities. The atmosphere plays a pivotal role as both a reservoir and a pathway for the interchange of MPs across diverse environments.

The MPs released into the atmosphere can settle and enter marine and terrestrial ecosystems, but they also have the potential to re-enter the atmosphere due to wind and waves. The main sources of MPs in the air are synthetic textiles, erosion from synthetic rubber tyres, and dust from cities. Additional sources, albeit with a lesser contribution, encompass plastic fragments from clothing and furniture, waste incineration, agricultural fields, industrial emissions, traffic emissions, plastics from agricultural activities (such as seedbeds), and sewage sludge used as fertilizers, among others [80]. These particles are transported through the atmosphere by wind and can act as vectors of chemicals, such as persistent pollutants and heavy metals, over short and long distances, given their ability to adsorb pollutants [81, 82].

Regarding sampling, there are two approaches to sample collection: passive sampling, which uses passive samplers (metallic pans and wooden brushes) to collect deposited MPs, and active sampling, which uses filters and air pumps to capture MPs in suspension, which may not be deposited [83]. To achieve a comprehensive global understanding, both sampling methods are essential, despite the limited number of studies addressing both approaches. Given the absence of a standardized method, the comparison of results becomes somewhat complicated [58].

Results have demonstrated variations in the composition and morphology between deposited and suspended MPs. Rao et al. [58] observed that fragments were predominant in deposited MPs, whereas both fragments and fibres were comparable in suspended MPs. Furthermore, the deposited MPs were primarily characterized by small black fragments (<100 μm), whereas the suspended MPs displayed a higher percentage of blue fibres, ranging in size from 100 to 500 μm [58]. Differences in concentrations are also observed among areas. In general, airborne MP concentrations tend to be higher in urbanized areas compared to rural areas, while being lower in wilderness areas [59]. The occurrence of airborne MPs can be influenced by several factors. For example, population density and economic development can shape MP characteristics, while meteorological factors such as precipitation and wind direction can impact their distribution [84].

Research on MPs in the atmosphere is in its early stages, and comparable studies, along with standardized sampling and measurement methods, are needed to ensure reproducibility and comparison of results.

1.4 Importance of Understanding the Impact of MPs: Future Research Policies and Perspectives

It is well evident that MPs are omnipresent, and they are accumulated in different compartments at increasing rates [11]. However, the variability in the concentration data with MPs and the big uncertainty are key factors hindering a realistic environmental risk assessment. NPs can enter living organisms, penetrating tissues and cells [85]. Up to 600 aquatic species have been harmfully affected by the ingestion of MPs, especially many kinds of marine sea turtle and seabird species [20]. Significant amounts of MPs have already been detected in the tissues of plants and both terrestrial and aquatic animals, and there is even evidence of their presence in human lungs, placenta, and blood [86–88], although this remains controversial mainly due to the ease of adventitious contamination [89]. MPs enter the organism mainly by ingestion, although other routes such as inhalation or skin contact may also contribute [90].

MPs and NPs are consumed directly or indirectly by a variety of fish, birds, and marine mammals due to their small size. This includes not only aquatic organisms such as filter-feeding animals, mussels, and whales, but also terrestrial animals, including invertebrates such as earthworms, dragonflies, and other insects [91], and vertebrates such as birds and herpetofauna [91–93]. MPs have been also detected in the faeces of ducks, geese, pigs, chickens, and sheep [94].

Some authors have recognized the trophic transfer of MPs [95–99], suggesting concomitant impacts on human health. Several studies have confirmed transfer across trophic levels, e.g. from algae to zooplankton to fish [100], or from invertebrates such as sandworms or crustaceous to fish [101]. In this context, humans at the top of the food chain may be the most vulnerable. MPs have been found in a variety of foods and beverages, including fish, edible crops, seafood, milk, and drinking water [94, 102]. The actual environmental risks of MPs are still unknown, and experts have only lately started to think about how those consequences can affect human health.

Regarding adverse effects caused by exposure to MPs, laboratory experiments have shown digestive, respiratory, endocrine, reproductive, and immune system disorders in aquatic organisms [18, 103–105]. Note that most species used in these studies (for ethical reasons) are marine organisms from the primary consumer group (like invertebrates), and an ecotoxicity data gap was revealed for freshwater and terrestrial species. Furthermore, scientists speculate that small particles are more likely to enter the organism due to their larger specific surface area, causing greater damage and more interaction with cellular components (Fig. 1.6). It is estimated that MPs with a size greater than 150 μm probably will not be absorbed. Those MPs with a size less than 150 μm may be transferred from the intestinal cavity to the lymph and circulatory system, while those MPs with a size ≤ 20 μm can penetrate organs and the smallest MPs (0.1 > 10 μm) to reach all organs, cross cell membranes, and the blood–brain barrier [104, 106]. *In vitro* studies with cerebral and epithelial human cells [107] supported the potential of small-sized MPs (40–250 nm) and NPs (10 μm) to cause cytotoxic effects, in terms of oxidative stress, at the cell level, strengthening scientific thoughts on possible consequences for human health [11].

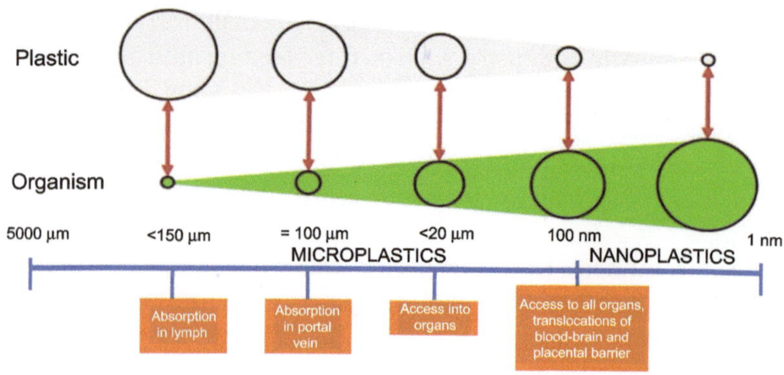

Figure 1.6 Relationship between the organism-to-plastic-size ratio and the dominant effects (adapted from Refs. [106, 108]).

All this evidence denotes that to properly assess and provide a basis for a sound risk assessment, more research on the effects of MPs is required. The increase in scientific interest and knowledge is usually complemented by important normative and policy markers [109]. In this sense, some organizations such as the US National Oceanographic and Atmospheric Administration (NOAA) or the European Marine Strategy Framework Directive (MSFD) have comprised "plastics litter" as an emerging form of pollution in their programmes. International experts committees (such as ICES or GESAMP) have worked to establish standardized protocols for monitoring plastic contamination in environmental matrices [110]. However, there is currently no single European law that comprehensively covers MPs. Thus, the establishment of monitoring protocols by researchers (e.g., method development, data collecting, thresholds) is a prerequisite so that they can be defined after and set later by governance or policy-making organizations (Fig. 1.7) [109, 111].

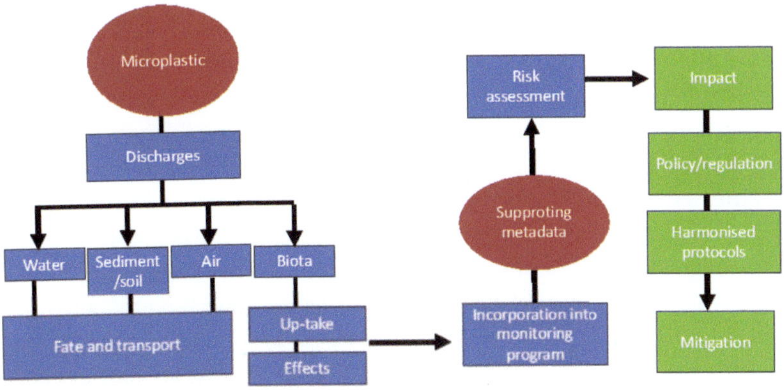

Figure 1.7 Role of monitoring in research and regulation related to microplastic contamination of the environment (adapted from Ref. [111]).

The following are some of the research gaps that should be addressed in the future:

- Standardized methods, monitoring techniques, and protocols for sampling, qualification, and quantification of MPs.
- A large database on the concentration levels of MPs in wastewater, lakes, rivers, effluents, estuaries, groundwater, soil, plants, atmosphere, and the human body.
- Role of MPs as a vector of hazardous organic pollutants and associated toxicological impacts at the field level.
- Impact of ingested MPs within the food web and trophic transfer.
- Long-term effects produced by the exposure to MPs at environmentally relevant concentrations.
- Regulatory framework and policies at the national and international levels.

1.5 Conclusion

This introductory chapter highlighted some diverse data on MPs associated with their pathways, classifications, occurrence, and movement in aquatic, terrestrial, and atmosphere ecosystems and recommended treatment scenarios to address the seriousness of the problem.

The current literature has demonstrated that MPs are widespread in the environment. MPs circulate and are dispersed between environmental compartments through rivers and the atmosphere, thus affecting areas far from their sources. However, there is a greater need for higher-quality occurrence data in more types of compartments, which are essential for accurate classification and chemical characterization. Monitoring must determine the complete size range of concentrations of MPs and must be expressed in meaningful units for comparison purposes. There exists limited data regarding the bioavailability of MPs in terrestrial systems, such as sludge, soil or plants, trophic transfer, and the human body. For all of that, research on standard analytical methods to characterize and quantify MPs at lower concentrations and sizes is required. Furthermore, there is still no consensus on what the principal sources of MPs in the environment are [diffuse (coatings, ware, tyres, etc.) or point sources (WWTP discharges, industrial emissions)]. Human activity is the primary cause, yet cross-cultural differences, resources, and lifestyles define the sources of MPs in each particular area.

Moving quickly toward evidence-based solutions necessitates open communication and knowledge exchange regarding MP contamination. The goal of interdisciplinary research dialogues should be to build and update action plans for a broad benefit and concentrate on harmonization, accurately expressing the status of knowledge, and prioritizing subjects that are crucial for both research and policy. To better understand the great concerns about MPs, the next chapters of this monograph include several case studies on worldwide geographical areas and scientific disciplines, including analytical methods, occurrence, fate, associated toxic chemicals potential exposed to the organisms, and risks of MPs in the environment, among others.

References

1. PlasticsEurope (2019). Plastics—The Facts 2019: An Analysis of European Plastics Production, Demand and Waste Data. http://www.plasticseurope.org.

2. Zhang, Y., Gao, T., Kang, S., Shi, H., Mai, L., Allen, D., Allen. S. (2022). Current status and future perspectives of microplastic pollution in typical cryospheric regions, *Earth-Sci. Rev.*, **226**, p. 103924.
3. GESAMP (2016). Sources, fate, and effects of microplastics in the marine environment: Part two of a global assessment. Eds. Kershaw, P.J., Rochmann, C.M., (IMO/FAO/ UNESCO-IOC/UNIDO/WMO/IAEA/ UN/UNEP/UNDP Joint Group of Experts on the Scientific Aspects of Marine Environmental Protection), *Rep. Stud. GESAMP* No. **93** (220 pp).
4. Tang, Y., Liu, Y., Chen, Y., Zhang, W., Zhao, J., He, S., et al. (2021). A review: Research progress on microplastic pollutants in aquatic environments, *Sci. Total Environ.*, **766**, p. 142572.
5. Prasad, M.N.V. (2023). *Microplastics in the Ecosphere: Air, Water, Soil, and Food*. Eds. Vithanage, M., Prasad, M.N.V., Chapter 3 "Microplastics: Global scenario" (John Wiley & Sons Ltd), pp. 29–64. https://doi.org/10.1002/9781119879534.ch3.
6. De Souza Machado, A.A., Kloas, W., Zarfl, C., Hempel, S., Rillig, M.C. (2018). Microplastics as an emerging threat to terrestrial ecosystems, *Glob. Chang. Biol.*, **24**, pp. 1405–1416.
7. Malizia, A., Monmany-Garzia, A.C. (2019). Terrestrial ecologists should stop ignoring plastic pollution in the Anthropocene time, *Sci. Total Environ.*, **668**, pp. 1025–1029.
8. Jamieson, A.J., Brooks, L.S.R., Reid, W.D.K. Piertney, S.B., Narayanaswamy, B. E., Linley, T. D. (2019). Microplastics and synthetic particles ingested by deep-sea amphipods in six of the deepest marine ecosystems on earth, *R. Soc. Open Sci.*, **6**, p. 180667.
9. Allen, S., Allen, D., Phoenix, V.R. Le Roux, G., Jiménez, P.D., Simonneau, A., et al. (2019). Atmospheric transport and deposition of microplastics in a remote mountain catchment, *Nat. Geosci.*, **12**, pp. 339–344.
10. Bergmann, M., Mützel, S., Primpke, S. Tekman, M.B., Trachsel, J., Gerdts, G. (2019). White and wonderful? Microplastics prevail in snow from the Alps to the Arctic, *Sci. Adv.*, 5, p. eaax1157.
11. Barboza, L.G.A., Vethaak, A.D., Lavorante, B.R.B.O., Lundebye, A.K., Guilhermino, L. (2018). Marine microplastic debris: An emerging issue for food security, food safety and human health, *Mar. Pollut. Bull.*, **133**, pp. 336–348.
12. Germanov, E.S., Marshall, A.D., Bejder, L., Fossi, M.C., Loneragan, N.R. (2018). Microplastics: No small problem for filter-feeding megafauna, *Trends Ecol. Evol.*, **33**, p. 227232.

13. Vandermeersch, G., Van Cauwenberghe, L., Janssen, C.R., Marques, A., Granby, K., Fait, G., et al. (2015). A critical view on microplastic quantification in aquatic organisms, *Environ. Res.*, **143**, p. 4655.

14. Cheung, L.T.O., Lui, C.Y., Fok, L. (2018). Microplastic contamination of wild and captive flathead Grey mullet (*Mugil cephalus*), *Int. J. Environ. Res. Public Health*, **15**, p. 597.

15. Renzi, M., Guerranti, C., Blašković, A. (2018). Microplastic contents from maricultured and natural mussels, *Mar. Pollut. Bull.*, **131**, pp. 248–251.

16. FAO (2016). *Contributing to Food Security and Nutrition for All*, "The State of World Fisheries and Aquaculture," Rome (200 pp). http://www.fao.org/3/a-i5555e.pdf.

17. Botterell, Z.L.R., Beaumont, N., Dorrington, T., Steinke, M., Thompson, R.C., Lindeque, P.K. (2019). Bioavailability and effects of microplastics on marine zooplankton: A review, *Environ. Pollut.*, **245**, p. 98110.

18. Foley, C.J., Feiner, Z.S., Malinich, T.D., Höök, T.O. (2018). A meta-analysis of the effects of exposure to microplastics on fish and aquatic invertebrates, *Sci. Total Environ.*, **631–632**, pp. 550–559.

19. Thompson, R.C., Olsen, Y., Mitchell, R.P., Davis, A., Rowland, S.J., John, A.W.J., et al. (2004). Lost at sea: Where is all the plastic? *Science*, **304**, p. 838.

20. Abeynayaka, A., Werellagama, I., Ngoc-Bao, P., Hengesbaugh, M., Gajanayake, P., Nallaperuma, B., et al. (2022). *Current Developments in Biotechnology and Bioengineering*. Eds. Bui, X.-Y., Nguyen, D.D., Nguyen, P.-D., Ngo, H.H., Pandey, A., Chapter 11 "Microplastics in wastewater treatment plants" (Elsevier Inc), pp. 311–337. https://doi.org/10.1016/B978-0-323-99874-1.00010-5.

21. Estahbanati, M.R.K., Rostami, S., Ghasemian, M., Kiendrebeogo, M., Drogui, P., Tyagi, R.D. (2023). *Current Developments in Biotechnology and Bioengineering*. Eds. Bui, X.T., Nguyen, D.D., Nguyen, P.D., Ngo, H.H., Pandey, A., Chapter 5 "Quantitative and qualitative identification, characterization, and analysis of microplastics and nanoplastics in water." 1st Ed. (Elsevier Inc) pp. 99–123. https://doi.org/10.1016/B978-0-323-99874-1.00010-5.

22. Martín, J., Santos, J.L., Aparicio, I., Alonso, E. (2022). Microplastics and associated emerging contaminants in the environment: Analysis, sorption mechanisms and effects of co-exposure, *TrAC Trends in Anal. Chem.*, **35**, p. e00170.

23. Mejías, C., Martín, J., Santos, J.L., Aparicio, I., Alonso, E. (2023). Adsorption of perfluoroalkyl substances on polyamide microplastics: Effect of sorbent and influence of environmental factors, *Environ. Res.*, **216**, p. 114834.

24. Mejías, C., Martín, J., Santos, J.L., Aparicio, I., Alonso, E. (2023). Role of polyamide microplastics as vector of parabens in the environment: An adsorption study, *Environ. Technol. Innov.*, **32**, p. 103276.

25. Burns, E.E., Boxall. A.B.A. (2018). Microplastics in the aquatic environment: Evidence for or against adverse impacts and major knowledge gaps, *Environ. Toxicol. Chem.*, **37**, pp. 2776–2796.

26. Cole, M., Lindeque, P., Halsband, C., Galloway, T.S. (2011). Microplastics as contaminants in the marine environment: A review, *Mar. Pollut. Bull.*, **62**, pp. 2588–2597.

27. Ni, B.-J., Xu, Q., Wei, W. (2023). *Microplastics in Urban Water Management* (John Wiley & Sons, Inc), ISBN 9781119759362.

28. Wright, S.L., Thompson, R.C., Galloway, T.S. (2013). The physical impacts of microplastics on marine organisms: A review, *Environ. Pollut.*, **178**, pp. 483–492.

29. Zhang, S.L., Wang, J.Q., Yan, P.K., Hao, X.H., Xu, B., Wang, W., Aurangzeib, M. (2021). Non-biodegradable microplastics in soils: A brief review and challenge, *J. Hazard. Mater.*, **409**, p. 124525.

30. Kim, J., Wager, Y.Z., Miller, C., Norton, J. (2023). *Microplastics in Urban Water Management*, Eds. Ni, B.-J., Xu, Q., Wei, W. Chapter 2 "Occurrence and removal of microplastics in drinking water systems" (John Wiley & Sons, Inc.) ISBN 9781119759362. https://doi.org/10.1002/9781119759379.ch2.

31. Long, Z., Pan, Z., Wang, W., Ren, J., Yu, X., Lin, L., et al. (2019). Microplastic abundance, characteristics, and removal in wastewater treatment plants in a coastal city of China, *Water Res.*,**155**, pp. 255–265.

32. Choi, J.S., Jung, Y.-J., Hong, N.-H., Hong, S.H., Park, J.W. (2018). Toxicological effects of irregularly shaped and spherical microplastics in a marine teleost, the sheepshead minnow (*Cyprinodon variegatus*), *Mar. Pollut. Bull.*, **129**, pp. 231–240.

33. Jia, Q.L., Chen, H., Zhao, X., Li, L., Nie, Y.-H., Ye, J-F. (2019). Removal of microplastics by different treatment processes in Shanghai large municipal wastewater treatment plants, *Huan Jing Ke Xue*, **40**, pp. 4105–4112.

34. Li, L., Geng, S., Wu, C., Song, K., Sun, F., Visvanathan, C., et al. (2019). Microplastics contamination in different trophic state lakes along the

middle and lower reaches of Yangtze River Basin, *Environ. Pollut.*, **254**, p. 112951.

35. Song, K., Li., L. (2023). *Microplastics in Urban Water Management*, Eds. Ni, B.-J., Xu, Q., Wei, W. Chapter 3 "Occurrence of microplastics in wastewater treatment plants" (John Wiley & Sons, Inc), ISBN 9781119759362. https://doi.org/10.1002/9781119759379.ch3.
36. Antunes, J., Frias, J., Micaelo, A., Sobral, P. (2013). Resin pellets from beaches of the Portuguese coast and adsorbed persistent organic pollutants, *Estuar. Coast. Shelf. Sci.*, **130**, pp. 62–69.
37. Chen, Q.Q., Li, Y., Li, B.W. (2020). Is color a matter of concern during microplastic exposure to *Scenedesmus obliquus* and *Daphnia magna*? *J. Hazard. Mater.*, **383**, p. 121224.
38. Frias, J., Sobral, P., Ferreira, A.M. (2010). Organic pollutants in microplastics from two beaches of the Portuguese coast, *Mar. Pollut. Bull.*, **60**, pp. 1988–1992.
39. Andrady, A.L. (2017). The plastic in microplastics: A review, *Mar. Pollut. Bull.*, **119**, pp. 12–22.
40. Brennecke, D., Duarte, B., Paiva, F., Caçador, I., Canning-Clode, J. (2016). Microplastics as vector for heavy metal contamination from the marine environment, *Estuar. Coast Shelf Sci.*, **178**, pp. 189–195.
41. Holmes, L.A., Turner, A., Thompson, R.C. (2014). Interactions between trace metals and plastic production pellets under estuarine conditions, *Mar. Chem.*, **167**, pp. 25–32.
42. Bhagwat, G., Tran, T.K.A., Lamb, D., Senathirajah, K., Grainge, I., O'Connor, W., et al. (2021). Biofilms enhance the adsorption of toxic contaminants on plastic microfibers under environmentally relevant conditions, *Environ. Sci. Technol.*, **55**, pp. 8877–8887.
43. Silvestrova, K., Stepanova, N. (2021). The distribution of microplastics in the surface layer of the Atlantic Ocean from the subtropics to the equator according to visual analysis, *Mar. Pollut. Bull.*, **162**, p. 111836.
44. Uurasjärvi, E., Pääkkönen, M., Setälä, O., Koistinen, A., Lehtiniemi, M. (2021). Microplastics accumulate to thin layers in the stratified Baltic Sea, *Environ. Pollut.*, **268**, p. 115700.
45. Kermenidou, M., Frydas, I.S., Moschoula, E., Kousis, D., Christofilos, D., Karakitsios, S., et al. (2023). Quantification and characterization of microplastics in the Thermaic Gulf, in the North Aegean Sea, *Sci. Total Environ.*, **892**, p. 164299.
46. Gong, X., Tian, L., Wang, P., Wang, Z., Zeng, L., Hu, J. (2023). Microplastic pollution in the groundwater under a bedrock island in the South China Sea, *Environ. Res.*, **239**, p. 117277.

47. Balestra, V., Vigna, B., De Costanzo, S., Bellopede, R. (2023). Preliminary investigations of microplastic pollution in karst systems, from surface watercourses to cave waters, *J. Contam. Hydrol.*, **252**, p. 104117.

48. Alvarado-Zambrano, D., Rivera-Hernández, J.R., Green-Ruiz, C., (2023). First insight into microplastic groundwater pollution in Latin America: The case of a coastal aquifer in Northwest Mexico, *Environ. Sci. Pollut. Res. Int.*, **30**, pp. 73600–73611.

49. Viitala, M., Steinmetz, Z., Sillanpää, M., Mänttäri, M., Sillanpää, M. (2022). Historical and current occurrence of microplastics in water and sediment of a Finnish lake affected by WWTP effluents, *Environ. Pollut.*, **314**, p. 120298.

50. Malla-Pradhan, R., Pradhan, B.L., Phoungthong, K., Joshi, T.P. (2022). Occurrence and distribution of microplastics from Nepal's second largest lake, *Water Air Soil Pollut.*, **233**, p. 423.

51. Kurki-Fox, J.J., Doll, B.A., Monteleone, B., West, K., Putnam, G., Kelleher, L., Krause, S., Schneidewind, U. (2023). Microplastic distribution and characteristics across a large river basin: Insights from the Neuse River in North Carolina, USA, *Sci. Total Environ.*, **878**, p. 162940.

52. Apetogbor, K., Pereao, O., Sparks, C., Opeolu, B. (2023). Spatio-temporal distribution of microplastics in water and sediment samples of the Plankenburg river, Western Cape, South Africa, *Environ. Pollut.*, **323**, p. 121303.

53. Simon, M., van Alst, N., Vollertsen, J. (2028). Quantification of microplastic mass and removal rates at wastewater treatment plants applying focal plane array (FPA)-based Fourier, *Water Res.*, **142**, pp. 1–9.

54. Franco, A.A., Arellano, J.M., Albendín, G., Rodríguez-Barroso, R., Quiroga, J.M., Coello, M.D. (2021). Microplastic pollution in wastewater treatment plants in the city of Cádiz: Abundance, removal efficiency and presence in receiving water body, *Sci. Total Environ.*, **776**, p. 145795.

55. Akca, M.O., Gündoğdu, S., Akca, H., Delialioğlu, R.A., Aksit, C., Turgay, O.C., et al. (2024). An evaluation on microplastic accumulations in Turkish soils under different land uses, *Sci. Total Environ.*, **911**, p. 168609.

56. Leitão, I.A., van Schaik, L., Ferreira, A.J.D., Alexandre, N., Geissen, V. (2023). The spatial distribution of microplastics in topsoils of an urban environment: Coimbra city case study, *Environ. Res.*, **218**, p. 114961.

57. Cohen, Q.M., Glaese, M., Meng, K., Geissen, V., Huerta-Lwanga, E. (2021). Parks and recreational areas as sinks of plastic debris in urban sites: The case of light-density microplastics in the City of Amsterdam, The Netherlands, *Environments*, **9**, p. 5.

58. Rao, W., Fan, Y., Li, H., Qian, X., Liu, T. (2024). New insights into the long-term dynamics and deposition-suspension distribution of atmospheric microplastics in an urban area, *J. Hazard. Mater.*, **463**, p. 132860.

59. Luo, D., Wang, Z., Liao, Z., Chen, G., Ji, X., Sang, Y., et al. (2024). Airborne microplastics in urban, rural and wildland environments on the Tibetan Plateau, *J. Hazard. Mater.*, **465**, p. 133177.

60. Klein, M., Bechtel, B., Brecht, T., Fischer, E.K. (2023). Spatial distribution of atmospheric microplastics in bulk-deposition of urban and rural environments: A one-year follow-up study in northern Germany, *Sci. Total Environ.*, **901**, p.165923.

61. Gaston, E., Woo, M., Steele, C., Sukumaran, S., Anderson, S. (2020). Microplastics differ between indoor and outdoor air masses: Insights from multiple microscopy methodologies, *Appl. Spectrosc.*, **74**, pp. 1079–1098.

62. Carpenter, E.J., Smith, K.L. (1972). Plastics on the Sargasso Sea surface, *Science*, **175**, pp. 1240–1241.

63. Carpenter, E.J., Anderson, S.J., Harvey, G.R., Miklas, H.P., Peck, B.B. (1972). Polystyrene spherules in coastal waters, *Science*, **178**, pp. 749–750.

64. Kye, H., Kim, J., Ju, S., Lee, J., Lim, C., Yoon, Y. (2023). Microplastics in water systems: A review of their impacts on the environment and their potential hazards, *Heliyon*, **9**, p. e14359.

65. Ali, N., Khan, M.H., Ali, M., Sidra, S. Ahmad, Khan, A., et al. (2024). Insight into microplastics in the aquatic ecosystem: Properties, sources, threats and mitigation strategies, *Sci. Total Environ.*, **913**, p. 169489.

66. Ma, P., Wang, Mu W., Liu, H., Chen, Yu, F., Xia, J. (2019). Research on ecotoxicology of microplastics on freshwater aquatic organisms, *Environ. Pollut. Bioavailab.*, **31**, pp. 131–137.

67. Dusaucy, J., Gateuille, D., Perrette, Y., Naffrechoux, E. (2021). Microplastic pollution of worldwide lakes, *Environ. Pollut.*, **284**, p. 117075.

68. Dey, U., Raj, D., Mondal, M., Roy, P., Mukherjee, A., Mondal, N.K., et al. (2023). Microplastics in groundwater: An overview of source, distribution, mobility constraints and potential health impacts during the Anthropocene, *Groundw. Sustain. Dev.*, **23**, p. 101036.

69. Parashar, N., Hait, S. (2023). Recent advances on microplastics pollution and removal from wastewater systems: A critical review, *J. Environ. Manage.*, **340**, p. 118014.

70. Kawecki, D., Nowack, B. (2019). Polymer-specific modelling of the environmental emissions of seven commodity plastics as macro- and microplastics, *Environ. Sci. Technol.*, **53**, pp. 9664–9676.

71. Nizzetto, L., Futter, M., Langaas, S. (2016). Are agricultural soils dumps for microplastics of urban origin? *Environ. Sci. Technol.*, **50**, pp. 10777–10779.

72. Surendran, U., Jayakumar, M., Raja, P., Gopinath, G., Chellam, P.V. (2023). Microplastics in terrestrial ecosystem: Sources and migration in soil environment, *Chemosphere*, **318**, p. 137946.

73. Amesho, K.T.T., Chinglenthoiba, C., Samsudin, M.S.A.B., Lani, M.N., Pandey, A., Desa, M.N.M., et al. (2023). Microplastics in the environment: An urgent need for coordinated waste management policies and strategies, *J. Environ. Manage.*, **344**, p. 118713.

74. Chae, Y., An, Y.J. (2018). Current research trends on plastic pollution and ecological impacts on the soil ecosystem: A review, *Environ Pollut.*, **240**, pp. 387–395.

75. Yang, H., Yumeng, Y., Yu, Y., Yinglin, H., Fu, B., Wang, J. (2022). Distribution, sources, migration, influence and analytical methods of microplastics in soil ecosystems, *Ecotoxicol. Environ. Saf.*, **243**, p. 114009.

76. Rillig, M.C., Ziersch, L., Hempel, S. (2017). Microplastic transport in soil by earthworms, *Sci Rep.*, **7**, p. 1362.

77. Dris, R., Gasperi, J., Tassin, B. (2018). Sources and fate of microplastics in urban areas: A focus on Paris megacity, *Handb. Environ. Chem.*, **58**, pp. 69–83.

78. Yin, M., Yan, B., Wang, H., Wu, Y., Wang, X., Wang, J., et al. (2023). Effects of microplastics on nitrogen and phosphorus cycles and microbial communities in sediments, *Environ. Pollut.*, **318**, p. 120852.

79. Zhang, Y., Kang, S., Allen, S., Allen, D., Gao, T., Sillampää, M. (2020). Atmospheric microplastics: A review on current status and perspectives, *Earth Sci. Rev.*, **203**, p. 103118.

80. Prata, J.C. (2018). Airborne microplastics: Consequences to human health? *Environ. Pollut.*, **234**, pp. 115–126.

81. Patidar, K., Ambade, B., Mohammad, F., Soleiman, A.A. (2023). Microplastics as heavy metal vectors in the freshwater environment: Distribution, variations, sources and health risk, *Phys. Chem. Earth*, **131**, p. 103448.

82. Su, X., Liu, M., Yuan, J., Huang, X., Lu, Z., Xu, J., He, Y. (2023). Potential risk of co-occurrence of microplastics and chlorinated persistent organic pollutants to coastal wetlands: Evidence from a case study, *Environ. Pollut.*, **320**, p. 121087.
83. Mbachu, O., Jenkins, G., Pratt, C., Kaparaju, P. (2020). A new contaminant superhighway? A review of sources, measurement techniques and fate of atmospheric microplastics, *Water Air Soil Pollut.*, **231**, p. 85.
84. Zhao, C., Liang, J., Zhu, M., Zheng, S., Zhao, Y., Sun, X. (2023). Occurrence, characteristics, and factors influencing the atmospheric microplastics around Jiaozhou Bay, the Yellow Sea, *Mar. Pollut. Bull.*, **196**, p. 115568.
85. Sun, N., Shi, H., Li, X., Gao, C., Liu, R. (2023). Combined toxicity of micro/nanoplastics loaded with environmental pollutants to organisms and cells: Role, effects, and mechanism, *Environ. Int.*, **171**, p. 107711.
86. Jenner, L.C., Rotchell, J.M., Bennett, R.T., Cowen, M., Tentzeris, V., Sadofsky, L.R. (2022). Detection of microplastics in human lung tissue using μFTIR spectroscopy, *Sci. Total Environ.*, **831**, p. 154907.
87. Leslie, H.A., van Velzen, M.J.M., Brandsma, S.H., Vethaak, A.D., Garcia-Vallejo, J.J., Lamoree, M.H. (2022). Discovery and quantification of plastic particle pollution in human blood, *Environ Int.*, **163**, p. 107199.
88. Ragusa, A., Svelato, A., Santacroce, C., Catalano, P., Notarstefano, V., Carnevali, O., et al. (2021). Plasticenta: First evidence of microplastics in human placenta, *Environ Int.*, **146**, p. 106274.
89. Kuhlman, R.L. (2022). Letter to the editor, discovery and quantification of plastic particle pollution in human blood, *Environ Int.*, **167**, p. 107400.
90. Rahman, A., Sarkar, A., Yadav, O.P., Achari, G., Slobodnik, J. (2021). Potential human health risks due to environmental exposure to nano- and microplastics and knowledge gaps: A scoping review, *Sci. Total Environ.*, **757**, p. 143872.
91. Li, J.Y., Yu, Y., Craig, N.J., He, W., Su, L. (2024). Interactions between microplastics and insects in terrestrial ecosystems: A systematic review and meta-analysis, *J. Hazard Mater.*, **462**, p. 132783.
92. Mackenzie, C.M., Vladimirova, V. (2023). Preliminary study and first evidence of presence of microplastics in terrestrial herpetofauna from Southwestern Paraguay, *Stud. Neotrop. Fauna Environ.*, **58**, pp. 16–24.
93. Mendez-Sanhueza, S., Torres, M., Pozo, K., Del Aguila, G., Hernandez, F., Jacobsen, C., et al. (2023). Microplastics in seabird feces from coastal areas of central Chile, *Animals*, **13**, p. 2840.

94. Wang, W., Do, A.T.N., Kwon, J.H. (2022). Ecotoxicological effects of micro- and nanoplastics on terrestrial food web from plants to human beings, *Sci. Total Environ.*, **834**, p. 155333.

95. Farrell, P., Nelson, K. (2013). Trophic level transfer of microplastic: *Mytilus edulis* (L.) to *Carcinus maenas* (L.), *Environ. Pollut.*, **177**, pp. 1–3.

96. Nelms, S.E., Galloway, T.S., Godley, B.J., Jarvis, D.S., Lindeque, P.K. (2018). Investigating microplastic trophic transfer in marine top predators, *Environ. Pollut.*, **238**, pp. 999–1007.

97. Rose, P.K., Yadav, S., Kataria, N., Khoo, K.S. (2023). Microplastics and nanoplastics in the terrestrial food chain: Uptake, translocation, trophic transfer, ecotoxicology, and human health risk, *TrAC Trends Anal. Chem.*, **167**, p. 117249.

98. Setälä, O., Fleming-Lehtinen, V., Lehtiniemi, M. (2014). Ingestion and transfer of microplastics in the planktonic food web, *Environ. Pollut.*, **185**, pp. 77–83.

99. Welden, N.A., Abylkhani, B., Howarth, L.M. (2018). The effects of trophic transfer and environmental factors on microplastic uptake by plaice, *Pleuronectes plastessa*, and spider crab, *Maja squinado*, *Environ. Pollut.*, **239**, pp. 351–358.

100. Mattson, K., Ekvall, M.T., Hansson, L.A., Linse, S., Malmendal, A., Cedervall, T. (2015). Altered behavior, physiology, and metabolism in fish exposed to polystyrene nanoparticles, *Environ. Sci. Technol.*, **49**, pp. 553–561.

101. Wang, F., Wu, H., Wu, W., Wang, L., Liu, J., An, L., et al. (2021). Microplastic characteristics in organisms of different trophic levels from Liaohe Estuary, China, *Sci Total Environ.*, **789**, p. 148027.

102. Madela, M. (2023). *Microplastics in the Ecosphere.* Eds. Vithanage, M., Prasad, M.N.V., "Microplastic toxicity to humans" (Wiley, USA), pp. 381–387.

103. Barboza, L.G.A., Vieira, L.R., Branco, V., Figueiredo, N., Carvalho, F., Carvalho, C., et al. (2018). Microplastics cause neurotoxicity, oxidative damage and energy-related changes and interact with the bioaccumulation of mercury in the European seabass, *Dicentrarchus labrax* (Linnaeus, 1758), *Aquat. Toxicol.*, **195**, pp. 49–57.

104. Barboza, L.G.A., Vieira, L.R., Guilhermino, L. (2018). Single and combined effects of microplastics and mercury on juveniles of the European seabass (*Dicentrarchus labrax*): Changes in behavioural responses and reduction of swimming velocity and resistance time, *Environ. Pollut.*, **236**, pp. 1014–1019.

105. Oliveira, M., Ribeiro, A., Hylland, K., Guilhermino, L. (2013). Single and combined effects of microplastics and pyrene on juveniles (0+ group) of the common goby *Pomatoschistus microps* (Teleostei, Gobiidae), *Ecol. Indic.*, **34**, pp. 641–647.

106. Lusher, A.L., Hollman, P.C.H., Mendoza-Hill, J.J. (2017). *FAO Fisheries and Aquaculture Technical Paper* **615** "Microplastics in fisheries and aquaculture: status of knowledge on their occurrence and implications for aquatic organisms and food safety" Rome, Italy.

107. Schirinzi, G.F., Pérez-Pomeda, I., Sanchís, J., Rossini, C., Farré, M., Barceló, D. (2017). Cytotoxic effects of commonly used nanomaterials and microplastics on cerebral and epithelial human cells, *Environ. Res.*, **159**, pp. 579–587.

108. Windsor, F.M., Durance, I., Horton, A.A., Thompson, R.C., Tyler, C.R., Ormerod, S.J. (2019). A catchment-scale perspective of plastic pollution, *Glob. Change Biol.*, **25**, p. 12071221.

109. Rotter, E. Grigalionyte-Bembic, K., Klun, Tiller., R. (2022) *Plastic Pollution and Marine Conservation.* Eds. Bonanno, G., Orlando-Bonaca, M. "Microplastics pollution: A thriller with many leading roles and an unknown ending" (Elsevier Inc). https://doi.org/10.1016/B978-0-12-822471-7.00004-3.

110. Bhattacharya, A., Roychoudhury, A. (2023). *Microplastics in the Ecosphere: Air, Water, Soil, and Food,* first edition. Eds. Vithanage, M., Prasad, M.N.V. Chapter 12 "Microplastics in the aquatic environment: Effects on ocean carbon sequestration and sustenance of marine life" (John Wiley & Sons Ltd.). https://doi.org/10.1002/9781119879534.ch12.

111. Lusher, A.L., Hurley, R., Arp, H.P.H., Booth, A.M., Bråte, I.L.N., Gabrielsen, G.W., et al. (2021). Moving forward in microplastic research: A Norwegian perspective, *Environ. Int.*, **157**, p. 106794.

Chapter 2

Analysis of Microplastics in Environmental Waters by Pyrolysis-GC/MS

Lúcia H.M.L.M. Santos[a,b]
[a]*Catalan Institute for Water Research (ICRA-CERCA), C. Emili Grahit 101, 17003 Girona, Spain*
[b]*University of Girona, Girona, Spain*
luciahelena.santos@sapo.pt

Microplastics (MPs) have been detected worldwide in a wide range of environmental compartments (e.g., wastewater, river and lake water, seawater, sediments, and biota), making the accumulation of plastics a growing environmental problem. Therefore, it is mandatory to gather data on their occurrence and environmental fate to better understand the effects of MPs on ecosystems and human health. Over the last years, pyrolysis coupled with gas chromatography and mass spectrometry (Py-GC/MS) appeared as a promising analytical method for the qualitative and quantitative analysis of MPs in the environment, especially because it has no particle size limitations.

Microplastic Pollution: Environmental Impact and Solutions
Edited by Julia Martín, Juan Luis Santos, Irene Aparicio, and Esteban Alonso

ISBN 978-981-5129-77-9 (Hardcover), 978-1-003-66006-4 (eBook)
www.jennystanford.com

DOI: 10.1201/9781003660064-2

This chapter presents a comprehensive overview of the sampling and sample preparation approaches as well as the instrumental analysis methods currently used for the analysis of MPs in environmental water using Py-GC/MS. Quality assurance and quality control (QA/QC) measures to ensure an accurate identification and quantification of plastic polymers are also described.

2.1 Introduction

Plastics are an essential material in our economy and daily lives. Consequently, global plastic production has drastically increased over the years. In 2022, it reached 400 million tons [1]. Unfortunately, a major drawback of plastic use is the high levels of waste generated, which ends up in the natural environment as plastic pollutants. It is estimated that in 2050, global plastic pollution will reach 34 billion tons [2]. The properties and characteristics that make plastics so versatile are also some of the reasons why the accumulation of plastics is a growing environmental concern, namely their persistence, durability, and capability of being transported through a wide range of environmental compartments [2, 3].

Although there is still some controversy related to the definition of microplastics (MPs), generally they have been defined as synthetic solid particles or polymeric matrices of any shape, ranging from 1 μm to 5 mm [4]. MPs can be divided into two categories relative to their origin: primary MPs, which correspond to the plastic pieces that are directly manufactured in the micrometric size for specific applications; and secondary MPs, which derive from the fragmentation of larger plastic pieces released in the environment and are subjected to photo- and thermal oxidation processes, mechanical abrasion, and biodegradation [5, 6]. MPs can enter the environment through various sources, such as the discharge of wastewater treatment plants (WWTPs), weathering and degradation of plastic waste in waterbodies, and terrestrial input from soil erosion or surface run-off [5]. Therefore, MPs have been ubiquitously detected in the environment (e.g., freshwater such as rivers and lakes [7–9], seawater [10, 11], sediments [12, 13], soils [14, 15], and biota [16, 17]). Some of the most common plastic polymers found in the environment are polyethylene (PE), polypropylene (PP), polystyrene

(PS), polyethylene terephthalate (PET), polyvinylchloride (PVC), polyester (PES), or polyamide (PA) [18]. Besides the size- and mass-related concentrations, knowing the chemical identity of MPs is very important, given that different polymers may act differently in the environment by adsorbing contaminants and/or leaching plastic additives from their composition, acting as vectors or carriers of contaminants to organisms and influencing the toxicity and impact of MPs to aquatic organisms [19, 20].

The environmental fate of MPs is a worrying problem, which in the last years has raised the attention of the scientific community, especially due to the possible risks of MPs to the ecosystems and human health. To better understand the environmental effects of MPs, researchers have focused on monitoring MPs in various environmental compartments, employing different techniques for sampling, sample processing, and identifying and quantifying MPs. Different analytical methods have been used in the analysis of MPs, namely visual identification, microscopy, spectroscopic methods, such as Fourier transform infrared (FTIR) and Raman spectroscopy, and thermal methods, including pyrolysis coupled to gas chromatography and mass spectrometry (Py-GC/MS) and thermal desorption coupled to gas chromatography and mass spectrometry (TED-GC/MS) [21–23]. Unfortunately, there are still no standardized methodologies for sampling and subsequent analysis of MPs in environmental samples, hampering the comparison of results among matrices and even within a matrix. Nevertheless, some progress has been made, and as the first attempt to harmonize sample preparation and MP analysis, the European Commission established a methodology to measure MPs in water intended for human consumption based on Fourier transform infrared micro-spectroscopy (μFTIR) and Raman micro-spectroscopy (μRaman) [24].

Although spectroscopic methods are the most widely used in the analysis of MPs, especially when combined with microscopy, thermal methods have gained increasing attention, especially Py-GC/MS. Spectroscopic methods have several advantages: They are non-destructive, require a minimum amount of sample, and in addition to the structural identification of MPs, they also give information about the total number of particles and particle size. However, their main drawbacks are that they require several steps of sample purification to isolate MPs as well as the spectra collection at each particle location, which implies a prolonged analysis time [25, 26]. They also have size limitations, since FTIR and Raman are only applicable in the identification of the polymeric composition of MP particles with a size greater than 10 μm and 1 μm, respectively [27].

Thermal methods are based on the thermal decomposition of polymers and their characterization through the analysis of the formed degradation products, providing molecular and mass spectral information on MPs composition [28]. Among thermal methods, Py-GC/MS has recently emerged as a powerful tool, since it can provide both qualitative and quantitative data, based on mass units, on MPs analysis in a single run. Its main advantages are that it allows the simultaneous analysis of MPs in the mixture, can reach levels of detection in the nanogram range, does not require an intensive purification and sometimes the sample pretreatment step, and does not have inherent size limitations, which makes it an attractive solution for the analysis of nanoplastics (plastic particles with size less than 1 μm). The main limitations of Py-GC/MS are its destructive nature, and the lack of information related to the morphological characteristics of MPs, such as particle size, shape, and colour, and its results may not be representative of heterogenous environmental samples [6, 25, 26]. Attending to their characteristics, thermal methods are generally recognised as complementary to spectroscopic approaches, such as FTIR and Raman. Despite their good performance, spectroscopic methods have two main limitations: a lack of selectivity, which does not allow the identification of specific monomers and co-monomers, mixtures, additives and degradation products; and the spectra are relative to the surface of MP particles. Py-GC/MS overcomes these limitations and characterizes MPs at a molecular level, analysing the whole MP particle, and identifying both synthetic polymers and organic plastic additives in a single analytical run [28].

This chapter will focus on the analysis of MPs in environmental water by Py-GC/MS, giving a comprehensive overview of (a) sampling, sample preparation, and instrumental analysis methods currently used; (b) quality assurance and quality control (QA/QC) measures applied; and (c) data reporting the presence of MPs in environmental water based on Py-GC/MS analysis. Figure 2.1 shows a schematic representation of the analytical procedure for the analysis of MPs in environmental water by Py-GC/MS, highlighting the principal techniques used in each step, which will be discussed in the following sections.

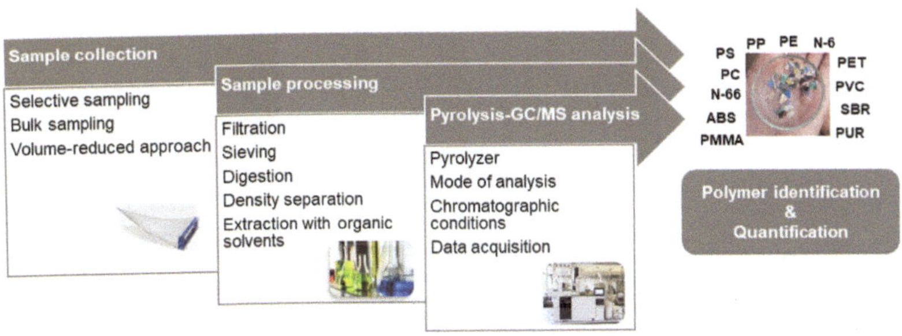

Figure 2.1 Overview of the different steps involved in the analysis of MPs in environmental water by Py-GC/MS.

2.2 Sampling

The sampling process is a crucial step regarding the collection of a representative environmental sample, and it should be done following the best practices to minimize undesirable contamination. Water samples for MPs analysis can be collected either from the water surface or the water column at different depths. Different approaches can be employed for sampling MPs in water, namely selective sampling, bulk sampling, and volume-reduced approach. The selection of the sampling technique will depend on the size of the MPs to be targeted.

Selective sampling is applicable in cases where the plastic items are large enough for identification with the naked eye, allowing their direct collection from environmental samples. This sampling technique is simple and straightforward but has a size limitation since only large MPs visible to the naked eye can be sampled [22]. This sampling strategy is less used for the analysis of MPs in environmental water by Py-GC/MS. However, there are some studies reported in literature describing the direct collection of unknown MPs samples from trawls [29] and Neuston nets [30, 31] used in the sampling of sea surface water. Usually, samples collected by selective sampling do not require further sample preparation and can be directly analysed by Py-GC/MS.

Bulk sampling implies collecting the entire sample without reducing the volume during the sampling process. This approach collects a relatively small amount of the sample, which may negatively affect the representativeness of the sample. In theory, bulk sampling allows the collection of all the MPs present in the sample independently of their size and visibility [22]. This sampling strategy has been used in the analysis of MPs and nanoplastics in wastewater [32–36], freshwater (e.g., rivers, lakes) [35–40], seawater [39, 41–43], and groundwater [38] by Py-GC/MS. Sample volume will depend on the type of environmental water, ranging from 1 L (influent) to 5 L (effluent) for wastewater, 1 L to 200 L for freshwater, and 1 L to 10 L for seawater. In the case of freshwater and seawater, sampling can be performed on the water surface or at different depths to evaluate the distribution of MPs in the water column. For instance, Li et al. [41] collected water samples from the Bohai Sea (China) at different depths to study both the distribution of small MPs and nanoplastics in the surface water and their vertical distribution in the top (0–2 m), middle (10 m), and bottom (>25 m) of the water column. Water samples are usually collected using stainless steel devices [32] or stainless steel or glass bottles [33, 34, 36, 43] that are submerged into the water for direct collection of the sample, but they can also be pumped directly to sample containers [38] or be collected using water sampling devices [37, 42]. For instance, Viitala et al. [37] used a Limnos water sampler operated from a motorboat to collect water samples from Lake Saimaa (Finland). Samples (10.4 L) were collected 0.5 m below the water surface and in some sampling points also at a depth of 5.0 m. Goβmann et al. [42] employed a sea surface scanner (S^3), an electric-powered and remotely controlled catamaran equipped with a sampling gear, to collect high-volume samples (10 L) from the sea surface microlayer and underlying water (1 m depth) of three fjord systems in the Swedish coast.

The volume-reduced approach involves reducing the entire volume of a bulk sample by fast filtration during sampling, preserving only a small fraction of the sample for further analysis. This approach has the advantage of covering a large quantity or area during the sampling process. The main disadvantage is that the size of the collected MPs is defined by the mesh size of the sampling tools used for the fast filtration, thus a significant amount of MPs may be lost, especially those of smaller size [22]. This sampling approach has been used in the analysis of MPs in wastewater [44], freshwater (e.g., canals, rivers, estuaries, lakes) [7, 9, 45–50], and seawater [27, 43, 51–53] by Py-GC/MS. There are different strategies to collect water samples by the volume-reduced approach. The most common is the use of Neuston nets (mesh size of 25 μm [43], or 300 μm [7, 43, 50, 53]), or manta trawl nets (mesh size of 330–333 μm [9, 47, 49]), operated from boats, or plankton nets with a mesh size of 333 μm, which are fixed in the middle of the river [46] for sampling surface water. A Neuston Catamaran with a net of 100 μm mesh size attached was also used for sampling North Sea surface water [27]. Normally, Neuston nets and manta trawl nets are conjugated with flowmeters to estimate the amount of sampled water. However, other approaches are also described in literature, such as the use of stainless steel sieve cascade [51, 52], combining different mesh sizes (e.g., 20 or 40, 63, and 125 μm [51], or 20 and 125 μm [52]), through which large volumes of water sample are filtered (e.g., between 44 and 336 L [51], or 56 L [52]); or using the sampling device Universal Filtering Object (UFO), a filtration system that comprises three interconnected stainless steel filters with 5000, 300, and 10 μm of pore size, working as a staged sieving apparatus [45]. Funck et al. [44] described the use of a cascadic MPs filtration plant using steel basket filters with mesh sizes of 100, 50, and 10 μm, which was installed in a WWTP, to sample MPs from WWTP effluent. This device allowed to filter a volume of 3500 L of effluent, of which 200 L was filtered through the 10 μm filter. Generally, after sample collection, MPs retained in the Neuston and manta trawl nets or the stainless steel sieves are transferred into metal sieves [9, 43, 47], membrane filters (e.g., 5 μm acetate cellulose membrane [43]), or glass bottles [46, 50–53] for further sample processing in the laboratory. Ultrapure water [46, 47, 50, 53] or a mixture of water/ethanol (10%) [52] can be used to facilitate the transfer of the collected particulate fraction. More detailed information on the sampling of environmental water using the described approaches can be found in Table 2.1.

Table 2.1 Examples of studies on the analysis of microplastics in environmental water using pyrolysis-GC/MS, showing the sampling and sample preparation approaches used as well as the microplastics detected and their respective concentrations

Location	Type of Sample	Sampling	Sample Preparation	Identified Polymers	Concentration	Ref.
			Wastewater			
Australia	WW influent	WW influent (1 L) was collected in pre-rinsed glass bottles.	Digestion of interfering organic materials by adding 60 mL of 30% H_2O_2 to 1 L of collected sample, mixed, and incubated at 60°C for 20–24 h in an orbital incubator shaker. After digestion, samples were vacuum filtered through a 1 µm glass fibre filter. The filter was stored in aluminium foil and dried at 60°C for 8 h. The filters were loaded into DCM precleaned 5 mL PLE extraction cell, and the void volume was filled with hydromatrix. PLE extraction consisted of a pre-extraction step with MeOH at 50°C and 1500 psi, a heat and a static time of 5 min, 1 cycle. Then samples were spiked with d_5-PS (IS) and extracted using DCM at 180°C and 1500 psi, a static time of 5 min and 2 extraction cycles. After extraction, 80 µL of the extract was transferred to a pyrolysis cup and left in a fume hood for 30 min to evaporate the solvent. After evaporation, 10 µL of TMAH, 25% in MeOH was added into the pyrolysis cup.	PLA PBAT	<MDL–0.15 mg/L <0.03–0.07 mg/L	[34]

Location	Type of Sample	Sampling	Sample Preparation	Identified Polymers	Concentration	Ref.
Australia	WWTP influent	WW influent (1 L) and treated effluent (5 L) were collected using a 1 L stainless steel bottle submerged into WWTP wastewater.	Digestion of interfering organic matter by adding 60 mL 30% H_2O_2 to 1 L of sample, mixing and incubating at 60°C for 20–24 h in an orbital incubator shaker. After digestion, samples were vacuum filtered through 1.0 μm glass fibre filters, and then were dried at 60°C for 20 h. Filters were loaded into DCM-precleaned 5 mL PLE extraction cells and the void volume was filled with hydromatrix. Then, samples were spiked with 40 μg of d_5-PS (IS) and PLE extracted using DCM at 180°C and 1500 psi, a static time of 5 min and 2 extraction cycles. After extraction, 160 μL of the extracted sample was transferred to a pyrolysis cup and left in a fume hood for 30 min to evaporate the solvent.	PE PVC PET PMMA	607–2340 μg/L 196–691 μg/L <5–67 μg/L <14–18 μg/L	[33]
	WWTP effluent			PP PET	<8–24 μg/L <5–7 μg/L	
Germany	WWTP effluent	WW effluent samples (~500 mL) were collected in two WWTPs.	WW effluent samples were size fractionated by sieving over a 500 μm stainless steel sieve. Fractions ≤ 500 μm were treated via an enzymatic-oxidative treatment. In a microplastics reactor, 10% SDS was added to the samples and incubated a 50°C for 24 h.	PE PVC PP PMMA/PUR PS PET	n.d.–2018 $\mu g/m^3$ 11–576 $\mu g/m^3$ n.d.–241 $\mu g/m^3$ n.d.–132 $\mu g/m^3$ n.d.–74 $\mu g/m^3$ n.d.–41 $\mu g/m^3$	[27]

(Continued)

Table 2.1 (*Continued*)

Location	Type of Sample	Sampling	Sample Preparation	Identified Polymers	Concentration	Ref.
			The technical enzyme purification steps with protease, cellulase, and chitinase as well as H_2O_2 a 35% were performed sequentially. The samples were incubated at optimal pH and temperature conditions for each step. Between treatments, the samples were flushed and rinsed several times with Milli-Q water. The samples were concentrated on a 0.2 µm Anodisc filter for FTIR analysis. Then, the filter was crushed and concentrated on a 1 µm glass fibre filter. The glass fibre was cut in half, and each half was folded and placed in a pyrolysis cup. 50 µL of IS mixture was added and evaporated, and 20 µL TMAH 25% in MeOH was added.			
–	WWTP effluent	Cascadic filtration plant using three basket filters with mesh sizes of 100, 50, and 10 µm. 3500 L of WWTP effluent were filtered through the 100 µm and the 50 µm filters. Through the 10 µm filter, 200 L was filtered. All basket filters were dried in an oven overnight at 60°C.	The filter surface was cleaned with a microspatula and ground, and an aliquot was taken for analysis. 5 µL of ethanol with internal standard styrene-d_8 was added.	PS	100 µm: 0.060 mg/m^3 50 µm: 0.009 mg/m^3 10 µm: 0.003–0.060 mg/m^3	[44]

Location	Type of Sample	Sampling	Sample Preparation	Identified Polymers	Concentration	Ref.
China	Liede WWTP influent	WW influent was sampled in glass bottles with glass lids.	WW influent sample was filtered through 1 μm glass microfibre filters. In a 50 mL long tapered centrifuge vial, 40 mL of the filtered sample was added. Then, the solution pH was adjusted to 3.0 with diluted HNO_3, and 40 μL of 20 mg/L BSA aqueous solution was added. The mixture was shaken on an orbital shaker at 300 rpm for 5 min under ambient temperature. In a water bath, 4 g of NaCl was added to the solution, followed by 15 min incubation at 80°C. After cooling to room temperature, the vial was centrifuged at 8000 rpm for 15 min. Then, the obtained protein with concentrated nanoplastics was redispersed in 200 μL of ultrapure water, transferred to a pyrolysis cup, and dried at 100°C for 10 min to remove water in three rounds (i.e., 70, 70, and 60 μL) before analysis.	PS	2.82 μg/L	[35]

(*Continued*)

Table 2.1 *(Continued)*

Location	Type of Sample	Sampling	Sample Preparation	Identified Polymers	Concentration	Ref.
			Freshwater			
Netherlands (Amsterdam)	Canal	Samples collected from a boat at a depth of 0.5 m using a plastic-free volume-reducing sampling device "Universal Filtering Object (UFO) system". Staged filtration: first down to 300 μm and then down to 10 μm.	Some 10 μm filters were cut into six pieces, transferred to a glass beaker, and covered with 10% H_2O_2 and left overnight in a shaking bath at 30°C and 55 rpm. Then, the filters were removed and rinsed with filtered water. The oxidized sample was filtered through a 10 μm stainless steel metal filter.	PE	n.d.–121.3 μg/m^3	[45]
				PP	2.5–116.3 μg/m^3	
				PET	8.0–87.1 μg/m^3	
				PVC	5.1–60.5 μg/m^3	
				PS	n.d.–19.7 μg/m^3	
				PMMA	n.d.–2.5 μg/m^3	
	Amstel River	Sample volumes ranged from 12–208 L and 69–419 L in summer and winter campaigns, respectively.	The filtered MPs were resuspended in NaI by sonication. The final separation of MPs was done by density separation using NaI. The supernatant was filtered through a 0.7 μm glass microfibre filter. Subsamples were punched out, placed into pyrolysis cups, and dried in the oven at 50°C.	PP	6.5–629.8 μg/m^3	
				PE	n.d.–156.8 μg/m^3	
				PET	9.2–82.9 μg/m^3	
				PVC	11.5–52.0 μg/m^3	
				PMMA	n.d.–0.1 μg/m^3	
	Estuary			PVC	3.1–15.0 μg/m^3	
				PP	0.2–9.8 μg/m^3	
				PET	5.2–9.4 μg/m^3	
				PE	n.d.–9.2 μg/m^3	
				PMMA	n.d.–0.2 μg/m^3	

Location	Type of Sample	Sampling	Sample Preparation	Identified Polymers	Concentration	Ref.
Finland	Lake Saimaa	Samples were collected 0.5 m below the surface (and in one sampling point also 5.0 m below the surface) using a Limnos water sampler operated from a motorboat. Water samples were filtered on a 20 μm stainless steel mesh. Five subsamples were collected for each composite sample (total sample volume = 10.4 L).	Filtered lake water samples were treated with the universal enzymatic purification protocol: 1–3 d incubations with SDS, protease, cellulase, 30% H_2O_2, and chitinase with proper buffers at varying temperatures. Then, samples were filtered on 20 μm metal meshes and stored in glass bottles with ultrapure water. Before analysis, samples were filtered on 1.2 μm glass microfibre filters, dried at 60°C, and 2.5 mL of 1,2,4-TCB containing 0.015% butylated hydroxytoluene and 2.5 mL of p-xylene were topped off. Polymers were extracted from the filters at 150°C for 60 min. After being cooled down, the supernatant was transferred to glass vials for analysis.	PP PE PS	<LOD–246.9 μg/L <LOD–39.8 μg/L <LOD–16.3 μg/L	[37]

(*Continued*)

Table 2.1 (*Continued*)

Location	Type of Sample	Sampling	Sample Preparation	Identified Polymers	Concentration	Ref.
Malaysia	Kelantan River basin	Samples were collected with a Neuston net with a 300 µm mesh size at a depth of 0.16 and 0.30 m, for 10–20 min per tow.	Wet peroxide oxidation of the sample by adding 20 mL 30% H_2O_2, heated at 60°C overnight and stirred at 180 rpm in a fume hood. Then, 20 mL 0.05M $FeSO_4$ were added and the cycle of heating and stirring was repeated.	PE	<72.9–3284.5 µg/L	[7]
				PP	<17.85–408 µg/L	
				PET	<14.30–142.6 µg/L	
	Langat River basin		After wet peroxidation, water samples were filtered through a 1.2 µm glass fibre filter and then dried in a desiccator for at least 24 h.	PE	116.3–31,370 µg/L	
				PP	<17.85–4355 µg/L	
				PET	<14.3–910.5 µg/L	
				N66	<10.35–485 µg/L	
				PS	<2.60–30.36 µg/L	
				N6	<2.45–28.37 µg/L	
				PMMA	<2.65–16.45 µg/L	
				PC	<2.55–2.99 µg/L	

Location	Type of Sample	Sampling	Sample Preparation	Identified Polymers	Concentration	Ref.
China	Fuhe River (surface water)	Surface water (200 L) was sampled with a pump from the shore at 0–30 cm deep. Groundwater (200 L) was drawn from local pumping wells with a stainless steel submergible pump.	Samples were filtered through 1 µm glass microfibre membrane filter and then concentrated by a crossflow ultrafiltration system with a molecular weight cut-off of 100 kDa. Then, the sample was digested by adding 5 mL H_2O_2 to 50 mL of concentrated water sample, held for 48 h in a 60°C water bath. After cooling, the sample was put in a centrifugal ultrafiltration tube (100 KDa) and centrifuged at 4000 rpm for 2 min. The retentate was transferred to a centrifuge tube and freeze-dried. Finally, the sample was added to 2 mL MeOH, vortexed for 20 s, transferred to a pyrolysis cup and dried in an oven at 60°C.	PP	0.117–0.389 µg/L	[38]
				PE	0.088–0.242 µg/L	
				PS	0.007–0.205 µg/L	
				PET	0.014–0.079 µg/L	
				PVC	0.006–0.057 µg/L	
				PMMA	n.d.–0.046 µg/L	
	Fuhe River (groundwater)			PP	0.015–0.108 µg/L	
				PE	n.d.–0.070 µg/L	
				PET	n.d.–0.024 µg/L	
				PS	n.d.–0.017 µg/L	
				PVC	n.d.–0.010 µg/L	
				PMMA	n.d.–0.003 µg/L	

(Continued)

Table 2.1 (*Continued*)

Location	Type of Sample	Sampling	Sample Preparation	Identified Polymers	Concentration	Ref.
China	Liao River	Water samples (8 L) were collected and stored in brown glass bottles.	Water samples were treated with Fenton reagent (Fe^{2+}, H_2O_2) and then filtered through 1 μm glass membranes to remove large particles. Alkylated Fe_3O_4 and 5 mL $CaCl_2$ (2 M) were added to 500 mL of filtered water for the extraction of NPs by the formation of Fe_3O_4-NPs aggregates. Formed aggregates were retained by filtration through 300 nm glass membranes. Loaded glass membranes were dried, folded, and then hung on a pyrolysis Eco-Stick of pyrolyser.	PS	0.24–0.31 μg/L	[39]
	Daliao River			PS	0.43–0.73 μg/L	
	Bohai sea			PS	n.d.–0.40 μg/L	
China	Liao River	Four samples (4 L each) were collected and stored in brown glass bottles.	Water samples were filtered through a 100 μm stainless steel screen and were treated with Fenton reagent (Fe^{2+}, H_2O_2) to eliminate interfering substances. For MP analysis: 500 mL water samples were filtered through a 1 μm glass membrane. After filtration, the membrane loaded with MPs was frozen in liquid N_2 for 5 min. Then, the membrane was grounded and transferred	PS MPs PMMA MPs	1.8–7.6 μg/L 0.6 μg/L	[40]

Location	Type of Sample	Sampling	Sample Preparation	Identified Polymers	Concentration	Ref.
			to a centrifuge tube. About 1 mL of water was added and mixed, and 50 μL of the mixture was transferred into a pyrolysis cup and further dried. For NP analysis: 100 μL of $MgSO_4$ (1 M) and 30 μL of 10% (m/v) TX-45 were added to 10 mL of filtrate, shaken and incubated in a water bath at 45°C for 15 min. Then it was centrifuged at 4000 rpm at 45°C for 10 min to obtain the surfactant-rich phase. 20 μL of TX-45-rich phase with NPs was transferred into a pyrolysis cup and subjected to thermal degradation at 190°C for 3 h before analysis.			
China	Pearl River	River water was sampled in glass bottles with glass lids.	The river water sample was filtered through 1 μm glass microfibre filters. Then, 40 mL of filtered sample was added in a 50 mL long tapered centrifuge vial. Then, the solution pH was adjusted to 3.0 with diluted HNO_3, and 40 μL of 20 mg/L BSA aqueous solution was added. The mixture was shaken on an orbital shaker at 300 rpm for 5 min under ambient temperature. About 4 g of NaCl	PS	1.92 μg/L	[35]

(*Continued*)

Table 2.1 (*Continued*)

Location	Type of Sample	Sampling	Sample Preparation	Identified Polymers	Concentration	Ref.
			was added into the solution, followed by 15 min incubation at 80°C in a water bath. After cooling to room temperature, the vial was centrifuged at 8000 rpm for 15 min. Then, the obtained protein with concentrated nanoplastics was redispersed in 200 μL of ultrapure water, transferred to a pyrolysis cup, and dried at 100°C for 10 min to remove water in three rounds (i.e., 70, 70, and 60 μL) before analysis.			
South Wales (UK)	River Tawe	Not indicated	Untreated river water (200 mL) was filtered through 0.45 μm PTFE membrane. Filter was dried, cut into small pieces (10 mm × 1 mm) with a laser cutter, and IS was added.	PS	241.8 μg/L	[54]
			Seawater			
China	Bohai Sea	Seawater (2 L) was collected and stored in brown glass bottles.	Oxidation of organic matter with Fenton reagent (Fe^{2+} and H_2O_2). Then, filtration through a 100 μm stainless steel screen. **MPs:** 500 mL of pretreated water was filtered by a 1 μm glass membrane. Then the glass membrane was freeze-dried in liquid N_2, grinding, and	PS (1–100 μm)	n.d.–0.41 μg/L	[41]

Location	Type of Sample	Sampling	Sample Preparation	Identified Polymers	Concentration	Ref.
			resuspending in 1 mL ultrapure water. About 50 μL of the suspension was transferred to a pyrolysis cup and dried.			
			NPs: In the filtrate of the 1 μm glass membrane, alkylated Fe_3O_4 (7.5 mg) and $CaCl_2$ (2 M, 5 mL) were added and shaken for 1 h. The mixture was filtrated trough a 300 nm glass membrane filter and dried at 80°C.	PS (1–999 nm)	n.d.–0.24 μg/L	
Sweden	Uddevalla Byfjord (seawater)	Seawater samples were collected with the Sea Surface Scanner (S^3), an electric-powered and remotely controlled catamaran used to collect high-volume samples. Seawater was collected at 1 m depth. Three samples of seawater (10 L each) was collected and then combined and treated as one sample.	Samples were filtered through 10 μm stainless steel filters and then treated with 20 mL 30% H_2O_2 and 20 mL HCl 2 M. Filter residues were transferred onto a 1 μm glass filter (Ø=15 mm). The glass fibre filter, together with the resulting filter cake, was folded and placed in a pyrolysis cup. About 20 μL deuterated PS (125 μg/mL in DCM) and 20 μL TMAH 12.5% in MeOH were added directly into the pyrolysis cup.	PMMA PVC PET PE PP PS PC MDI-PUR	0.36–8.11 μg/L 1.39–6.19 μg/L 0.55–1.44 μg/L 0.12–0.41 μg/L 0.02–0.11 μg/L 0.01–0.05 μg/L <LOQ–0.01 μg/L n.d.–<LOQ	[42]

(*Continued*)

Table 2.1 (*Continued*)

Location	Type of Sample	Sampling	Sample Preparation	Identified Polymers	Concentration	Ref.
	Askeröfjorden (seawater)			PMMA	0.23–9.21 µg/L	
				PVC	1.87–5.34 µg/L	
				PE	0.13–0.52 µg/L	
				PET	0.15–0.31 µg/L	
				PA6	n.d.–0.27 µg/L	
				MDI-PUR	n.d–0.18 µg/L	
				PP	0.01–0.11 µg/L	
				PC	<LOQ–0.01 µg/L	
				PS	<LOQ	
	Gullmar Fjord (seawater)			PVC	1.55–4.72 µg/L	
				PMMA	<LOQ–2.05 µg/L	
				PET	0.03–1.10 µg/L	
				PE	0.05–0.17 µg/L	
				PP	0.06–0.10 µg/L	
				PA6	n.d.–0.11 µg/L	
				PS	<LOQ–0.03 µg/L	
				PC	<LOQ–0.02 µg/L	
				MDI-PUR	n.d.–0.02 µg/L	

Location	Type of Sample	Sampling	Sample Preparation	Identified Polymers	Concentration	Ref.
Germany	North Sea (German Bight)	Between 44 and 336 L of water from 2.5 m depth was pumped through an onboard deck-wash system. MPs were concentrated using a stainless steel sieve cascade (20 or 40, 63, and 125 μm mesh size).	Samples were filtered through a 20 μm stainless steel sieve and transferred onto 1 μm glass fibre filters underplayed by 8–12 μm cellulose filters. In the case of noticeable organic matter, 2 mL of 30% H_2O_2 was applied to sample filters. Glass fibre filters were wetted with ethanol, folded, and transferred into pyrolysis cups. 50 μL of pyrolysis IS and 20 μL of TMAH 25% in MeOH were added to the cups.	PMMA PVC PE PET PP PS	n.d.–1004 ng/L n.d.–416 ng/L n.d.–275 ng/L n.d.–240 ng/L n.d.–60 ng/L n.d.–29 ng/L	[51]
Germany	North Sea	Seawater sample (56 L) was taken through an onboard deck-wash system and filtered through 125 μm and 20 μm stainless steel sieves.	Preconcentrated 20 μm and 125 μm size fractions were transferred onto 1 μm glass fibre filters. The MPs-loaded glass fibre filter was directly introduced in the pyrolysis cup. Before analysis, 50 μL of a mixture of pyrolysis IS and 20 μL of TMAH 25% in MeOH were added to the pyrolysis cup.	PVC PMMA PE PP PET	6.5 μg 5.9 μg 5.1 μg 2.3 μg 1.8 μg	[52]

(*Continued*)

Table 2.1 (*Continued*)

Location	Type of Sample	Sampling	Sample Preparation	Identified Polymers	Concentration	Ref.
—	North Sea (southern)	Seawater samples were collected with a Neuston Catamaran with a mesh size of 100 µm towed alongside the vessel for up to 20 min.	Seawater samples were size fractionated by sieving through a 500 µm stainless steel sieve. The fraction of size ≤ 500 µm was treated via an enzymatic-oxidative treatment. In a microplastics reactor, 10% SDS was added to the samples and incubated a 50°C for 24 h. The technical enzyme purification steps with protease, cellulase, and chitinase as well as 35% H_2O_2 were performed sequentially. The samples were incubated at optimal pH and temperature conditions for each step. Between treatments, the samples were flushed and rinsed several times with Milli-Q water. The samples were concentrated on a 0.2 µm Anodisc filter for FTIR analysis. Then, the filter was crushed and concentrated on a 1 µm glass fibre filter. The glass fibre was cut in half, and each half was folded and placed in a pyrolysis cup. 50 µL of IS mixture was added and evaporated, and 20 µL TMAH 25% in MeOH was added.	PVC PE PP PMMA/PUR	n.d.–6 µg/m^3 n.d.–2 µg/m^3 n.d.–1 µg/m^3 n.d.–1 µg/m^3	[27]

Location	Type of Sample	Sampling	Sample Preparation	Identified Polymers	Concentration	Ref.
—	North Atlantic subtropical gyre	Mesoplastics (5 mm to 20 cm) were collected by a sailing vessel. Large MPs (1–5 mm) were collected using Neuston net (0.5 × 0.4 m rectangular frame fitted with a 2 m long net) with a mesh size of 300 μm from the surface layer at a depth of 0–20 cm. Tow duration was set to 30 min. Contents of the tows were filtered through 300 μm sieves. Small MPs (20–999 μm) were collected using a Neuston net (0.3 × 0.1 m rectangular frame fitted with 3 m long net) with a mesh size of 25 μm from the surface layer at a depth of 0–11 cm. The tow duration was set to 10 min.	**Large MPs (1–5 mm)** On the boat, the contents of the tows were filtered through 300 μm sieves and the plastic debris (large MPs) were removed with tweezers. 10 μg of plastic was introduced into the pyrolysis cup. **Small MPs (20–999 μm)** In the laboratory, the 5 μm cellulose acetate membrane was immersed in 80 mL ultrapure water at ambient temperature under gentle agitation for 1 h. After the removal of the filter, 20 mL of a 10 M NaOH solution and 300 μL of a 50 g/L SDS solution were added. The solution was gently agitated for 4 h and stored for 1 week at ambient temperature. Then, the solution was filtered through 0.7 μm glass fibre filters. 0.7 μm glass fibre membranes were cut into eight pieces, and one piece was rolled and introduced in the pyrolysis cup. **Nanoplastics (1–999 nm)** 1 L of seawater was filtered through a 1.2 μm poly(ether sulfone) membrane.	**Mesoplastics** PE PP PS PVC **Large MPs** PE PP **Small MPs** PE PP PVC PS PET **Nanoplastics** PVC PET PS PE	 59% 17% 12% 6% 90% 10% 73% 13% 8% 2% 1% 70% 17% 9% 4%	[43]

(Continued)

Table 2.1 (*Continued*)

Location	Type of Sample	Sampling	Sample Preparation	Identified Polymers	Concentration	Ref.
		The content of the tows was filtered through a 5 µm cellulose acetate membrane in a closed filtration unit.	Then, the filtered seawater was concentrated by ultrafiltration with a 10 kDa poly(ether sulfone) membrane. The final retentate volume was 10 ± 2 mL.			
		Nanoplastics (1–999 nm) Seawater (1 L) was collected directly by plunging the glass bottle in the sea. The seawater sampling was taken on the tender at 300 m from the boat after turning off the engine.	Concentrated seawater was freeze-dried and then ground in an agate mortar for homogenization. About 25 mg of lyophilizate was introduced into the pyrolysis cup.			

PLA: polylactic acid; PBAT: polybutylene adipate terephthalate; PE: polyethylene; PVC: polyvinylchloride; PET: polyethylene terephthalate; PMMA: polymethylmethacrylate; PP: polypropylene; PS: polystyrene; PUR: polyurethane; N6 or PA6: polyamide 6; N66: polyamide 66; PC: polycarbonate; MDI-PUR: methylene diphenyl diisocyanate-polyurethane; H_2O_2: hydrogen peroxide; NaI: sodium iodide; SDS: sodium dodecyl sulphate; DCM: dichloromethane; PLE: pressurized liquid extraction; MeOH: methanol; d5-PS: deuterated polystyrene; TMAH: tetramethylammonium hydroxide; NPs: nanoplastics; HCl: hydrochloric acid; $FeSO_4$: iron (II) sulphate; NaOH: sodium hydroxide; HNO_3: nitric acid; BSA: bovine serum albumin; 1,2,4-TCB: 1,2,4 – trichlorobenzene; Fe_3O_4: ferroferric oxide; $CaCl_2$: calcium chloride; TX-45: Triton X-45; NaCl: sodium chloride; $MgSO_4$: magnesium chloride

Both bulk sampling and volume-reduced approach are the most common sampling strategies used in the analysis of MPs in environmental water by Py-GC/MS (Table 2.1). These two approaches can also be used in a complementary way, depending on the size range of MPs, which pretends to be covered during the sampling campaign. For instance, for the characterization of the presence of different fractions of MPs, namely large MPs (1–5 mm), small MPs (20 μm to 1 mm), and nanoplastics (1–999 nm), in the North Atlantic subtropical gyre, Ter Halle et al. [43] collected large and small MPs using Neuston nets with mesh sizes of 300 μm and 25 μm, respectively, while the fraction of nanoplastics was directly collected by plunging a 1 L glass bottle in the sea followed by further concentration by ultrafiltration.

2.3 Sample Preparation

After sample collection, it is necessary to extract, isolate, and concentrate the MPs present in the water samples for subsequent analysis by Py-GC/MS. Sample preparation is a crucial step that demands an appropriate design to reach the limits of detection and quantification of the instrumental method. Furthermore, several techniques are needed to cover a wide size range of MPs. Various techniques can be used for the extraction of MPs and the elimination of interfering compounds from the matrix, such as organic matter. The most used for the extraction of MPs from water samples are filtration, sieving, digestion, and density separation. Examples of sample preparation procedures used for the analysis of MPs in environmental water by Py-GC/MS are described in Table 2.1.

Filtration is one of the techniques most frequently used in the separation of MPs from water samples. Usually, filtration is applied after a sample pretreatment step, such as digestion or density separation [33, 41, 42]. Different types of membrane filters can be used. The most employed in the preparation of water samples for MPs analysis by Py-GC/MS are the glass fibre membrane filters [7, 27, 33, 34, 37, 40–43, 45, 46, 50–52]. However, other types of filters are also described in the literature, such as Anodisc (aluminium oxide) inorganic filter membranes [32], stainless steel filters [37, 42, 45], PTFE membranes [54], or nylon filters [47]. The fraction of MPs analysed is limited by the pore size of the filter used. Filters of various pore sizes have been used, namely 0.2 μm [32], 0.45 μm [54], 0.7 μm [43, 45], 1 μm [27, 33, 34, 40–42, 51, 52], 1.2 μm [7, 37, 46, 50], 10 μm [42, 45], 20 μm [37], and 180 μm [47].

Sieving is another technique frequently used for the separation of MPs from water matrices. It is similar to filtration, but instead of membrane filters, stainless steel sieves are used with different mesh sizes. The sieves retain the MP particles and other particulate material present in the water samples that are greater than the mesh size, allowing the removal of water and smaller particles from the sample. The main advantage of sieving is that it can extract MPs from large volumes of water to get representative experimental data [41]. The mesh size of the sieves is selected based on the targeted size range of MPs to be collected. For the analysis of MPs in environmental water by Py-GC/MS, several mesh sizes have been used, namely 5 mm [7, 46], 1 mm [7], 500 μm [27], 300 μm [7, 46, 50], 100 μm [41, 50], and 20 μm [51]. Water samples can be sieved directly [7, 27, 46, 51], sometimes using sequential sieves with different mesh sizes [7, 46], or be previously digested [41], as in the case of water samples with a high organic content. Normally, sieving is used as a sample pretreatment [7, 27, 41, 46, 50], followed by other sample preparation techniques for the concentration of MPs.

Digestion is a process aiming at removing interfering organic materials present in the environmental matrices. Common digestion protocols apply oxidizers, acids, or alkaline solutions as well as enzymatic digestion, which can be used alone or in combination, depending on the type of matrix and its organic content. Oxidative digestion using hydrogen peroxide (H_2O_2) is one of the most common digestion protocols applied to the removal of natural organic matter in water samples [7, 32–34, 45]. A typical wet peroxidation protocol consists of adding 30% H_2O_2 to the sample, which is then incubated at 60°C. Different incubation times are described in the literature, namely 3 h [32], 20–24 h [33, 34], or overnight [7]. However, the use of 10% H_2O_2 and overnight agitation at 30°C [45] have also been reported. Fenton's reagent (Fe^{2+} and H_2O_2) is often used as well [7, 9, 41, 46, 47, 50, 53]. To a lesser extent, enzymatic digestion [27, 37] and alkaline digestion using a 10 M sodium hydroxide (NaOH) solution [43] have also been described to remove organic matter from surface water, wastewater, and seawater. For instance, Viitala et al. [37] applied an enzymatic digestion protocol, which included sodium dodecyl sulphate (SDS), protease, cellulase, 30% H_2O_2, and chitinase with proper buffers and incubation times for the treatment of lake water. Moreover, the combination of different digestion protocols was proposed by some authors, for example the use of 30% H_2O_2 and hydrochloric acid (HCl) 2 M [42]. Finally, it should be highlighted that the digestion solution and time are selected according to the type of sample in order to ensure the complete removal of organic matter without affecting the integrity of MPs [33].

Density separation is a technique commonly applied to isolate MPs from environmental samples. It is based on the density differences between MPs and other interfering compounds present in the sample and uses the buoyant force of a liquid with an intermediate density to separate the lighter materials from the denser ones after a thorough shaking and settling of the materials in the liquid. The specific densities of MPs range from 0.8 to 1.70 g/cm^3 [22], and this technique takes advantage of the density of MPs to separate them from interfering materials present in the sample. The salt solutions described in the literature for the density separation of MPs from environmental water for further analysis by Py-GC/MS are saturated sodium chloride (NaCl) solution [9, 46, 47, 50, 53] (density = 1.2 g/cm^3), and sodium iodide (NaI) solution (density = 1.65 g/cm^3) [32, 45]. After shaking for resuspending the MP particles, different settlement times to allow the MPs to float to the surface are indicated, namely 3 h [32] or overnight [45, 47, 53]. It should be highlighted that although NaCl solution is most often used for density separation of MPs, as it is inexpensive, nontoxic, and environmentally friendly compared to other salt solutions, it is inefficient for separating more dense MPs from environmental matrices [47].

Besides the abovementioned sample preparation techniques, organic solvents can also be used to extract MPs from membrane filters. A few procedures using pressurized liquid extraction (PLE) are described for this purpose [33, 34]. Dichloromethane is a commonly used extraction solvent [33, 34] since it dissolves most of the plastic polymers present in the environment as MPs [28]. A pre-extraction clean-up step with methanol to remove interfering organic matrix compounds can also be included as it was shown by Okoffo et al.

[34] for the analysis of micro-bioplastics in environmental samples. A limitation of this sample preparation technique is that the analysis of MPs is limited to those polymers that are soluble in the selected extraction organic solvent.

Generally, the sample preparation of environmental water for the analysis of MPs by Py-GC/MS implies the sequential application of various techniques described earlier to extract and concentrate MPs and eliminate interfering organic and inorganic matrix components as much as possible. For example, for the preparation of wastewater samples, Lee et al. [32] used wet peroxidation, density separation with NaI solution, and filtration by 0.2 μm Anodisc inorganic filters, while Okoffo et al. [33] applied wet peroxidation, filtration with glass fibre filters, and PLE.

Given that Py-GC/MS is not able to provide information on MP particle size, the sequential fractioning of collected MPs during sample preparation is described in the literature [40, 41, 43] to separate MPs in different range sizes, allowing to obtain more reliable information on their environmental distribution. Depending on the MPs size fraction considered, different sample preparation protocols can be applied. For instance, after water sample pretreatment using Fenton oxidation followed by sieving through a 100 μm stainless steel sieve, Li et al. [41] combined filtration through a 1 μm glass fibre membrane and alkylated ferrosoferric oxide (Fe_3O_4) adsorption to separate and extract small MPs (<100 μm) and nanoplastics, respectively, in seawater samples from the Bohai Sea (China), while for the analysis of small MPs (<5 μm) and nanoplastics in water samples from Liao River (China), the same authors combined filtration trough a 1 μm glass fibre membrane with cloud-point extraction, respectively [40]. Ter Halle et al. [43] collected four distinct fractions of MPs in the North Atlantic subtropical gyre, which were processed following different procedures. Mesoplastics (5 mm to 20 cm) and large MPs (1–5 mm) were directly analysed by Py-GC/MS by introducing 10 μg of sample in the pyrolysis cup, while small MPs (20 μm to 1 mm) required an alkaline digestion followed by filtration through a 0.7 μm glass fibre membrane, and nanoplastics (1–999 nm) were concentrated from seawater by ultrafiltration.

Typically, filtration is the last step in sample preparation, and then the filters loaded with MPs are introduced in the pyrolysis cup before Py-GC/MS analysis. For that, there are different approaches described in the literature that can be used: (a) filters are dried and then cut into small pieces [43, 54] or punched [45], milled using an agate mortar [32], or folded [42, 51] to be transferred to the pyrolysis cup; or (b) filters are frozen in liquid nitrogen, ground, resuspended in ultrapure water, and then a small aliquot (e.g., 50 μL) of the uniformly mixed suspension is transferred to the pyrolysis cup and dried before Py-GC/MS analysis [40, 41]. The surface of the filter can also be scratched with tweezers and the obtained powder transferred to the pyrolysis cup [32]. In the case that extraction with organic solvents is used, an aliquot of the extracted sample is directly transferred into the pyrolysis cup and the solvent is evaporated before Py-GC/MS analysis [33, 34].

Although sample preparation is desirable for eliminating the interference of organic matter and lowering the limits of detection and quantification, sometimes this step can be excluded when Py-GC/MS is used for MPs analysis. For example, Funck et al. [44] used a cascadic filtration plant to sample MPs from WWTP effluent that were analysed by Py-GC/MS without further sample preparation. After sample collection, the extracted samples were ground with a mortar and then weighted to a 70 μL ceramic crucible before analysis. Large MPs that are collected by selective sampling are also analysed without sample preparation. Nevertheless, due to limitations on the amount of sample that can be introduced in the pyrolyser, before Py-GC/MS analysis, each MP piece can be cut into small pieces and directly introduced into a pyrolysis cup [29] or a large number of MPs can be polled and then cryo-milled into a powered mixture, which is weighted into the pyrolysis cup [30, 31].

Finally, it should be highlighted that analytical methods using FTIR or Raman and Py-GC/MS as complementary tools for the analysis of MPs in water samples have recently appeared in the literature. A common sample preparation for both analytical methods is suggested to enable the comparison of qualitative and quantitative data. In this way, after FTIR analysis, filters loaded with MPs are transferred to the pyrolysis cup for further Py-GC/MS analysis. For that, different preparation methods are proposed. Lee et al. [32] tested two preparation methods for the determination of MPs in raw wastewater, namely whole filter collection, which consisted of gently crushing the whole Anodisc filter in an agate mortar, homogenizing and weighing 1.5–2.5 mg of the crushed sample to the pyrofoil; and the surface scratching method, which consisted of carefully scraping with tweezers the aluminium oxide surface of the Anodisc filter membrane and then directly transferring the entire sample collected in powdered form to the pyrofoil. Better results were obtained by scratching the MP particles from the surface of the Anodisc filter, because when the entire filter was crushed before analysis, there was a significant dilution of the sample, increasing the limits of detection [32]. In turn, Primpke et al. [27] crushed the aluminium oxide (Anodisc) filter and concentrated it on a 1 μm glass fibre filter, which was then cut into half, and each half was folded and introduced in a pyrolysis cup. Alternatively, some authors proposed complementary sample preparation procedures after FTIR and Raman analysis. For instance, Viitala et al. [37] combined Raman and Py-GC/MS for the MPs analysis in Lake Saimaa water samples. After Raman analysis, the glass filters loaded with MPs were dried at 60°C and then MPs were extracted using 1,2,4-trichlorobenzene containing 0.015% butylated hydroxytoluene and p-xylene before Py-GC/MS analysis. Sefiloglu et al. [45] also used two different sample preparation approaches for μFTIR and Py-GC/MS analysis. After density separation, water samples for μFTIR analysis were filtered through a 10 μm stainless steel filter, and the particles retained in the filter were resuspended by sonication in ultrapure water and then filtered through a 0.2 μm alumina filter (Anodisc), while for Py-GC/MS analysis, the supernatant of the density separation was directly filtered through a 0.7 μm glass fibre filter, which was then punched and introduced in the pyrolysis cup.

2.3.1 Sample Preparation for Nanoplastics Analysis

Since Py-GC/MS does not have particle size limit restrictions, it appears as a powerful and promising tool for the identification and quantification of nanoplastics (plastic particles smaller than 1 μm). However, sample preparation for nanoplastics analysis is still a very challenging task. In recent years, various sample preparation approaches for the extraction of nanoplastics from environmental water samples have been reported, including ultrafiltration [43] or crossflow ultrafiltration [38, 48], agglomeration [39, 41], cloud-point extraction [36, 40], and protein corona–mediated extraction [35].

Generally, water samples are first filtered through glass fibre membrane filters, typically with a pore size of 1 µm [35, 36, 38–41], to separate the MPs (loaded in the filter) and nanoplastics (in the filtrate) fractions. Sometimes, a digestion step can be applied before filtration to remove most of the organic matter [39–41]. The use of crossflow ultrafiltration for the sample preparation of nanoplastics can remove small molecules of natural organic matter and has high efficiency for concentrating large volumes of water. Ter Halle et al. [43] used ultrafiltration with a 10 kDa poly(ether sulfone) membrane to concentrate nanoplastics from seawater, while Xu et al. [38] proposed using double ultrafiltration, which consisted of crossflow ultrafiltration (molecular weight cut-off of 100 kDa, equivalent to approximately 10 nm) followed by H_2O_2 oxidation and second centrifugal ultrafiltration (100 kDa) to extract nanoplastics from river and groundwater samples. Finally, in both ultrafiltration approaches, the treated sample was freeze-dried to obtain a powder containing nanoplastics, which was crushed in an agate mortar for homogenization, and 25 mg of lyophilizate was weighted to the pyrolysis cup [43], or the freeze-dried powder was resuspended in methanol and this suspension was transferred to the pyrolysis cup and dried at 60°C before Py-GC/MS analysis [38].

Combining alkylated Fe_3O_4 agglomeration with membrane filtration was suggested by Li et al. [39] to extract nanoplastics from different environmental water matrices. The authors added alkylated Fe_3O_4 and $CaCl_2$ to filtered water to form Fe_3O_4-NP agglomerates, which were retained by filtration through a 300 nm glass filter membrane. The glass membrane was then dried, carefully folded, and hung in the pyrolysis Eco-Stick of the pyrolyser for Py-GC/MS analysis.

Cloud-point extraction is a preconcentration technique based on the formation of micelles when a non-ionic surfactant is heated above its cloud-point temperature [36]. Cloud-point extraction with Triton X-45 (TX-45) was described for the extraction of nanoplastics from WWTP influent and effluent, river water, and seawater. It consisted of adding the surfactant TX-45 to the water sample, which was incubated at 45°C for 15 min and then centrifuged to separate the surfactant-rich phase, where the nanoplastics were trapped. An aliquot of this phase was transferred to a pyrolysis cup and thermally degraded at 190°C for 3 h to degrade the surfactant TX-45 before Py-GC/MS analysis [36, 40].

Lastly, nanoplastics extraction from environmental waters mediated by protein corona was proposed by Zhou et al. [35]. This sample preparation procedure consisted of promoting the formation of protein corona in the surface of nanoplastics by adding bovine serum albumin (BSA) to water samples followed by protein precipitation via salting out. The highest recoveries of nanoplastics were obtained using the following experimental conditions: BSA concentration of 20 mg/L, equilibration time of 5 min, pH 3.0, 10% (w/v) NaCl, incubation temperature of 80°C, and incubation time of 15 min. At the end, proteins loaded with nanoplastics were separated by centrifugation, redispersed in 200 μL of ultrapure water, transferred to a pyrolysis cup and dried at 100°C for 10 min to evaporate the water in three rounds (i.e., 70, 70, and 60 μL) to transfer the total volume of protein resuspension into the pyrolysis cup.

2.4 Analysis by Pyrolysis-GC/MS

Pyrolysis coupled with gas chromatography and mass spectrometry (Py-GC/MS) has recently gained increasing attention for the analysis of MPs in the environment because it allows both the identification and quantification of MPs in complex environmental matrices. It is based on the thermal decomposition of polymers into smaller fragments under an inert atmosphere. Then, the pyrolysis fragments are separated by gas chromatography as a function of time and characterized by mass spectrometry, producing a pyrogram. The identification of the degradation products is achieved by comparison of their MS spectra with mass spectral libraries or by selecting distinctive indicator compounds and their respective ions. After pyrolysis, each polymer presents characteristic degradation products and their respective indicator ions, which form the polymer's molecular profile, and can be chosen as marker compounds for its identification and quantification [6, 26, 55].

Table 2.2 gives an overview of instrumental methods of Py-GC/MS reported in the literature for the qualitative and quantitative analysis of MPs in environmental water. Besides conventional plastic polymers, Py-GC/MS has also been used for the analysis of bioplastics in wastewater samples [34].

Table 2.2 Examples of instrumental methods for the qualitative and quantitative analysis of microplastics in environmental water by pyrolysis-GC/MS

Water Sample Type	Polymer	Indicator Compound	Indicator Ion (m/z)	Type of Pyrolyser	Pyrolysis T (°C) and (Time)	Split Ratio	GC Column	GC Gradient	MS Mode (m/z Range)	Internal Standard	LOD (μg)	LOQ (μg)	Ref.
Wastewater (influent)	HDPE	1-octadecene	55	Curie Point	590 (15 s)	Splitless	HP-5ms UI	50°C held for 1 min, 50 to 320°C at 10°C/min, 320°C held for 10 min	Full scan mode (50–650)	Antracene-d_{10}	—	—	[32]
		1,15-octadecadiene	**55**										
		1-octadecane	57										
	PP	2,4-dimethyl-1-heptene	70								—	—	
		2,4,6-trimethyl-1-nonene	**69**										
		2,4,6,8-tetramethyl-10-undecene	69										
		2,4,6,8,10-pentamethyl-12-dodecene	69										
	PS	Styrene	107								—	—	
		2,4-dimethyl-1-butene	91										
		2,4,6-trimethyl-1-hexene	**91**										
	PET	**Dimethyl terephthalate***	**163**								—	—	
	PA6	**N-methyl-caprolactame***	**113**								—	—	
Wastewater (influent and effluent)	PMMA	Methyl acrylate	55, 85	Microfurnace	650 (12 s)	50:1	Ultra-Alloy® 5 (30 m × 0.25 mm × 0.25 μm)	40°C held for 2 min, 40 to 320°C at 20°C/min, 320°C held for 14 min	Full scan mode (40–600)	d_5-PS	0.09	—	[33]
		Methyl methacrylate	69, **100**, 89										
		(Z)-trimethyl 4,6-dimethylhept-2-ene-2,4,6-tricarboxylate	121, 149										
		(Z)-dimethyl 2,4-dimethylpent-2-enedioate	67, 95, 127, 111, 154										
		dimethyl 2, 2-dimethyl-4-methylenepentanedioate	81, 101, 109, 125, 140										
		$C_{11}H_{18}O_4$	81, 95, 109, 123										

(Continued)

Table 2.2 (*Continued*)

Water Sample Type	Polymer	Indicator Compound	Indicator Ion (m/z)	Type of Pyrolyser	Pyrolysis T (°C) and (Time)	Split Ratio	GC Column	GC Gradient	MS Mode (m/z Range)	Internal Standard	LOD (µg)	LOQ (µg)	Ref.
	PS	Styrene	104, 78								0.08	—	
		3-butene-1,3-diyldibenzene (styrene dimer)	91, 130, 193, 208										
		5-hexene-1,3,5-triyltribenzene (styrene trimer)	**91**, 117, 194, 312										
	PP	n-pentane	72, 55								0.05	—	
		2-methyl-1-pentene	84, 69, 56										
		2,4-dimethyl-1-heptene	70, 83, 126, **111**										
	PVC	Benzene	78, 74, 52, 39								0.37	—	
		Toluene	91, 89, 65, 51										
		Indene	116, 119										
		Styrene	104, 78										
		Indene	116, 19, 132, 90										
		Mil1-methylindene	130, 115, 132, 91, 77, 64, 51										
		Naphthalene	**128**, 132, 146, 116, 102, 77										
		1 methylnaphthalene	142, 91, 115, 129, 158										
		Acenaphthene	154, 76, 128, 102, 143, 128										

Water Sample Type	Polymer	Indicator Compound	Indicator Ion (m/z)	Type of Pyrolyser	Pyrolysis T (°C) and (Time)	Split Ratio	GC Column	GC Gradient	MS Mode (m/z Range)	Internal Standard	LOD (µg)	LOQ (µg)	Ref.
	PE	Alkadiene ($C_{10, 12, 14}$)	81, 109, 139								0.93	—	
		n-alkene ($C_{10, 12, 14}$)	**83**, 111, 140										
		n-alkane ($C_{10, 12, 14}$)	85, 114, 141										
	PET	**dimethyl terephthalate***	194, **163**								0.03	—	
	PC	p-methoxy-tert-butylbenzene*	164, 149								0.07	—	
		2,2-bis(4′-methoxy-phenyl)propane*	256, **241**										
	d5-PS	**Styrene monomer**	**109**, 82, 54, 107								—	—	
		Styrene dimer	96, 109, 65, 213, 218, 68										
		Styrene trimer	96, 121, 147, 207, 73, 44										
Wastewater (effluent)	PS	Styrene	51, 78, 104	Filament	600 (2 s)	10:1	OPTIMA 1 MS (30 m × 0.25 mm × 0.25 µm)	40°C held for 4 min, 40 to 200°C at 5°C/ min, 200 to 300°C at 10°C/min	Full scan mode (40–450)	Styrene-d_8	0.0004	0.001	[44]
		2,4-Diphenyl-1-butene	91, 104, 130, 208								0.009	0.03	
		2,4,6-Triphenyl-1-hexene	91, 117, 194, 207								0.009	0.03	

(Continued)

Table 2.2 (*Continued*)

Water Sample Type	Polymer	Indicator Compound	Indicator Ion (m/z)	Type of Pyrolyser	Pyrolysis T (°C) and (Time)	Split Ratio	GC Column	GC Gradient	MS Mode (m/z Range)	Internal Standard	LOD (μg)	LOQ (μg)	Ref.
	PE	1,12-Tridecadiene	55, 81, 67, 95								0.4	1.2	
		1,13-Tetradecadiene	81, 95, 109								0.3	1.0	
		1,14-Pentadecadiene	55, 81, 95, 109								0.3	1.1	
Wastewater (influent)	PLA	2,3-pentadione	43, 57, 74, 100	Microfurnace	600 (12 s)	50:1	Ultra Alloy 5 (30 m × 0.25 mm × 0.25 μm)	40°C held for 2 min, 40 to 320°C at 20°C/min, 320°C held for 14 min	Full scan mode (40–600)	d_5-PS	0.17	—	[34]
		meso-lactide (3,6-dimethyl-1,4-dioxane-2,5-dion)	43, 45, 56										
		DL-lactide (3,6-dimethyl-1,4-dioxane-2,5-dion)	43, 45, 56										
		methyl 2-hydroxypropanoate*	46, 61, 89										
		methyl 2-methoxypropanoate*	**59**, 88, 75, 43, 103										
	PCL	*ε*-caprolactone	55, 84, 114								0.10	—	
		dimer of *ε*-caprolactone	115, 97, 55										
		trimer of *ε*-caprolactone	115, 97, 55										
		pentyl pent-4-enoate	114, 97, 55, 68										
		5-oxo-5-(pentyloxy)pentyl pent-4-enoate	115, 97, 69, 185										
		5-hexenoic acid	60, 68, 96, 114										

Water Sample Type	Polymer	Indicator Compound	Indicator Ion (m/z)	Type of Pyrolyser	Pyrolysis T (°C) and (Time)	Split Ratio	GC Column	GC Gradient	MS Mode (m/z Range)	Internal Standard	LOD (μg)	LOQ (μg)	Ref.
		6-(5-hexenoyloxy)hexanoic acid	55, 69, 97, 115, 128										
		methyl 6-methoxyhexanoate*	**45**, 55, 74, 87, 98, 113, 128, 145										
	PBAT	tetrahydrofuran	71, 42, 53								0.09	—	
		cyclopentanone	41, 55, 84, 69, 95										
		benzoic acid	51, 77, 105, 122										
		3-butenyl 4-hydroxylbutyl adipate	55, 83, 111, 129, 183, 201										
		2-((but-3-enyloxy)carbonyl)benzoic acid	41, 54, 104, 121, 149										
		dimethyl terephthalate*	163, 194, 135, 103, 120, 76, 50										
		dimethyl hexanedioate*	**114**, 143, 101, 74, 83, 59, 41										

(*Continued*)

Table 2.2 *(Continued)*

Water Sample Type	Polymer	Indicator Compound	Indicator Ion (m/z)	Type of Pyrolyser	Pyrolysis T (°C) and (Time)	Split Ratio	GC Column	GC Gradient	MS Mode (m/z Range)	Internal Standard	LOD (µg)	LOQ (µg)	Ref.
	PBS	tetrahydrofuran	71, 42, 53								0.14	—	
		1,4-butanediol	71, 42										
		3-dibutenyl succinate	55, 101, 155										
		3-butenyl-4-hydroxybutyl succinate	55, 73, 101, 155, 173										
		bis(4-hydroxybutyl) succinate	55, 73, 101, 119, 142, 160										
		dibut-3-enyl-butane-1,4-diyl disuccinate	55, 101, 155, 173, 227										
		dimethyl succinate*	**115**, 87, 55, 101										
		butylene glycol methyl ether*	45, 58, 71, 89										
	PHAs	2-butenoic acid (*cis*)	57, 68, 86								0.06	—	
		2-butenoic acid (*trans*)	57, 68, 86										
		1H-indene,1-chloro-2,3-dihydromethyl	43, 63, 89, 115, 144,										
		2-butenoate (*cis*)*	59, 69, 85, 100										
		methyl 2-butenoate (*trans*)*	41, 59, 69, 100, 85										

Water Sample Type	Polymer	Indicator Compound	Indicator Ion (m/z)	Type of Pyrolyser	Pyrolysis T (°C) and (Time)	Split Ratio	GC Column	GC Gradient	MS Mode (m/z Range)	Internal Standard	LOD (µg)	LOQ (µg)	Ref.
Wastewater (influent and effluent) and surface water (river water, seawater)	PS	Styrene	104	Microfurnace	590 (18 s)	30:1	HP-5MS (30 m × 0.25 mm × 0.25 µm)	50°C held for 2 min, 50 to 320°C at 10°C/min, 320°C held for 3 min	Full scan mode (10–550)	No	1.1†		[36]
		Styrene trimer	**91**, 312										
	PMMA	**Methyl methacrylate**	**100**								0.6†		
Wastewater (influent and effluent) and river water	PS	Styrene	51, 78, 104	Microfurnace	590 (18 s)	10:1	HP-5MS (30 m × 0.25 mm × 0.25 µm)	50°C held for 2 min, 50 to 320°C at 10°C/min, 320°C held for 3 min	Full scan mode (10–550)	No	0.08†	—	[35]
		3-butene-1,3-diyldibenzene (styrene dimer)	91, 104, 115, 130, 193, 208										
		5-hexene-1,3,5-triylbenzene (styrene trimer)	**91**, 117, 194, 207, 312										
	PMMA	**Methyl methacrylate**	32, 41, 59, 69, 85, **100**								0.03†	—	
Wastewater (effluent) and seawater	PE	Alkanes (e.g., C_{20})	85	Microfurnace	590 (60 s)	15:1	DB-5MS (30 m × 0.25 mm × 0.25 µm)	35°C held for 2 min, 35 to 310°C at 3°C/min, 310°C held for 30 min	Full scan mode (50–650)	Mixture of TOHA, DOHA, and cholanic acid	0.7	—	[27]
		α-Alkenes (e.g., C_{20})	83										
		α,ω-Alkenes (e.g., $\mathbf{C_{20}}$)	**82**										

(Continued)

Table 2.2 (*Continued*)

Water Sample Type	Polymer	Indicator Compound	Indicator Ion (m/z)	Type of Pyrolyser	Pyrolysis T (°C) and (Time)	Split Ratio	GC Column	GC Gradient	MS Mode (m/z Range)	Internal Standard	LOD (µg)	LOQ (µg)	Ref.
	PP	**2,4-Dimethyl-1-heptene**	126, **70**								0.8	—	
		2,4,6,8-Tetramethyl-1-undecene (isotactic)	100, 69										
		2,4,6,8-Tetramethyl-1-undecene (heterotactic)	100, 69										
		2,4,6,8-Tetramethyl-1-undecene (syndiotactic)	100, 69										
	PS	Styrene	104								0.9	—	
		2,4-Diphenyl-1-butene	91										
		2,4,6-Triphenyl-1-hexene	**91**										
	PVC	**Benzene**	**78**								0.8	—	
		Chlorobenzene	112										
	PMMA	Methylacrylate	55								0.8	—	
		Methyl methacrylate	**100**, 69										
	PA6	**ε-Caprolactam**	**113**								1	—	
		N*-methyl-caprolactam	**127**										
	PET	**Dimethyl terephthalate***	**163**								0.7	—	
	PC	*p*-Methoxy-tert-butylbenzene*	164, 149								0.9	—	
		2,2-Bis(4'-methoxy-phenyl)propane*	256, **241**										
		4,4'-Methylenbis(N-methylaniline)*	226										
	MDI-PUR	*N,N*-Dimethyl-4-(4-methylamino) benzylanilin*	240								0.9	—	
		4,4'-Methylenbis(*N,N*-dimethylaniline)*	253, **254**										

Water Sample Type	Polymer	Indicator Compound	Indicator Ion (m/z)	Type of Pyrolyser	Pyrolysis T (°C) and (Time)	Split Ratio	GC Column	GC Gradient	MS Mode (m/z Range)	Internal Standard	LOD (µg)	LOQ (µg)	Ref.
Surface water (canal, Amstel River, estuary)	PE	**1-hexadecene ($C_{16}H_{32}$)**	**111**, 83, 97	Microfurnace	600 (18 s)	10:1	Agilent DB-5HT (30 m × 0.25 mm × 0.25 µm)	40°C held for 2 min, 40 to 360°C at 20°C/min, 360°C held for 1 min	SIM	Poly(4-fluoro styrene)	0.7065	2.3549	[45]
		1-undecene ($C_{11}H_{22}$)	111, 83, 97										
		1-triacontene ($C_{30}H_{60}$)	83, 97, 111										
	PP	**2,4-dimethyl-1-heptene**	**70**, 83, 126								0.0456	0.1519	
		2,4,6,8-tetramethyl-1-undecene	125, 111, 97										
	PS	**5-hexene-1,3,5-triyltribenzene (styrene trimer)**	**91**, 117, 194								0.3298	1.0992	
		Butene-1,3-diyldibenzene (styrene dimer)	91, 130, 208										
	PVC	**1,2-dihydronaphthalene**	**130**, 129, 115								0.2417	0.8058	
		1-methylnaphthalene	142, 141, 115										
		2-methylnaphthalene	142, 141, 115										
		1-methylindene	130, 129, 115										
	PET	**Benzoic acid**	**105**, 77, 122								0.0402	0.1341	
		Benzophenone	105, 182, 77										

(*Continued*)

Table 2.2 *(Continued)*

Water Sample Type	Polymer	Indicator Compound	Indicator Ion (m/z)	Type of Pyrolyser	Pyrolysis T (°C) and (Time)	Split Ratio	GC Column	GC Gradient	MS Mode (m/z Range)	Internal Standard	LOD (µg)	LOQ (µg)	Ref.
	PMMA	**Methyl methacrylate**	**100**, 69, 41								0.0003	0.0010	
River water and groundwater	PVC	**Benzene**	**78**	Microfurnace	650 (12 s)	50:1	DB5 (30 m × 0.25 mm × 0.25 µm)	40°C held for 2 min, 40 to 320°C at 20°C/min	Full scan mode (10–600)	No	0.001†	0.004†	[38]
	PMMA	**Methyl methacrylate**	**100**								0.001†	0.002†	
	PP	**2,4-dimethyl-1-heptene**	**126**								0.001†	0.002†	
	PS	**5-hexene-1,3,5-triyltribenzene**	**91**								0.002†	0.006†	
	PE	**1-decene**	**83**								0.001†	0.002†	
	PET	**Benzoic acid**	**105**								0.002†	0.002†	
Surface water (lake, river) and seawater	PS	Styrene **Styrene trimer**	104 **91**	Microfurnace	590 (18 s)	30:1	HP-5MS (15 m × 0.25 mm × 0.25 µm)	50°C held for 2 min, 50 to 280°C at 10°C/min, 280°C held for 5 min, 280 to 320°C at 30°C/min	Full scan mode (50–550)	No	0.02†	0.07†	[39]
	PMMA	**Methyl methacrylate**	**100**								0.03†	0.10†	
River water	PS	**Styrene trimer**	**91**	Microfurnace	590 (18 s)	30:1	HP-5MS (30 m × 0.25 mm × 0.25 µm)	50°C held for 2 min, 50 to 320°C at 10°C/min, 320°C held for 3 min	Full scan mode (10–550)	No	MPs = 0.05† NPs = 1.9†	MPs = 0.2† NPs = 6.4†	[40]

Water Sample Type	Polymer	Indicator Compound	Indicator Ion (m/z)	Type of Pyrolyser	Pyrolysis T (°C) and (Time)	Split Ratio	GC Column	GC Gradient	MS Mode (m/z Range)	Internal Standard	LOD (μg)	LOQ (μg)	Ref.
	PMMA	**Methyl methacrylate**	100								MPs = 0.05† NPs = 1.1†	MPs = 0.2† NPs = 3.6†	
River water	PE	α.ω-alkenes (e.g., C_{10})	55, 67,81, 95	Microfurnace	600 (—)	—	Ultra ALLOY-5 (30 m × 0.25 mm × 0.25 μm)	40°C held for 2 min, 40 to 320°C at 20°C/min, 320°C held for 13 min	Full scan mode# (50–500)	No	—	—	[46]
		α-alkenes (e.g., C_{10})	55, 70, 83, 97										
		Alkanes (e.g., C_{10})	55, 71, 85, 98										
	PP	2.4-dimethylhept-1-ene	55, 70, 83, 126								—	—	
		2.4.6.8-tetramethyl-1-undecene (isotactic)	55, 69, 83, 111										
		2.4.6.8-tetramethyl-1-undecene (heterotactic)	55, 69, 83, 111										
		2.4.6.8-tetramethyl-1-undecene (syndiotactic)	69, 111, 125, 154										
	PS	Styrene	104, 78								—	—	
		3-butene-1.3-diyldibenzene	208, 91										
		5-hexene-1.3.5-triyltribenzene	207, 91										
	PVC	Benzene	78								—	—	
		Toluene	91										
		Naphthalene	128										
		Acenaphthene	142										
		Methylnaphthalene	154										

(*Continued*)

Table 2.2 (*Continued*)

Water Sample Type	Polymer	Indicator Compound	Indicator Ion (m/z)	Type of Pyrolyser	Pyrolysis T (°C) and (Time)	Split Ratio	GC Column	GC Gradient	MS Mode (m/z Range)	Internal Standard	LOD (µg)	LOQ (µg)	Ref.
	TDI-PU	2,4-totylene diisocyanate	174, 145, 132, 118, 91								—	—	
		2,6-tolylene diisocyanate	174, 145, 132, 118, 91										
River water	PVC	**Toluene**	**91**	Filament	500 (20 s)	75:1	5% diphenyl dimethyl polysiloxane (Restek) (30 m × 0.25 mm × 0.25 µm)	40 to 300°C at 10°C/min	Full scan mode[#] (50–500)	Chlorobenzene-d5	48.9[†]	—	[54]
		Ethyl benzene	91										
		Styrene	104										
		Indane	**117**										
		Indene	116										
		1,4-Dihydronaphthalene	**130**										
		Naphthalene	**128**										
		1-Methyl-naphthalene	**142**										
		Biphenyl	154										
	PP	**7-Methyl-decene**	**111**, 97								—	—	
		3-Dodecene	**111**, 97										
		2,4-Dimethyl-1-decene	**111**, 97										
		5-Methyl-undecene	**111**, 97										
		7-Methyl-1-undecene	**111**, 97										
		7-Methyl-2-decene	**111**, 97										
		1-Tridecene	**111**, 97										
		2,6,8-Trimethyl-octene	**111**, 97										
		2,4,6-Trimethyl-nonene	**111**, 97										
		2,4,6,8-Tetramethyl-1-decene	**111**, 97										

Water Sample Type	Polymer	Indicator Compound	Indicator Ion (m/z)	Type of Pyrolyser	Pyrolysis T (°C) and (Time)	Split Ratio	GC Column	GC Gradient	MS Mode (m/z Range)	Internal Standard	LOD (μg)	LOQ (μg)	Ref.
	PS	Toluene	91								42.9†	—	
		Styrene	**104**										
		1-Ethyl-2-methyl benzene	**117, 105**										
		1,2,3-Trimethyl-benzene	120, **105**										
		2-Propenyl-benzene	**117, 91**										
		3-Butenyl-benzene	132, **91**										
		3-Phenyl-1-propyne	115										
		1-Methylenepropyl-benzene	132, 117, 91										
		Bibenzyl	182, 104, 91										
		1,1'-(1-Methyl-1,2-ethanediyl)bis-benzene	105, 91										
		Stilbene	**180**, 165										
		1,2-Dihydro-3-phenylnaphthalene	191, 128, 91										
		2,5-Diphenyl-1,5-hexadiene	143, 130, 104, 91										
Seawater	PS	Styrene	104	Microfurnace	590 (18 s)	30:1	HP-1MS (15 m × 0.25 mm × 0.25 μm)	50°C held for 2 min, 50 to 280°C at 10°C/min, 280°C held for 5 min, 280 to 320°C at 30°C/min	SIM	No	MPs = 12 ng/L NPs = 4.6 ng/L	MPs = 40 ng/L NPs = 15 ng/L	[41]
		Styrene trimer	**91**, 312										

(*Continued*)

Table 2.2 *(Continued)*

Water Sample Type	Polymer	Indicator Compound	Indicator Ion (m/z)	Type of Pyrolyser	Pyrolysis T (°C) and (Time)	Split Ratio	GC Column	GC Gradient	MS Mode (m/z Range)	Internal Standard	LOD (µg)	LOQ (µg)	Ref.
Seawater	PE	Alkanes (e.g., C_{20})	282 [M], 85	Microfurnace	590 (60 s)	12.5:1	DB5 (30 m × 0.25 mm × 0.25 µm) + pre-column Trajan P/N 064062 (10 m × 250 µm/363 µm)	35°C held for 2 min, 35 to 310°C at 4°C/min, 310°C held for 60 min	Full scan mode (50–650)	dPS	0.1–0.3	0.1–0.3	[42]
		α-Alkenes (e.g., C_{20})	280 [M], 83										
		α,ω-Alkenes[a] (e.g., C_{20})	278 [M], 95, **82**										
	PP	**2,4-dimethylhept-1-ene**	126 [M], **70**								<0.1	0.5	
		2,4,6,8-tetramethyl-1-undecene (isotactic)	210 [M], 100, 69										
		2,4,6,8-tetramethyl-1-undecene (heterotactic)	210 [M], 100, 69										
		2,4,6,8-tetramethyl-1-undecene (syndiotactic)	210 [M], 100, 69										
	PS	Styrene	104 [M]								0.001	0.002	
		2,4-diphenyl-1-butene	208 [M], 91										
		2,4,6-triphenyl-1-hexene	312 [M], **91**										
	PVC	Benzene	78 [M]								0.010–0.050	0.050	
		Chlorobenzene	112 [M]										
		Naphthalene	**128** [M]										

Water Sample Type	Polymer	Indicator Compound	Indicator Ion (m/z)	Type of Pyrolyser	Pyrolysis T (°C) and (Time)	Split Ratio	GC Column	GC Gradient	MS Mode (m/z Range)	Internal Standard	LOD (µg)	LOQ (µg)	Ref.
	PET	Dimethyl terephthalate*	194 [M], **163**								0.010–0.050	0.050	
	PMMA	Methacrylate	86 [M], 55								<0.1	0.1–0.3	
		Methyl methacrylate	**100** [M], 69										
	PC	*p*-Methoxy-*tert*-butylbenzene*	242 [M], 164, 149								0.001	0.002	
		2,2-Bis(4'-methoxy-phenyl)propane*	256 [M], **241**										
	PA6	**ε-Caprolactam**	**113** [M]								0.1–0.3	0.5	
		N-methyl caprolactam*	**127** [M]										
	MDI-PUR	4,4'-Methylenbis(N-methylaniline)*	226 [M]								0.1–0.3	0.5	
		N-N-Dimethyl-4-(4-methylamino) benzylanilin*	240 [M]										
		4,4'-Methylenbis(N,N-dimethylaniline)*	**254** [M], 253, 210, 134										
	dPS	Styrene (deuterated)	112 [M]								—	—	
		2,4-diphenyl-1-butene (deuterated)	224 [M], 98										
		2,4,6-triphenyl-1-hexene (deuterated)	336 [M], **98**										
Seawater	PE	*n*-alkanes (e.g., C_{16}-C_{26})	85	Microfurnace	590 (60 s)	15:1	DB-5MS (30 m × 0.25 mm × 0.25 µm)	35°C held for 2 min, 35 to 310°C at 3°C/min, 310°C held for 30 min	Full scan mode (50–650)	Mixture of DOHA, TOHA and Cholan	9.2	—	[51]
		n-alkenes (e.g., C_{16}-C_{26})	83										
		***n*-alkadienes** (e.g., $\mathbf{C_{16}}$-$\mathbf{C_{26}}$)	**82**										

(*Continued*)

Table 2.2 *(Continued)*

Water Sample Type	Polymer	Indicator Compound	Indicator Ion (m/z)	Type of Pyrolyser	Pyrolysis T (°C) and (Time)	Split Ratio	GC Column	GC Gradient	MS Mode (m/z Range)	Internal Standard	LOD (µg)	LOQ (µg)	Ref.
	PP	**2,4-dimethylhept-1-ene**	126, **70**								0.8	—	
		2,4,6,8-tetramethyl-1-undecene	100, 69										
	PET	**Dimethyl terephthalate***	**163**								0.9	—	
	PS	Styrene	104								0.9	—	
		2,4-diphenyl-1-butene	91										
		2,4,6-triphenyl-1-hexene	**91**										
	PVC	**Benzene**	**78**								0.7	—	
		Chlorobenzene	112										
		Naphthalene	128										
	PC	*p*-methoxy-*tert*-butylbenzene*	164, 149								0.9	—	
		2,2-bis(4'-methoxy-phenyl)propane (dimethyl bisphenol A)*	256, **241**										
	PMMA	Methyl acrylate	55								0.8	—	
		Methyl methacrylate	**100**, 69										
	PA6	**ε-caprolactam**	**113**								1.0	—	
		N-methylcaprolactam*	**127**										
	PA66	**1,8-diazacyclotetradecane-2,7-dione**	**128, 226**								1.2	—	
	MDI-PUR	4,4'-methylenbis(*N*-methylaniline)*	226								1.2	—	
		N,N-dimethyl-4-(4-methylamino) benzylaniline*	240										
		4,4'-methylenbis(*N,N*-dimethylaniline)*	253, **254**										
Seawater	PE	Alkanes (e.g., C_{20})	85	Cure-point vs Microfurnace	590 (—)	15:1	DB-5 MS (30 m × 0.25 mm × 0.25 µm)	35°C held for 2 min, 35 to 310°C at 3°C/min, 310°C held for 30 min	Full scan mode (50–650)	Mixture of DOHA, antracene-d^{10}, androstane and cholinic acid	0.5	—	[52]
		α-Alkenes (e.g., C_{20})	83										
		α,ω-Alkenes (e.g., C_{20})	**82**										
		***n*-C_{16-26}-alkadienes**	**82**										

Water Sample Type	Polymer	Indicator Compound	Indicator Ion (m/z)	Type of Pyrolyser	Pyrolysis T (°C) and (Time)	Split Ratio	GC Column	GC Gradient	MS Mode (m/z Range)	Internal Standard	LOD (µg)	LOQ (µg)	Ref.
	PP	**2,4-dimethylhept-1-ene**	126, **70**								0.3	—	
		2,4,6,8-tetramethyl-1-undecene (isotactic)	100, 69										
		2,4,6,8-tetramethyl-1-undecene (heterotactic)	100, 69										
		2,4,6,8-tetramethyl-1-undecene (syndiotactic)	100, 69										
	PS	Styrene	104								0.8	—	
		2,4-diphenyl-1-butene	91										
		2,4,6-triphenyl-1-hexene	**91**										
	PVC	**Benzene**	**78**								0.3	—	
		Chlorobenzene	112										
	PMMA	Methylacrylate	55								0.8	—	
		Methyl methacrylate	**100**, 69										
	PA6	**ε-caprolactam**	**113**								0.5	—	
		N-methylcaprolactam*	**127**										
	PET	**Dimethyl terephthalate***	**163**								0.6	—	
	PC	*p*-methoxy-*tert*-butylbenzene*	149, 164								0.9	—	
		2,2-bis(4'-methoxy-phenyl)propane (dimethyl bisphenol A)*	256, **241**										
	MDI-PUR	4,4'-methylenbis(*N*-methylaniline)*	226								1.4	—	
		N,N-dimethyl-4-(4-methylamino) benzylaniline*	240										
		4,4'-methylenbis(*N,N*-dimethylaniline)*	253, **254**										

(*Continued*)

Table 2.2 (*Continued*)

Water Sample Type	Polymer	Indicator Compound	Indicator Ion (m/z)	Type of Pyrolyser	Pyrolysis T (°C) and (Time)	Split Ratio	GC Column	GC Gradient	MS Mode (m/z Range)	Internal Standard	LOD (μg)	LOQ (μg)	Ref.
Seawater	PE	*n*-alkadiene (C_7-C_{38})	55	Microfurnace	700 (60 s)	Meso- and large MPs: 100:1 Small MPs: 10:1 Nanoplastics: 5:1	SLB 5MS (60 m × 0.25 mm × 25 μm)	50°C held for 2 min, 50 to 180°C at 15°C/min, 180 to 310°C at 15°C/min, 310°C held for 44 min	Full scan mode (50–600)	No	—	—	[43]
		n-alkene (C_7-C_{38})	55										
		n-alkane (C_7-C_{38})	55										
	PP	2,4-dimethylhept-1-ene	—								—	—	
		Benzene	78										
		Toluene	91										
		Dimethylbenzene	91										
	PS	Styrene	104								—	—	
		Toluene	91										
		Methylstyrene	117, 118										
		Benzene	78										
	PVC	Benzene	78								—	—	
		Toluene	91										
		Naphthalene	128										
		Styrene	104										
		Indene	116										
	PET	Benzene	78								—	—	
		Biphenyl	154										
		Toluene	91										

Water Sample Type	Polymer	Indicator Compound	Indicator Ion (m/z)	Type of Pyrolyser	Pyrolysis T (°C) and (Time)	Split Ratio	GC Column	GC Gradient	MS Mode (m/z Range)	Internal Standard	LOD (µg)	LOQ (µg)	Ref.
Particles collected in seawater	PE	1-Nonene (C9)	83, 97	Microfurnace	700 (60 s)	Particle <5 µg – 5:1 Particle >5 µg – 20:1	RXi-5 ms® (60 m × 0.25 mm × 25 µm)	40°C held for 2 min, 40 to 261°C at 13°C/min, 261 to 300°C at 6°C/min, 300°C held for 2 min	Full scan mode (33–500)	No	0.070	—	[29]
		1 Decene (C10)	83, 97										
		1-Undecene (C11)	83, 97										
		1-Dodecene (C12)	83, 97										
		1-Tridecene (C13)	83, 97										
		1-Tetradecene (C14)	83, 97										
		1-Pentadecen (C15)	83, 97										
		1-Hexadecene (C16)	83, 97										
	PS	**Styrene**	78, 104								0.003	—	
		3-butene-1,3-diyldibenzene (styrene dimer)	91, 208										
	PMMA	**Methyl methacrylate**	41, 69, 100								0.029	—	
	PP	**2,4-dimethyl-1-heptene**	70								0.027	—	
	PA-6	**ε-caprolactam**	113								0.110	—	
	PC	Phenol	66, 94								0.116	—	
		p-Cresol	77, 107										
		p-Ethylphenol	107, 122										
		p-Vinylphenol	91, 120										
		p-Isopropenylphenol	119, 134										
	PET	Benzene	52, 78								0.015	—	
		Acetophenone	51, 77, 105										
		Vinyl benzoate	52, 77, 105										
		Benzoic acid	77, 105, 122										
		Divinyl terephthalate	104, 175										

(*Continued*)

Table 2.2 (*Continued*)

Water Sample Type	Polymer	Indicator Compound	Indicator Ion (m/z)	Type of Pyrolyser	Pyrolysis T (°C) and (Time)	Split Ratio	GC Column	GC Gradient	MS Mode (m/z Range)	Internal Standard	LOD (µg)	LOQ (µg)	Ref.
	uPVC	Benzene	52, 78								0.592	—	
		Toluene	91										
		Styrene	78, 104										
		Indene	116										
		Naphthalene	128										
		2-methylnaphthalene	115, 142										
		1-methylnaphthalene	115, 142										

Note: The indicator compound and its respective indicator ion used for quantification are indicated in bold.

†Data expressed in µg/L.

— Data not indicated.

#Analysed using a QToF mass spectrometer detector.

*Only after TMAH treatment (thermochemolysis); aMean of n-C16-C26-alkadienes used for quantification; HDPE: high density polyethylene; PP: polypropylene; PS: polystyrene; PET: polyethylene terephthalate; PA6: polyamide 6; PMMA: polymethylmethacrylate; PVC: polyvinylchloride; PC: polycarbonate; PLA: polylactic acid; PCL: polycaprolactone; PBAT: polybutylene adipate terephthalate; PBS: polybutylene succinate; PHAs: polyhydroxyalkanoate; MDI-PUR: methylene diphenyl diisocyanate-polyurethane; TDI-PU: polyurethane; PA 66: polyamide 66; SIM: selected ion monitoring; dPS: deuterated polystyrene; DOHA: 9-Dodecyl-1,2,3,4,5,6,7,8-octahydro anthracene; TOHA: 9-Tetradecyl-1,2.3,4,5,6,7,8-Octahydro anthracene; Cholan: cholanic acid/cholanic acid methyl esters.

The analysis of MPs in environmental water by Py-GC/MS can be done following different working modes. The most common is the single-shot mode (or flash pyrolysis) [27, 34–41, 44, 46, 51], which consists of performing the pyrolysis at a single temperature, usually higher than 500°C [56]. Various pyrolysis temperatures are reported in the literature for the analysis of MPs in environmental water (Table 2.2), namely 400°C [7], 500°C [54], 560°C [48], 590°C [27, 32, 35, 36, 39–42, 51, 52], 600°C [34, 44–46], 650°C [33, 38], 700°C [29, 37, 43, 50, 53], and 750°C [49]. The double-shot mode is another working mode that can be used [33, 45]. It involves the analysis of the sample employing two temperature programs: In the first one, a low-temperature ramp is applied to the sample and the volatile compounds (e.g., plastic additives) are analysed through a thermal desorption step; and in the second one, pyrolysis of the non-volatile macromolecules (e.g., polymers) occurs at a single temperature, according to single-shot parameters [56, 57]. The first temperature program (thermal desorption step) can also be used as a clean-up step as described by Okoffo et al. [33]. These authors used a ramp from 100 to 300°C to thermally desorb/remove potentially interfering volatile and semi-volatile organic compounds co-extracted with MPs from wastewater and water samples [33]. To improve the analysis of certain polymers, thermochemolysis can be used as well [27, 32–34, 42, 43, 51, 52]. It consists of adding a derivatizing agent to the sample, usually tetramethyl ammonium hydroxide (TMAH), to chemically derivatize the polymer in the pyrolysis chamber by esterification, transesterification, and methylation reactions [56, 57]. This working mode allows to improve the sensitivity of the analysis by the Py-GC/MS of various plastic polymers, such as PET, polycarbonate (PC), polyurethane (PUR), and bioplastics (e.g., polylactic acid (PLA), polycaprolactone (PCL), polybutylene succinate (PBS), polybutylene adipate terephthalate (PBAT), and polyhydroxyalkanoate (PHA)) (Table 2.2).

There are different types of pyrolysis systems, which differ in terms of temperature generation and mode of operation, heat transfer, and sample amount introduced in the pyrolyser. Depending on the heating mechanism, pyrolysis systems can be classified into pulse-mode pyrolysers, where the sample is introduced cold and receives a thermic pulse (e.g., filament and Curie-point pyrolysers), and continuous-mode pyrolysers (e.g., microfurnace pyrolyser) [56,

57]. For the analysis of MPs in environmental water, different types of pyrolysers have been used, namely filament pyrolysers [37, 44, 48, 49, 54], Curie-point pyrolysers [32], and microfurnace pyrolysers [27, 29, 33–36, 38–41, 43, 45, 46, 51, 52] (Table 2.2):

- Filament pyrolysers have a coil made of a very thin wire of resistive metal (commonly platinum). The samples are introduced in open or semiclosed quartz tubes that are placed in the heated platinum coil and are heated isothermally or with a temperature program [21, 56].
- In the Curie-point pyrolysers, the samples are introduced on a ferromagnetic alloy wire or in semiclosed ferromagnetic targets having the desired Curie-point, which is the exact temperature at which the ferromagnetic material loses its magnetism. The alloy defines the exact pyrolysis temperature, which is almost automatically reached when placed in a high-frequency coil chamber. Using this type of pyrolyser, pyrolysis is only performed in the isothermal mode [21, 56].
- Microfurnace pyrolysers are the most common type of pyrolysers used for the analysis of MPs in environmental water (Table 2.2). They are vertical microfurnaces that have a hollow quartz tube covered by a heater device for rapid heating and cooling. Samples are transferred to inert deactivated cups, usually stainless steel cups, which are then dropped by free fall due to gravity into the centre of the vertical furnace. Using this type of pyrolyser, the samples can be isothermally pyrolysed (at a given temperature) (single-shot mode) or heated with a temperature programme (double-shot mode) [21, 56]. The injection volume is defined by the capacity of the pyrolysis cup, which is 80 μL, restricting the amount of MPs that can be analysed in an analytical run [41].

Mass spectrometry (MS) is used for the detection of the degradation compounds separated by gas chromatography. The most used MS system to obtain the molecular profile of the generated degradation products is the single quadrupole mass analyser. However, recently high-resolution mass spectrometry (HRMS) has also been coupled with GC, and there are some instrumental methods using time of flight (TOF) or quadrupole-time of flight (Q-ToF) detectors for the analysis of MPs in environmental water reported in the literature

[46, 50, 53, 54, 58]. Generally, mass spectrometry detectors are operated in full-scan mode, covering a range of m/z between 10 and 650 (Table 2.2). The generated degradation products detected by MS are then identified using commercial mass spectra libraries (e.g., F-Search, NIST) [38, 43, 46, 48], in-house libraries created based on the molecular profile of plastic standards [27, 34, 43, 48–50, 53, 58] or literature data [46, 49]. Nevertheless, studies using single ion monitoring (SIM) mode were recently reported in the literature [41, 45]. SIM mode has a higher sensitivity and specificity, allowing lower limits of detection, but there is a lack of information since only specific indicator compounds and their respective ions are acquired by the MS detector, not allowing a future retrospective analysis of the acquired results.

Recently, analytical methods combining μFTIR or μRaman and Py-GC/MS as complementary tools for the analysis of MPs have appeared in the literature, enabling the physical characterization of MPs (e.g., size, colour, shape), the identification of the polymer type, and the quantification of the number and mass of MPs in environmental water [32, 37, 45].

Finally, it should be noted that besides Py-GC/MS, TED-GC/MS is another thermoanalytical technique that has been applied for the analysis of MPs in environmental water [9, 47, 59].

2.4.1 Polymer Identification

The degradation products formed from the pyrolysis of plastic polymers at a defined temperature and under an inert atmosphere are characteristics of a certain polymer. Therefore, these pyrolyzates are used to identify the polymer type of unknown MPs. For that, specific indicator compounds and their respective indicator ions are chosen for each polymer, which normally are mono- or oligomeric components of the polymer. It should be noted that the specific marker compounds for a reliable identification of a polymer are in general minor pyrolytic products. Furthermore, the selection of the appropriate indicator compounds is a crucial step to get reliable results, and it should take into account that some pyrolyzates can be shared by different plastic polymers and/or can exist in natural polymers, such as humic substances, lignin, chitin, cellulose, and wood [6, 55, 57]. A typical example of the critical choice of indicator compounds is the case of PS. When PS undergoes thermal degradation, it produces styrene at a higher yield. Nevertheless, styrene is also a pyrolytic product of other polymers (e.g., acrylonitrile-butadiene-styrene (ABS) and styrene butadiene rubbers) as well as of natural polymers, such as humic substances, vegetables, degraded lignin from wood and paper [6, 37]. Therefore, the identification of PS by Py-GC/MS is based on the simultaneous identification of three compounds, namely styrene, styrene dimer (3-butene-1,3-diyldibenzene), and styrene trimer (5-hexene-1,3,5-triyltribenzene). Although these last two compounds are less intense, they are specific indicator compounds of PS [30]. Styrene trimer is often selected for PS quantification (Table 2.2). Pyrolyzates of natural polymers, such as chitin, wool, and wood, can also cause some interference in the analysis of PVC, while the remaining natural, long aliphatic chains, such as fats or waxes, may interfere in the analysis of PE [37]. Due to interferences in the analysis of PE, Xu et al. [38] proposed the validation of its identification based on three criteria: (a) the presence of a homologous series of the characteristic PE triplets (alkadiene, *n*-alkene, and *n*-alkane); (b) the presence of a homologous series of more than five triplets within C_7-C_{41} of the PE standard; and (c) the standard deviation of the peak areas of the individual C_{10} triplets within two times the standard deviation of the PE standard.

Indicator compounds can be selected from the molecular profile of a polymer obtained by direct pyrolysis or by derivatization with TMAH (thermochemolysis), taking into account the most abundant and/or polymer-specific compounds. Generally, thermochemolysis gives less pyrolysis reaction products, but is more specific, mainly consisting of methyl derivatives of the constituents of the polymer chain. It also improves the sensitivity and the peak shape of the pyrolytic products [30, 34]. Table 2.2 shows an overview of the most used indicator compounds for the analysis of several MPs by Py-GC/MS reported in the literature. In short, PE has a very characteristic pyrogram, showing multiple triplets composed of *n*-alkadiene followed by *n*-alkene and *n*-alkane, usually in the indicated order, and having an increasing number of carbons in the molecule backbone throughout the pyrogram. These triplets are very specific to PE. Although the alkenes show the highest response, the dialkenes are more specific to PE and are recommended as indicator compounds together with *n*-alkenes and *n*-alkanes above C_{16}, given that molecules with a higher number of carbons are less prone to organic matrix interferences [25, 60]. PP typically decomposes into six compounds with relatively high abundance, namely 2,4-dimethylhept-1-ene (the typical marker compound of PP), 2,4,6-trimethylnon-1-ene (two diastereoisomers), and 2,4,6,8-tetramethylundec-1-ene (four diastereoisomers) [25, 60]. As mentioned earlier, the pyrolysis of PS forms styrene and its oligomers (styrene dimer and styrene trimer), which are preferred as indicator compounds due to their specificity [25, 30]. After pyrolysis, PET shows multiple signals of benzoate and terephthalate derivatives and oligomers. These compounds are polar and produce broad peaks with tailing. Thermochemolysis helps to improve the analysis of PET by forming dimethyl terephthalate, a more specific compound with a higher response [25]. PVC decomposes in several small aromatic molecules such as acenaphthene, anthracene, benzene, naphthalene, toluene, and xylenes; however, none of them are specific to this polymer. Benzene and naphthalene have been proposed as indicator compounds of PVC, though they are not specific to this polymer [25, 60]. This can cause limitations in the quantification of PVC as shown by Goβmann et al. [42]. Naphthalene is normally selected for the quantification of PVC, but it is suspected to be a pyrolytic product of different polymeric soot residues, which are ubiquitously present in the environment. Thus, an overestimation of the quantification of PVC in environmental water samples might occur when naphthalene is selected as an indicator compound [42].

The identification of plastic polymers is performed based on two parameters: (a) retention time and (b) spectral matching, by comparing the MS spectrum of the characteristic indicator compounds of the sample with the MS spectrum of reference from commercial libraries (e.g., F-Search, NIST) or in-house libraries created based on the molecular profile of individual reference plastic standards analysed either by direct pyrolysis or by thermochemolysis adding TMAH. A positive identification of MPs is based on a minimum of 80% similarity between both MS spectra [7, 29–31, 55]. To minimize false positive results, Primpke et al. [27] suggested using a third identification parameter, the retention index (RI). The authors checked the RI of the identified indicator compounds against the respective RI from the library. A RI window was set based on the RI of the respective reference target compound ± 40, and a positive identification was considered if the RI of the identified indicator compound of the sample felt in the established RI window.

Finally, it should be highlighted that the Py-GC/MS of environmental samples consists of a measurement of the whole sample. The generated pyrograms sum up all the formed pyrolytic products independently of their original precursor polymer. Thus, any potential interference of natural organic materials should be excluded by an adequate sample preparation to get more accurate results [27].

2.4.2 Polymer Quantification

Although the identification of MPs in a sample relies on identifying multiple indicator compounds, the quantification is based on a single selected characteristic indicator compound for each polymer, which is also used to build the calibration curve. Therefore, sample preparation is highly recommended in the analysis of environmental samples to avoid matrix effects that could hinder an accurate estimation of MPs concentration.

The quantification of MPs can be performed by either external calibration or internal calibration, which is related to an internal standard that is added before pyrolysis. Calibration curves are built by plotting the mass of a standard polymer versus its instrumental signal. For that, an indicator ion of the selected characteristic indicator compound of the polymer is extracted from the pyrogram, and the peak area of the selected characteristic indicator compound obtained in the extracted ion chromatogram is used to build the calibration curve. Generally, $r^2 > 0.9$ is considered an acceptable linearity [6, 57]. The lowest point of the calibration curve is restricted by the minimum amount of standard polymer that can be weighted by the analytical balance. The use of an inert diluent (e.g., calcium carbonate ($CaCO_3$), kaolin, deactivated silica) [31, 32, 61] is indicated to facilitate weighting the standard polymers, lowering the limits of detection, and avoiding overloading of the GC column. The preparation of the different concentrations of the calibration curve by serial dilutions of a stock solution prepared in a mixture of dichloromethane and methyl alcohol is also proposed [38]. Dibke et al. [51] compared external and internal calibration curves to determine which had the best linear fit for the quantification of MPs. For each polymer, linear regressions were performed for standard polymer mass and respective peak area (external calibration curves) and standard polymer mass and ratios of standard polymer peak area to internal standard peak area (internal calibration curves). The calibration curves that showed the highest average r^2 were selected for the quantification of each polymer [51]. The quantification of MPs in environmental water samples is done using the same selected characteristic indicator compound of each polymer that is used to build the calibration curve.

The use of internal standards is a common and preferable practice concerning calibration since they can help overcome some of the challenges associated with calibrant/sample preparation and instrument variability [57]. Different compounds have been used as internal standards in the analysis of MPs by Py-GC/MS (Table 2.2), including poly(4-fluoro styrene) [45], a mixture of 9-dodecyl-1,2,3,4,5,6,7,8-octahydro anthracene (DOHA), 9-tetradecyl-1,2,3,4,5,6,7,8-octahydro anthracene (TOHA), and cholanic acid/cholanic acid methyl esters [27, 51], a mixture of 9-dodecyl-1,2,3,4,5,6,7,8-octahydro anthracene, anthracene-d_{10}, androstane, and cholanic acid [52], deuterated compounds, such as anthracene-d_{10} [30, 32], chlorobenzene-d_5 [54], styrene-d_8 [44], and even deuterated polymers such as deuterated PS (e.g., d_5-PS) [33, 34, 42].

For a reliable quantification of MPs in real samples, the calibration curve and samples should be prepared in the same experimental conditions, so the sample preparation protocol applied to the water samples should also be employed for the preparation of the calibration curves. For instance, Okoffo et al. [33] used PLE-extracted plastic polymer standards to prepare the calibration curves, while Ishimura et al. [31] recommended the use of the same diluent (e.g., $CaCO_3$) in the calibration curve and the samples for the quantification of MPs. Fischer and Scholz-Böttcher [52] introduced a glass fibre membrane loaded with MPs directly in the pyrolysis cup for MPs analysis and the same type of filter was used for calibration and blanks since it does not alter pyrolysis and offers a constant pyrolysis surface.

2.5 Quality Assurance and Quality Control (QA/QC)

Quality assurance and quality control (QA/QC) are fundamental for MPs analysis. QA/QC embraces measures that cover all steps of the analytical procedure, i.e. sampling, sample preparation, and Py-GC/MS analysis. Each step has its own challenges for QA/QC, since MPs are everywhere in the environment, and contamination is a critical issue in MPs analysis, which can interfere with both qualitative and quantitative determination.

During the sampling step, samples may be contaminated by the atmosphere, the containers, and/or instruments used for sample collection and other factors inherent to the sampling location. To minimize potential contamination of the samples during the sampling, Okoffo et al. [33] washed/rinsed the sampling devices with Milli-Q water filtered through 0.7 μm glass fibre filters and used them to collect the samples at the next site or between samples to avoid cross-contamination. Plastic items (e.g., bottles, packaging, containers, *etc.*) should be removed from the vicinity of the sampling site, and plastic equipment should be avoided as much as possible to prevent contamination from degradation [33]. Stainless steel filters used in the sampling step should be pre-rinsed with filtered water and heated at 525°C in a muffle oven before use [45]. To assess the potential presence of MPs in the air around the sampling area, Hendrickson et al. [47] placed two clean, empty and dry Petri dishes on the boat while collecting the samples. After sampling, the Petri dishes were rinsed three times with ultrapure water and the rinses were collected for further analysis. After sample collection, the collected water samples should be immediately sealed to avoid contamination from the sources and during transport to the laboratory [39, 41]. Field blanks are performed during the sampling campaign in the same sampling conditions by replacing the water sample by prefiltered water [7, 37, 43, 51]. Ideally, only natural fibre clothing (e.g., 100% cotton or wool) and cotton lab coats should be worn during the sampling to avoid contamination with synthetic fibres [9, 33, 39, 45, 47].

It is desirable to use metal or glass laboratory ware during the entire process of sample preparation and Py-GC/MS analysis to prevent possible contamination by MPs. Before use, all the glassware material and equipment should be rinsed with filtered water [35, 38, 43, 46, 50], filtered water and ethanol (96% or 99%) [32, 42], acetone and dichloromethane [33], or only dichloromethane [34]. Alternatively, it is also suggested that glass material should be pre-heated at 525°C in a muffle oven [45]. Minor et al. [9] suggested cleaning all non-metal materials with soap and water, followed by acid leaching in 10% HCl, rinsed thoroughly with deionized water before use, and finally the glass material was combusted for 4 h at 450°C. All materials must be covered with aluminium foil until use [33, 46, 50]. Also, the working areas should be wiped down with 70% ethanol [33, 34], prefiltered water and 70% ethanol [45], or only prefiltered water [46] prior to starting the experiments. Ideally, all steps of sample preparation should be done in a clean room. However, not all laboratories have this facility, so sample pretreatment must be performed inside a laminar flow cabinet or, at least, in a fume hood to prevent indoor airborne MPs contamination [32–34, 37, 43, 45, 46]. Even so, during the sample preparation, samples present in the glass material (e.g., beakers), filtration systems, etc. should be always covered with aluminium foil or a glass plate when not being processed [32–35, 42, 43, 45, 46], and principally when they are moved outside the laminar flow cabinet [32]. Okoffo et al. [34] recommended the use of aluminium foil precleaned with dichloromethane. All solutions and reagents used in the sample preparation should be freshly prepared and filtered before use through metal filters (pore size 20 or 5 µm) [32, 37], glass fibre filters (pore size 0.7 µm or 1.2 µm) [37, 45], or 0.3 µm filters [42]. All the glass fibre filters used in the sample preparation, including those used to filter the solvents, must be previously heated in a muffle oven at 400°C for 4 h [51], or 500°C [45], while in the case of stainless steel filters, they should be precleaned with filtered water and heated in a muffle oven at 525°C before use to remove any contamination from the material [45]. When a drying process is needed, samples should be covered with aluminium foil to prevent airborne contamination [41]. Procedural blanks, consisting of performing the entire sample preparation protocol without the matrix using the same material, reagents and experimental conditions applied to real samples, are needed and are treated in parallel to real samples. Normally, a procedural blank is included in each batch of processed samples (usually consisting of 5–6 samples, or even less (e.g., 1–3 samples)), to monitor processing and extraction contamination, allowing the estimation of secondary contamination [9, 28, 33, 34, 38–41, 43, 45, 51]. Recovery experiments should be included to assess the quality of the applied sample preparation protocol. The use of natural fibre clothing (e.g., cotton lab coats) and nitrile gloves by the operators during all steps of the sample processing and analysis is recommended to minimize contamination [7, 9, 32, 34, 35, 37, 38, 42, 43, 45–47]. Some authors mention that synthetic fibres are even prohibited in the laboratory [42] and that only one person is present in the laboratory during the sample processing [46].

Airborne contamination is a critical aspect of MPs analysis during all the steps of the analytical procedure: sampling, sample preparation, and instrumental analysis. Different strategies are reported in the literature to evaluate the possible risks of airborne contamination of the samples during sample preparation, the so-called laboratory blanks. For instance, open beakers filled with 250 mL of water filtered through 0.7 μm glass fibre filters are placed in the laminar flow cabinet or in the fume hood as well as in the laboratory bench where samples are processed, and then they are treated and extracted in the same batch of the real samples [33]. Zhou et al. [35] applied the same strategy, but using vials containing 40 mL of filtered water, which were placed on the laboratory bench where experiments were carried on. These were then treated and extracted using the same procedure as real samples. Another strategy is to place glass fibre filters (usually with a pore size of 1.2 μm or 2.7 μm) in open Petri dishes next to the working space, which are then treated and analysed following the same procedure of real samples [34, 37, 50]. Exposing clean filter papers to the air [7] or leaving empty Petri dishes open to the lab environment [9] during sample preparation is also suggested to evaluate potential airborne contamination. Alternatively, Li et al. [39] exposed 300 nm glass fibre membranes to the laboratory air for 1 h and then analysed them by Py-GC/MS, while Li et al. [40] exposed an empty pyrolysis cup to air for 30 min (the expected time to process a sample), covered it with aluminium foil, and heated at 190°C for 3 h and then analysed by Py-GC/MS to evaluate the impact of the air deposition of MPs in the laboratory.

Prior analysis, Py-GC/MS should be checked by doing an instrumental blank (a blank run without a pyrolysis cup) [33, 34] to guarantee that there is no plastic contamination of the system. To avoid any possible contamination, some authors recommend the use of pyrolysis cups new and previously washed with dichloromethane [33, 34], while others suggest burning the pyrolysis cup or the pyrolysis filament and glass cell with a torch before use [44, 45] to remove any external contamination. Zhou et al. [35] burned the pyrolysis cups for more than 3 min in an alcohol lamp and then ran them on Py-GC/MS before adding the sample. Instrumental blanks should be analysed in each batch of samples, normally after every 7 or 10 samples, to demonstrate the lack of cross-contamination, carryover of target plastic polymers, secondary contamination, background response, or potential instrumental contamination [33–35]. Ter Halle et al. [43] went further and ran an instrumental blank between each sample to avoid cross-contamination. Additionally, the injection of a midpoint calibration, added to an internal standard if applicable to the analysis procedure, at regular intervals (for instance, every 10 samples) throughout the analytical sequence was proposed to check any instrumental shift in the sensitivity of accuracy and performance of the instrumental method [34, 35].

When considered appropriate, all data should be blank corrected before further data processing by subtracting the average blank value from the data of real samples [33, 45, 51].

Finally, a detailed report of QA/QC measures should be included in the studies to better understand the possible pitfalls and limitations of the analytical procedure used for the analysis of MPs, contributing to improving the quality of the entire analytical procedure and of the generated data [21].

2.6 Occurrence of Microplastics in Environmental Waters

Understanding the occurrence and distribution of MPs in environmental waters is of great importance to assess the potential risks of MPs to the environment and human health. Nevertheless, just a few studies reported the mass concentration of MPs, especially those of small range size (<10–20 μm), and concerning nanoplastics, data is almost unknown, mainly due to analytical limitations in sampling, sample preparation, and analysis. In this context, Py-GC/MS appeared as a powerful tool to overcome this gap of knowledge, especially in what concerns nanoplastics analysis, since it overcomes the size limitations of spectroscopic techniques, such as FTIR and Raman, given that mass concentrations reported by Py-GC/MS are not dependent on particle size, making these measurements more robust [33]. However, sample preparation protocols for the concentration of nanoplastics are still sparse, as it was described in Section 2.3.1, hindering the knowledge of the occurrence and distribution of nanoplastics in the environment. The lack of standardized protocols for the sampling, processing, and analysis of MPs and nanoplastics in the environment also hampers the comparison of published data.

The analysis of MPs by Py-GC/MS allowed their detection and quantification in different environmental water samples all over the world, namely wastewater, freshwater (e.g., rivers, canals, estuaries, lakes), seawater, and even groundwater. Table 2.1 summarizes some examples of studies reporting the presence of MPs in environmental water using Py-GC/MS. Usually, PE, PVC, PET, PP, and PS are among the most common polymers identified in wastewater, freshwater, and seawater (Table 2.1). The presence of MPs of bioplastics, such as polylactic acid (PLA) and polybutylene adipate terephthalate (PBAT), has also been reported in wastewater [34]. According to Table 2.1, PE showed the highest mass concentration in wastewater and freshwater (Kelantan River basin) (up to 2340 μg/L [33] and 31,370 μg/L [7], respectively), while PMMA had the highest reported concentration in seawater (9.21 μg/L at Askeröfjorden [42]). These variations in the predominance of polymer type in different environmental waters are also seen within a type of water sample when different range sizes of MPs are analysed. For instance, Ter Halle et al. [43] analysed different size fractions of MPs, namely mesoplastics (5 mm to 20 cm), large MPs (1–5 mm), small MPs (20–999 μm), and nanoplastics (1–999 nm), in the North Atlantic subtropical gyre, showing that PE was the most abundant polymer in the mesoplastics, large and small MP fractions, while in the case of nanoplastics, there was a predominance of PVC. In fact, PE was the less abundant polymer identified in the nanoplastics fraction, only accounting for 4% of the total mass of detected MPs. Thus, it can be concluded that data comparison on MPs analysis is a hard and challenging task and should be done with caution, due to the broad range of methods used for sampling and sample processing, which limits the particle size of the MPs reported [33, 37]. Studies that go lower in the size range fraction (for instance, MPs >1 μm) often report higher concentrations than those that focus on MPs >10–20 μm or even >300–333 μm (depending on the sampling and sample preparation approaches used) [33], which hamper data comparison. For instance, Campanale et al. [46] stated that when 300 μm mesh size nets are used for sampling, the MPs concentration can be underestimated, since only particles >300 μm are captured, neglecting the smaller fraction of MPs. Therefore, harmonization of criteria for sampling, sample pretreatment, and analysis as well as the reporting of MPs concentration is needed to facilitate data comparison on MPs analysis in environmental water.

2.7 Conclusions and Future Perspectives

Microplastics are ubiquitously distributed in the environment, representing a critical problem of environmental pollution, given that their concentration is globally increasing, threatening ecosystems and human health. Thus, fast and reliable analytical methods are required for the identification and quantification of MPs in various environmental compartments. This chapter highlights the scientific efforts made in the sampling, sample preparation, and instrumental methods for the analysis of MPs in environmental water by Py-GC/MS. Although spectroscopic methods have been the most widely used in the analysis of MPs, in recent years Py-GC/MS has gained increasing attention due to its potential to identify and quantify, in a mass base, MPs in environmental samples, including the smallest MPs and nanoplastics, overcoming the size limitation inherent to spectroscopic methods. Besides plastic polymers, Py-GC/MS can also analyse plastic additives in a single analytical run.

Sampling is a critical step to get a representative water sample and should be carefully planned. Although different approaches can be employed for the sampling of MPs in water, bulk sampling and volume-reduced approaches are most commonly used in the analysis of MPs in environmental water by Py-GC/MS. It was highlighted that the selection of the sampling method, and especially the choice of the mesh size nets or sieves used for sampling, will limit the size of the analysed MPs.

The extraction of MPs from environmental water samples is a crucial aspect of getting reliable and accurate data by Py-GC/MS, since sometimes under- or overestimation of MPs concentration may occur. Herein, the most used analytical techniques for the extraction of MPs from water samples were described, giving examples of their different applications. Normally, sample preparation implies the sequential application of various techniques to extract and concentrate MPs and eliminate interfering organic and inorganic matrix components as much as possible. As Py-GC/MS recently appeared as a powerful and promising tool for the analysis of nanoplastics, examples of sample preparation approaches for the extraction of nanoplastics from environmental water samples were also described. Sample preparation for either the smallest MPs and nanoplastics is still a very challenging analytical field, and further research is needed for their optimization and to explore the high potential of Py-GC/MS to identify and quantify plastic particles of small range size, i.e., smallest MPs (1–20 μm) and nanoplastics (<1 μm), in environmental water samples.

A critical aspect of Py-GC/MS analysis of MPs is the selection of characteristic indicator compounds for each polymer. Some pyrolyzates can be shared by different plastic polymers and/or can exist in natural polymers, such as humic substances, lignin, chitin, cellulose, and wood. Thus, choosing the most polymer-specific marker compounds is essential for reliable identification of MPs. Adding a derivatizing agent, such as TMAH, improves the Py-GC/MS analysis of various plastic polymers such as PET and PC. Furthermore, new polymeric products have been developed, such as bioplastics, and research on their pyrolytic behaviour should take place to allow a more complete understanding of environmental contamination by MPs.

The analysis of MPs obligated the redefinition of QA/QC protocols to ensure an accurate identification and quantification of plastic polymers. Several QA/QC measures that have been applied were reviewed in this chapter. These cover the entire analytical procedure (sampling, sample preparation, and Py-GC/MS analysis) since MPs are everywhere in the environment and best practices are mandatory to limit plastic contamination in the field and laboratory.

The hazards posed by MPs can be dependent on the concentration, size, and even polymer type, so these data are of high interest. Nowadays, any of the analytical methods available for the analysis of MPs can provide all these data, thus combining Py-GC/MS with spectroscopical methods, such as μFTIR and μRaman, as complementary tools would be a good solution to overcome analytical limitations and get a more complete overview of the problem of MPs contamination by providing data on size, mass-based concentration, and polymer type.

Finally, it should be highlighted that future research should move towards standardization or, at least, harmonization of methodologies for the sampling, processing, and analysis of MPs in environmental water to facilitate the data comparison on MPs distribution and environmental fate, allowing a better assessment of their potential risks to the environment and human health.

Acknowledgements

The author acknowledges the support of the Economy and Knowledge Department of the Catalan Government through a Consolidated Research Group (ICRA-ENV-2021 SGR 01282 and ICRA-TECH-2021 SGR 01283).

References

1. Plastics Europe (2024). Plastics: The *fast* Facts 2023. [Available from: https://plasticseurope.org/knowledge-hub/plastics-the-fast-facts-2023/].
2. Briassoulis, D. (2023). Agricultural plastics as a potential threat to food security, health, and environment through soil pollution by microplastics: Problem definition, *Sci. Total Environ.*, **892**, pp. 164533.
3. Cusworth, S.J., Davies, W.J., McAinsh, M.R., Gregory, A.S., Storkey, J., and Stevens, C.J. (2024). Agricultural fertilisers contribute substantially to microplastic concentrations in UK soils, *Commun. Earth Environ.*, **5**, pp. 7.
4. Hartmann, N.B., Hüffer, T., Thompson, R.C., Hassellöv, M., Verschoor, A., Daugaard, A.E., Rist, S., Karlsson, T., Brennholt, N., Cole, M., Herrling, M.P., Hess, M.C., Ivleva, N.P., Lusher, A.L., and Wagner, M. (2019). Are we

speaking the same language? Recommendations for a definition and categorization framework for plastic debris, *Environ. Sci. Technol.*, **53**, pp. 1039–1047.

5. Li, C., Busquets, R., and Campos, L.C. (2020). Assessment of microplastics in freshwater systems: A review, *Sci. Total Environ.*, **707**, pp. 135578.
6. Ainali, N.M., Kalaronis, D., Kontogiannis, A., Evgenidou, E., Kyzas, G.Z., Yang, X., Bikiaris, D.N., and Lambropoulou, D.A. (2021). Microplastics in the environment: Sampling, pretreatment, analysis and occurrence based on current and newly-exploited chromatographic approaches, *Sci. Total Environ.*, **794**, pp. 148725.
7. Anuar, S.T., Abdullah, N.S., Yahya, N.K.E.M., Chin, T.T., Yusof, K.M.K.K., Mohamad, Y., Azmi, A.A., Jaafar, M., Mohamad, N., Khalik, W.M.A.W.M., and Ibrahim, Y.S. (2023). A multidimensional approach for microplastics monitoring in two major tropical river basins, Malaysia, *Environ. Res.*, **227**, pp. 115717.
8. Di, M., and Wang, J. (2018). Microplastics in surface waters and sediments of the Three Gorges Reservoir, China, *Sci. Total Environ.*, **616–617**, pp. 1620–1627.
9. Minor, E.C., Lin, R., Burrows, A., Cooney, E.M., Grosshuesch, S., and Lafrancois, B. (2020). An analysis of microlitter and microplastics from Lake Superior beach sand and surface-water, *Sci. Total Environ.*, **744**, pp. 140824.
10. Lorenz, C., Roscher, L., Meyer, M.S., Hildebrandt, L., Prume, J., Löder, M.G.J., Primpke, S., and Gerdts, G. (2019). Spatial distribution of microplastics in sediments and surface waters of the southern North Sea, *Environ. Pollut.*, **252**, pp. 1719–1729.
11. Leistenschneider, C., Wu, F., Primpke, S., Gerdts, G., and Burkhardt-Holm, P. (2024). Unveiling high concentrations of small microplastics (11–500 μm) in surface water samples from the southern Weddell Sea off Antarctica, *Sci. Total Environ.*, **927**, pp. 172124.
12. Dierkes, G., Lauschke, T., Becher, S., Schumacher, H., Földi, C., and Ternes, T. (2019). Quantification of microplastics in environmental samples via pressurized liquid extraction and pyrolysis-gas chromatography, *Anal. Bioanal. Chem.*, **411**, pp. 6959–6968.
13. Scherer, C., Weber, A., Stock, F., Vurusic, S., Egerci, H., Kochleus, C., Arendt, N., Foeldi, C., Dierkes, G., Wagner, M., Brennholt, N., and Reifferscheid, G. (2020). Comparative assessment of microplastics in water and sediment of a large European river, *Sci. Total Environ.*, **738**, pp. 139866.

14. Rolf, M., Laermanns, H., Horn, J., Kienzler, L., Pohl, C., Dierkes, G., Kernchen, S., Laforsch, C., Löder, M.G.J., and Bogner, C. (2024). Multi-method analysis of microplastic distribution by flood frequency and local topography in Rhine floodplains, *Sci. Total Environ.*, **927**, pp. 171927.

15. Cusworth, S.J., Davies, W.J., McAinsh, M.R., and Stevens, C.J. (2023). A nationwide assessment of microplastic abundance in agricultural soils: The influence of plastic crop covers within the United Kingdom, *Plants, People, Planet*, 1–11.

16. Fischer, M., and Scholz-Böttcher, B.M. (2017). Simultaneous trace identification and quantification of common types of microplastics in environmental samples by pyrolysis-gas chromatography–mass spectrometry, *Environ. Sci. Technol.*, **51**, pp. 5052–5060.

17. Ribeiro, F., Okoffo, E.D., O'Brien, J.W., Fraissinet-Tachet, S., O'Brien, S., Gallen, M., Samanipour, S., Kaserzon, S., Mueller, J.F., Galloway, T., and Thomas, K.V. (2020). Quantitative analysis of selected plastics in high-commercial-value Australian seafood by pyrolysis gas chromatography mass spectrometry, *Environ. Sci. Technol.*, **54**, pp. 9408–9417.

18. Picó, Y., and Barceló, D. (2019). Analysis and prevention of microplastics pollution in water: Current perspectives and future directions, *ACS Omega*, **4**, pp. 6709–6719.

19. Santos, L.H.M.L.M., Rodríguez-Mozaz, S., and Barceló, D. (2021). Microplastics as vectors of pharmaceuticals in aquatic organisms: An overview of their environmental implications, *Case Stud. Chem. Environ. Eng.*, **3**, pp. 100079.

20. Llorca, M., Álvarez-Muñoz, D., Ábalos, M., Rodríguez-Mozaz, S., Santos, L.H.M.L.M., León, V.M., Campillo, J.A., Martínez-Gómez, C., Abad, E., and Farré, M. (2020). Microplastics in Mediterranean coastal area: Toxicity and impact for the environment and human health, *Trends Environ. Anal. Chem.*, **27**, pp. e00090.

21. Primpke, S., Booth, A.M., Gerdts, G., Gomiero, A., Kögel, T., Lusher, A., Strand, J., Scholz-Böttcher, B.M., Galgani, F., Provencher, J.F., Aliani, S., Patankar, S., and Vorkamp, K. (2023). Monitoring of microplastic pollution in the Arctic: Recent developments in polymer identification, quality assurance and control, and data reporting, *Arct. Sci.*, **9**, pp. 176–197.

22. Wang, W., and Wang, J. (2018). Investigation of microplastics in aquatic environments: An overview of the methods used, from field sampling to laboratory analysis, *TrAC Trends Anal. Chem.*, **108**, pp. 195–202.

23. Caldwell, J., Taladriz-Blanco, P., Lehner, R., Lubskyy, A., Ortuso, R.D., Rothen-Rutishauser, B., and Petri-Fink, A. (2022). The micro-, submicron-, and nanoplastic hunt: A review of detection methods for plastic particles, *Chemosphere*, **293**, pp. 133514.

24. European Commission (2024). Commission Delegated Decision (EU) 2024/1441 of 11 March 2024 supplementing Directive (EU) 2020/2184 of the European Parliament and of the Council by laying down a methodology to measure microplastics in water intended for human consumption, *Official Journal of the European Union*, pp. 1–7.

25. Yakovenko, N., Carvalho, A., and Ter Halle, A. (2020). Emerging use thermo-analytical method coupled with mass spectrometry for the quantification of micro(nano)plastics in environmental samples, *TrAC Trends Anal. Chem.*, **131**, pp. 115979.

26. Peñalver, R., Arroyo-Manzanares, N., López-García, I., and Hernández-Córdoba, M. (2020). An overview of microplastics characterization by thermal analysis, *Chemosphere*, **242**, pp. 125170.

27. Primpke, S., Fischer, M., Lorenz, C., Gerdts, G., and Scholz-Böttcher, B.M. (2020). Comparison of pyrolysis gas chromatography/mass spectrometry and hyperspectral FTIR imaging spectroscopy for the analysis of microplastics, *Anal. Bioanal. Chem.*, **412**, pp. 8283–8298.

28. La Nasa, J., Biale, G., Fabbri, D., and Modugno, F. (2020). A review on challenges and developments of analytical pyrolysis and other thermoanalytical techniques for the quali-quantitative determination of microplastics, *J. Anal. Appl. Pyrolysis*, **149**, pp. 104841.

29. Hermabessiere, L., Himber, C., Boricaud, B., Kazour, M., Amara, R., Cassone, A.-L., Laurentie, M., Paul-Pont, I., Soudant, P., Dehaut, A., and Duflos, G. (2018). Optimization, performance, and application of a pyrolysis-GC/MS method for the identification of microplastics, *Anal. Bioanal. Chem.*, **410**, pp. 6663–6676.

30. Matsui, K., Ishimura, T., Mattonai, M., Iwai, I., Watanabe, A., Teramae, N., Ohtani, H., and Watanabe, C. (2020). Identification algorithm for polymer mixtures based on Py-GC/MS and its application for microplastic analysis in environmental samples, *J. Anal. Appl. Pyrolysis*, **149**, pp. 104834.

31. Ishimura, T., Iwai, I., Matsui, K., Mattonai, M., Watanabe, A., Robberson, W., Cook, A.-M., Allen, H.L., Pipkin, W., Teramae, N., Ohtani, H., and Watanabe, C. (2021). Qualitative and quantitative analysis of mixtures of microplastics in the presence of calcium carbonate by pyrolysis-GC/MS, *J. Anal. Appl. Pyrolysis*, **157**, pp. 105188.

32. Lee, S.-Y., An, J., and Kwon, J.-H. (2023). Sequential quantification of number and mass of microplastics in municipal wastewater using Fourier-transform infrared spectroscopy and pyrolysis gas chromatography-mass spectrometry, *Environ. Pollut.*, **336**, pp. 122452.
33. Okoffo, E.D., Rauert, C., and Thomas, K.V. (2023). Mass quantification of microplastic at wastewater treatment plants by pyrolysis-gas chromatography–mass spectrometry, *Sci. Total Environ.*, **856**, pp. 159251.
34. Okoffo, E.D., Chan, C.M., Rauert, C., Kaserzon, S., and Thomas, K.V. (2022). Identification and quantification of micro-bioplastics in environmental samples by pyrolysis–gas chromatography–mass spectrometry, *Environ. Sci. Technol.*, **56**, pp. 13774–13785.
35. Zhou, X.-X., He, S., Gao, Y., Li, Z.-C., Chi, H.-Y., Li, C.-J., Wang, D.-J., and Yan, B. (2021). Protein corona-mediated extraction for quantitative analysis of nanoplastics in environmental waters by pyrolysis gas chromatography/mass spectrometry, *Anal. Chem.*, **93**, pp. 6698–6705.
36. Zhou, X.-X., Hao, L.-T., Wang, H.-y.-Z., Li, Y.-J., and Liu, J.-F. (2019). Cloud-point extraction combined with thermal degradation for nanoplastic analysis using pyrolysis gas chromatography–mass spectrometry, *Anal. Chem.*, **91**, pp. 1785–1790.
37. Viitala, M., Steinmetz, Z., Sillanpää, M., Mänttäri, M., and Sillanpää, M. (2022). Historical and current occurrence of microplastics in water and sediment of a Finnish lake affected by WWTP effluents, *Environ. Pollut.*, **314**, pp. 120298.
38. Xu, Y., Ou, Q., Jiao, M., Liu, G., and van der Hoek, J.P. (2022). Identification and quantification of nanoplastics in surface water and groundwater by pyrolysis gas chromatography–mass spectrometry, *Environ. Sci. Technol.*, **56**, pp. 4988–4997.
39. Li, Q., Lai, Y., Li, P., Liu, X., Yao, Z., Liu, J., and Yu, S. (2022). Evaluating the occurrence of polystyrene nanoparticles in environmental waters by agglomeration with alkylated ferroferric oxide followed by micropore membrane filtration collection and Py-GC/MS analysis, *Environ. Sci. Technol.*, **56**, pp. 8255–8265.
40. Li, Q.-c., Lai, Y.-j., Yu, S.-j., Li, P., Zhou, X.-x., Dong, L.-j., Liu, X., Yao, Z.-w., and Liu, J.-f. (2021). Sequential isolation of microplastics and nanoplastics in environmental waters by membrane filtration, followed by cloud-point extraction, *Anal. Chem.*, **93**, pp. 4559–4566.
41. Li, Q., Sun, H., Bai, Q., Li, P., Lai, Y., Yu, S., Liu, X., Yao, Z., Cai, Y., and Liu, J. (2023). Spatial distribution of polystyrene nanoplastics and small microplastics in the Bohai Sea, China, *Sci. Total Environ.*, **881**, pp. 163222.

42. Goßmann, I., Mattsson, K., Hassellöv, M., Crazzolara, C., Held, A., Robinson, T.-B., Wurl, O., and Scholz-Böttcher, B.M. (2023). Unraveling the marine microplastic cycle: The first simultaneous data set for air, sea surface microlayer, and underlying water, *Environ. Sci. Technol.*, **57**, pp. 16541–16551.
43. Ter Halle, A., Jeanneau, L., Martignac, M., Jardé, E., Pedrono, B., Brach, L., and Gigault, J. (2017). Nanoplastic in the North Atlantic subtropical gyre, *Environ. Sci. Technol.*, **51**, pp. 13689–13697.
44. Funck, M., Yildirim, A., Nickel, C., Schram, J., Schmidt, T.C., and Tuerk, J. (2020). Identification of microplastics in wastewater after cascade filtration using Pyrolysis-GC–MS, *MethodsX*, **7**, pp. 100778.
45. Sefiloglu, F.Ö., Stratmann, C.N., Brits, M., van Velzen, M.J.M., Groenewoud, Q., Vethaak, A.D., Dris, R., Gasperi, J., and Lamoree, M.H. (2024). Comparative microplastic analysis in urban waters using μ-FTIR and Py-GC-MS: A case study in Amsterdam, *Environ. Pollut.*, **351**, pp. 124088.
46. Campanale, C., Stock, F., Massarelli, C., Kochleus, C., Bagnuolo, G., Reifferscheid, G., and Uricchio, V.F. (2020). Microplastics and their possible sources: The example of Ofanto river in southeast Italy, *Environ. Pollut.*, **258**, pp. 113284.
47. Hendrickson, E., Minor, E.C., and Schreiner, K. (2018). Microplastic abundance and composition in Western Lake superior as determined via microscopy, Pyr-GC/MS, and FTIR, *Environ. Sci. Technol.*, **52**, pp. 1787–1796.
48. Mintenig, S.M., Bäuerlein, P.S., Koelmans, A.A., Dekker, S.C., and van Wezel, A.P. (2018). Closing the gap between small and smaller: Towards a framework to analyse nano- and microplastics in aqueous environmental samples, *Environ. Sci. Nano*, **5**, pp. 1640–1649.
49. Ravit, B., Cooper, K., Buckley, B., Yang, I., and Deshpande, A. (2019). Organic compounds associated with microplastic pollutants in New Jersey, U.S.A. surface waters, *AIMS Environ. Sci.*, **6**, pp. 445–459.
50. Vilakati, B., Venkataraman, S., Nyoni, H., Mamba, B.B., Omine, K., and Msagati, T.A.M. (2022). Qualitative characterisation and identification of microplastics in a freshwater dam at Gauteng Province, South Africa, using pyrolysis–gas chromatography–time of flight–mass spectrometry (Py–GC–ToF–MS), *Environ. Sci. Pollut. Res.*, **29**, pp. 83452–83468.
51. Dibke, C., Fischer, M., and Scholz-Böttcher, B.M. (2021). Microplastic mass concentrations and distribution in German Bight waters by pyrolysis–gas chromatography–mass spectrometry/thermochemolysis reveal potential impact of marine coatings: Do ships leave skid marks?, *Environ. Sci. Technol.*, **55**, pp. 2285–2295.

52. Fischer, M., and Scholz-Böttcher, B.M. (2019). Microplastics analysis in environmental samples: Recent pyrolysis–gas chromatography–mass spectrometry method improvements to increase the reliability of mass-related data, *Anal. Methods*, **11**, pp. 2489–2497.

53. Vilakati, B., Sivasankar, V., Mamba, B.B., Omine, K., and Msagati, T.A.M. (2020). Characterization of plastic micro particles in the Atlantic Ocean seashore of Cape Town, South Africa and mass spectrometry analysis of pyrolyzate products, *Environ. Pollut.*, **265**, pp. 114859.

54. Sullivan, G.L., Gallardo, J.D., Jones, E.W., Hollliman, P.J., Watson, T.M., and Sarp, S. (2020). Detection of trace sub-micron (nano) plastics in water samples using pyrolysis-gas chromatography time of flight mass spectrometry (PY-GCToF), *Chemosphere*, **249**, pp. 126179.

55. Santos, L.H.M.L.M., Insa, S., Arxé, M., Buttiglieri, G., Rodríguez-Mozaz, S., and Barceló, D. (2023). Analysis of microplastics in the environment: Identification and quantification of trace levels of common types of plastic polymers using pyrolysis-GC/MS, *MethodsX*, **10**, pp. 102143.

56. Picó, Y., and Barceló, D. (2020). Pyrolysis gas chromatography–mass spectrometry in environmental analysis: Focus on organic matter and microplastics, *TrAC Trends Anal. Chem.*, **130**, pp. 115964.

57. Seeley, M.E., and Lynch, J.M. (2023). Previous successes and untapped potential of pyrolysis–GC/MS for the analysis of plastic pollution, *Anal. Bioanal. Chem.*, **415**, pp. 2873–2890.

58. Vilakati, B., Sivasankar, V., Nyoni, H., Mamba, B.B., Omine, K., and Msagati, T.A.M. (2021). The Py–GC-TOF-MS analysis and characterization of microplastics (MPs) in a wastewater treatment plant in Gauteng Province, South Africa, *Ecotoxicol. Environ. Saf.*, **222**, pp. 112478.

59. Dümichen, E., Eisentraut, P., Bannick, C.G., Barthel, A.-K., Senz, R., and Braun, U. (2017). Fast identification of microplastics in complex environmental samples by a thermal degradation method, *Chemosphere*, **174**, pp. 572–584.

60. Dehaut, A., Hermabessiere, L., and Duflos, G., (2020) *Handbook of Microplastics in the Environment*, eds. Rocha-Santos, T., Costa, M., and Mouneyrac, C., "Microplastics detection using Pyrolysis-GC/MS-based methods", (Springer Nature Switzerland), pp. 1–35.

61. Matsueda, M., Mattonai, M., Iwai, I., Watanabe, A., Teramae, N., Robberson, W., Ohtani, H., Kim, Y.-M., and Watanabe, C. (2021). Preparation and test of a reference mixture of eleven polymers with deactivated inorganic diluent for microplastics analysis by pyrolysis-GC–MS, *J. Anal. Appl. Pyrolysis*, **154**, pp. 104993.

Chapter 3

Analysis of Microplastics and Small Microplastics (<100 µm) in Natural Waters via Micro-FTIR

Fabiana Corami
Fabiana Corami: Institute of Polar Sciences, CNR-ISP, Campus Scientifico - Ca' Foscari University of Venice, Via Torino, 155, 30172, Venezia-Mestre, Italy
fabiana.corami@cnr.it

3.1 Introduction

Microplastics (MPs) are present in different environmental compartments (abiotic and biotic, Fig. 3.1); after the atmospheric compartment [1–9], natural waters are a substantially significant reservoir for MPs [10–21], which through different transport pathways can move from mid to high latitudes via marine currents.

The assessment of the abundance of MPs in natural waters, marine or fresh, is of paramount importance precisely for assessing the degree of plastic pollution, identifying sources of contamination,

Microplastic Pollution: Environmental Impact and Solutions
Edited by Julia Martín, Juan Luis Santos, Irene Aparicio, and Esteban Alonso

ISBN 978-981-5129-77-9 (Hardcover), 978-1-003-66006-4 (eBook)
www.jennystanford.com

DOI: 10.1201/9781003660064-3

and assessing potential risks to all aquatic inhabitants. Besides, in the One Health Approach [22], investigating how plastic pollution affects good environmental status and quality of organisms would also allow for an integrated response intended to achieve sustainable balance and optimize people's health.

When designing a study regarding MPs in natural waters, it is essential to establish clear objectives for the research (Fig. 3.1). It is essential to indicate in which matrix (e.g., marine waters, transitional waters, deltas, and mouths, rivers, or lakes) the abundance and distribution of MPs will be studied, as much as establishing when sampling campaigns will be performed (e.g., assessment of seasonal variations, or as a function of depth). Assessing the pathways of MPs from freshwaters to marine waters is crucial; for instance, it is essential to plan upstream and downstream sampling in rivers, as much as in lakes; it is critical to consider the circulation and stratification (thermal or chemical) of the water that could influence the pathways to the sea.

Another critical major issue concerns effluents from wastewater treatment plants (WWTPs), which may be discharged directly into the sea, along coastlines, rivers, or other freshwater basins [23–32]. Indeed, WWTPs can play an impactful role in the dispersal of MPs into the environment, especially for those particles of such size as to bypass all the treatment units in the plant [33–40]; these plastic particles are usually below 100 μm (small microplastics, SMPs), and they are relevantly overlooked in natural waters.

Indeed, size is a key factor in evaluating the occurrence and threats of MPs in natural waters. The European Chemical Agency [41] provided a definition of the composition of plastic polymers and their size: "MPs are solid polymer-containing particles, to which additives or other substances have been added, with particle dimensions ranging from 1 nm to 5 mm, fibre lengths ranging from 3 nm to 15 mm, and a length-to-diameter ratio > 3" (Fig. 3.2).

This definition in size came after that one proposed by the Technical Subgroup on Marine Litter for the implementation of the Marine Strategy Framework Directive (MSFD) [42]; the group recommended a subdivision of the microlitter into "visible size fraction" (>1 mm) and "invisible fraction" (<1 mm). Besides, the "visible size fraction" was further categorized into macroplastics (>25 mm), mesoplastics (5–25 mm), and large microplastics (1–5 mm). At first, the minimum size of microplastics was defined according to the sampling and pretreatment procedures (this will be explored in more detail later), but the European Chemicals Agency (ECHA) definition also shows a maximum and a minimum size, including nanoplastics.

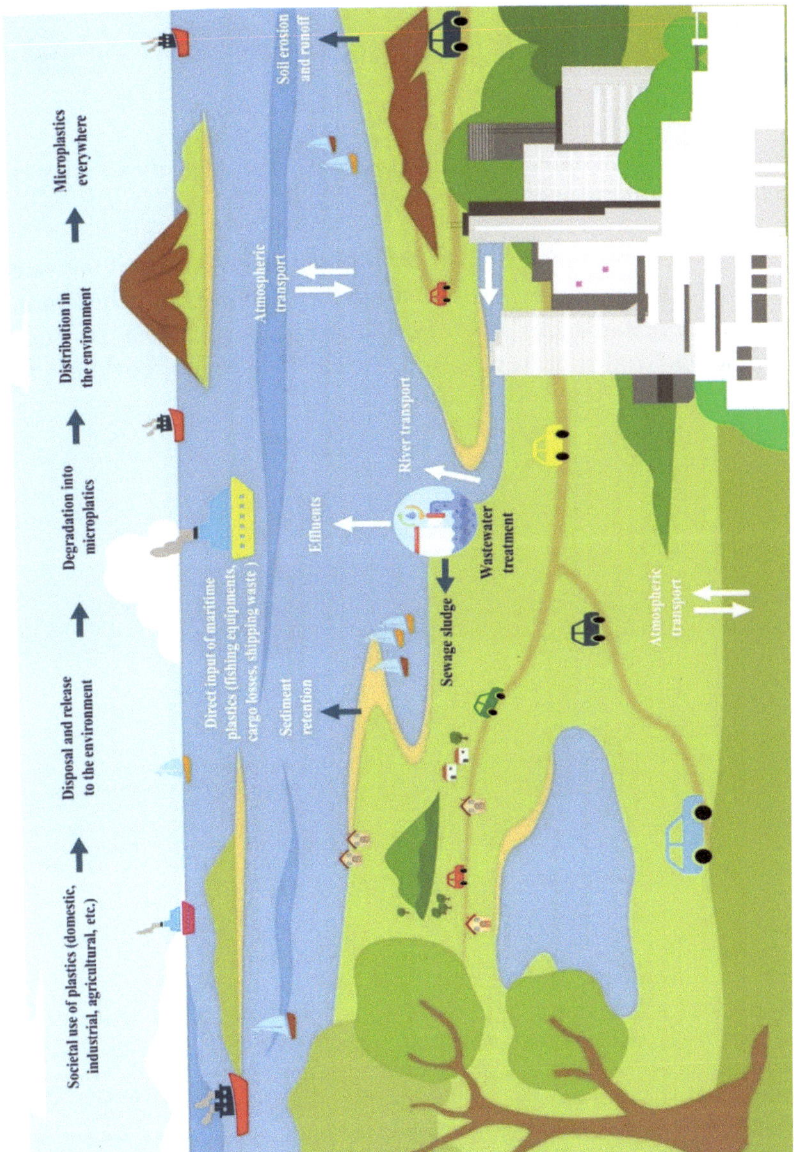

Figure 3.1 Microplastics' cycle in the natural waters: sources and transport pathways in the environment.

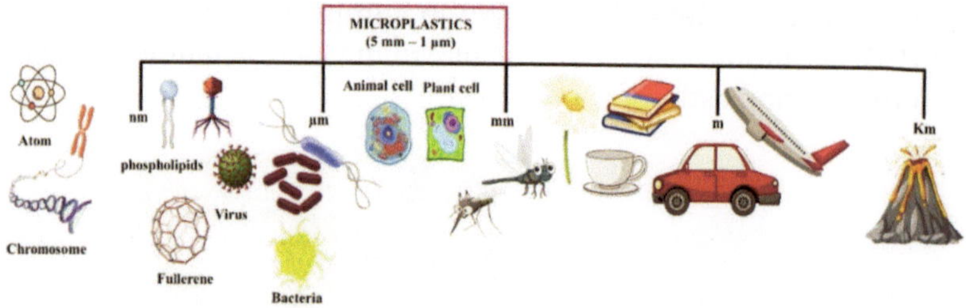

Figure 3.2 The km to nm size scale depicts the size range of microplastics (adapted from Mostri di Plastica, Ref. [12]).

However, this umbrella-like definition was revised, and the size range of microplastics goes from 1 μm to 5 mm [43–46]; therefore, MPs and nanoplastics turn out to be separate, also because interactions, transport routes, and the fate of these may be different.

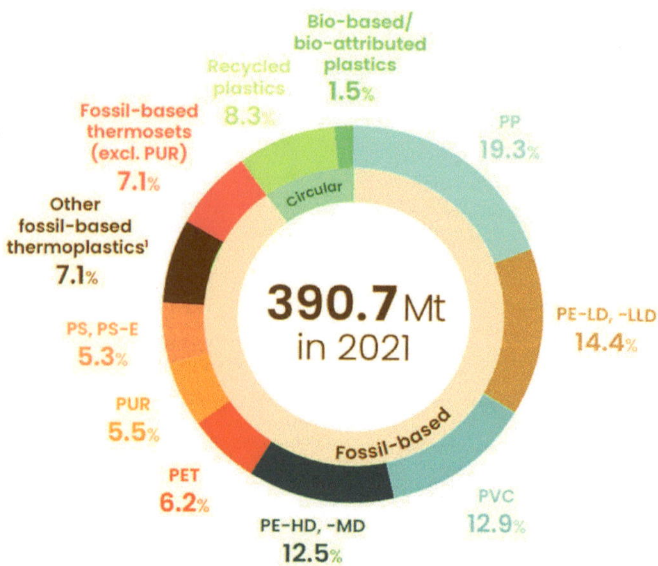

Figure 3.3 Rounded estimations of the distribution of plastic production by type (2021, adapted from *Plastics—the Facts 2022*, Plastic Europe, 2022). Plastic production from polymerization and production of mechanically recycled plastics are included, while polymers not used in converting plastic parts and products (e.g., for textiles, adhesives, sealants, coatings, etc.) were excluded. [1]PBT, PEEK, PEI, POM, PPA, PSU/PES/PPSU, PTFE, PVDF, and other thermoplastics not listed separately are included here. Sources: Conversio Market & Strategy GmbH and nova-Institute.

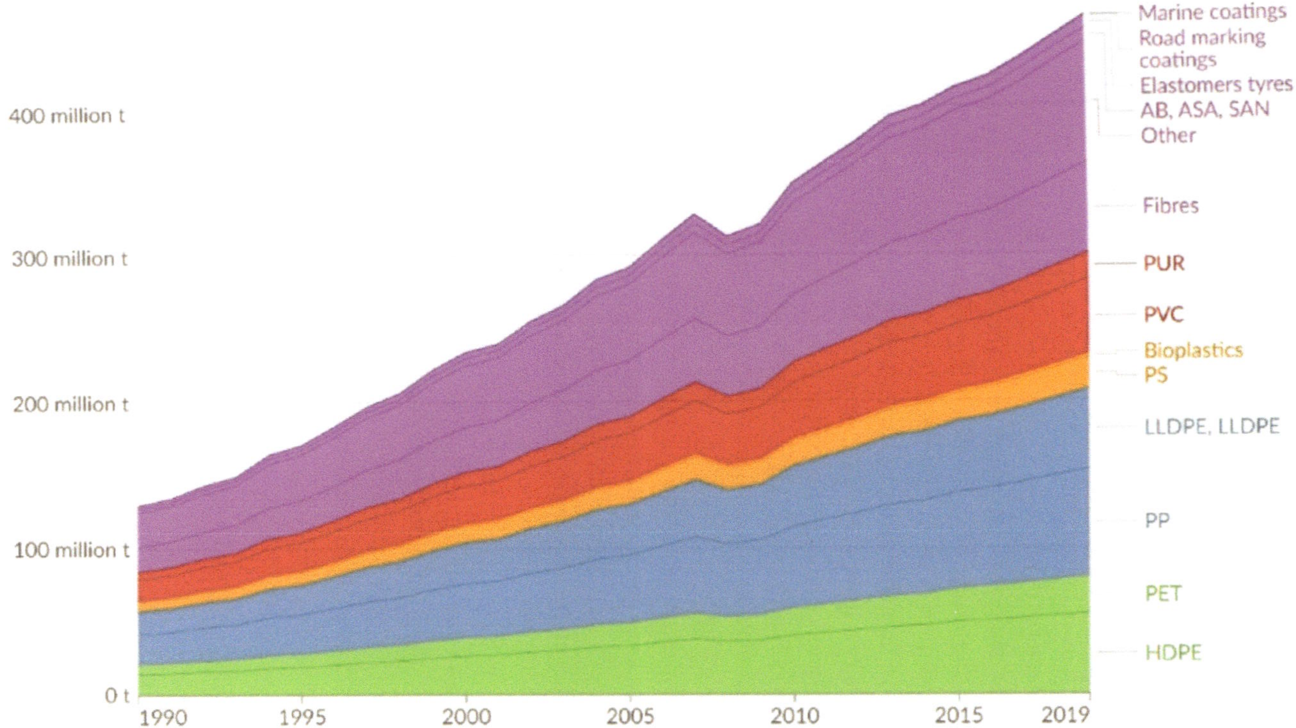

Figure 3.4 Global primary production by polymer (1999–2021). Many polymers are grouped, e.g., marine coatings, bioplastics, coatings, etc. Besides, polymers are grouped according to their recyclability and are colour coded: green for widely recycled, blue for moderately recycled, orange for limited recyclability, red for usually non-recycled, and violet for unknown recyclability. Source: OECD 2022.

Although the term "small microplastics" was employed with different upper limits, e.g., <500 µm [22], between 0.1 and 1000 µm [47], <300 µm [48], <100 µm [6, 11, 49–60], the latter is the most widely accepted. Therefore, when planning a study regarding MPs in natural waters, it is of utmost importance to consider which size is to be studied to be then able to fully understand the transport pathways, possible interactions with biota, transport pathways, and fate of these particles.

Regarding plastic polymers, they are produced and employed in an extensive range of applications (Figs. 3.3 and 3.4) and can have many different plastic additives (e.g., plasticizers, antioxidants, lubricants, flame retardants, curing agents, vulcanizers, etc). Plastic polymers can have a wide range of density; plastic polymers (e.g., polyethylene and polypropylene) are considered low-density polymers (density <1 g/cm^3), while fluorinated polymers, such as polytetrafluoroethylene, are considered high-density polymers (density ≥2 g/cm^3). Therefore, when investigating the occurrence of microplastics in natural waters, it is crucial to consider which polymers are to be investigated, and then the pretreatment method should be chosen accordingly.

Furthermore, it should be noted that plastic objects contain plastic additives; while some of these may be released in water, others may be present in the environment as particles following the break-up of the object, leading to the formation of microplastics. Thus, all the microlitter components should be unambiguously identified because their identification is an essential prerequisite for a precise, robust, accurate analysis, which should represent the bulk [6, 11, 61–66].

3.2 Sampling Procedures

As for any other environmental matrix, sampling microplastics in natural waters is one of the most delicate steps of the whole analytical procedure, as it highly affects the representativity of the obtained results. Considering the size range of MPs (1 µm to 5 mm), it is paramount to identify which size range is the focus of the study so that the most appropriate sampling method can be chosen.

It should be noted that depths, currents, the closeness of the coastline or shores (also in the case of rivers or lakes), the proximity of estuaries and deltas (in the case of the sea and rivers), and the presence of tributaries (inlets and outlets, in lakes) are all variables that must be considered when designing a sampling, which is then representative of the environment under investigation.

Sampling MPs in marine waters can be different from sampling them in fresh waters, especially in relation to the surface area of the waterbody and the flow rate of the streams. Manta trawls are commonly employed samplers for MPs (Fig. 3.5a) [67–77], and the minimum size of particles collected is given by the mesh size of the net, which is usually 300–330 μm.

It should be noted, however, that below 300 μm, microplastics may still be present in the environment. Other mesh sizes were employed for the manta net, ranging from 500 μm to 100 μm [78–82].

Also, neuston net (Fig. 3.5c) [48, 74, 83–89], bongo net (Fig. 3.5d) [74, 85, 90–94], and plankton net (Fig. 3.5e) [73, 74, 88, 95–98] were employed to sample MPs in natural waters. Trawls and nets need a boat and suitable equipment to be deployed in the water.

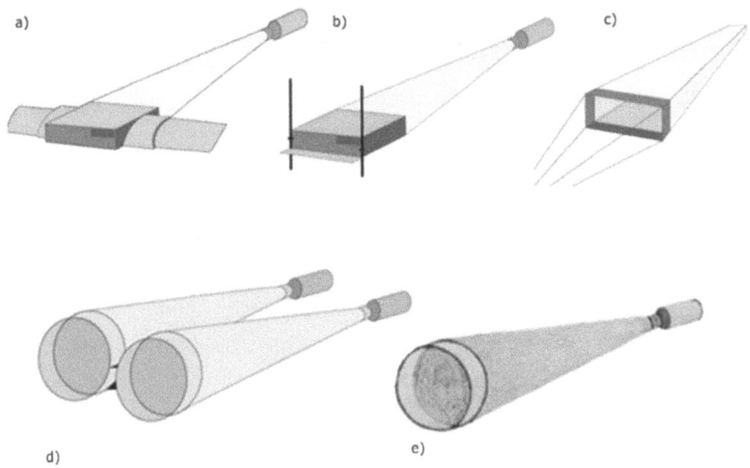

Figure 3.5 Examples of nets and trawls employed to collect microplastics > 100 μm in natural waters: (a) manta trawl, (b) microplastic trap, (c) neuston net, (d) bongo net, (e) plankton net.

While bongo and plankton nets are used to collect water at different depths, manta trawls and neuston nets should be submerged and towed on a linear course, and to do so, they need to be equipped with a flow meter to evaluate the volume of water filtrated and to minimize the mishaps due to waves, winds, and boat movements, which can affect the estimation of the filtered water volume.

A shortcoming of sampling with manta nets, especially with those whose mesh size is 300 μm, is that microfibres are not retrieved, as well as all the particles are less than 300 μm; this will lead to a potential underestimation of all the microplastics present in the waters since SMPs and microfibres account for a highly relevant portion of the microplastics [11, 99, 100]. A flaw of sampling with the nets is the clogging due to phytoplankton and zooplankton, mainly when their blooms occur.

Microplastic traps (Fig. 3.5b) are employed for sampling MPs in small rivers and shallow watercourses. They are provided with an aluminium frame, stakes for anchoring to the riverbed, and a nylon filter net, whose mesh size will represent the minimum size for the collected particles. The traps can be placed in several locations to cover the width of the watercourse and have a more representative sampling [101–103].

Other sampling systems involve the use of pumps and cascade filters, characterized by different mesh sizes [102–107]; the pumps can be positioned at the desired depth, and the water is pumped and passed over the different filters, separating the microplastics.

A major drawback of these samplers is that the filters may clog in the presence of a significant amount of particulate organic material and planktonic blooms (phytoplankton and zooplankton). Hence, smaller particles might get trapped in the larger fraction; this accidental entrapment could also happen due to possible electrostatic interactions, which cause microplastics to stick together or to stick with the other particles in the particulate matter, increasing the size of microplastics.

A further sampling option is to collect discrete samples via Niskin bottles, rosettes, integrating water samplers [76, 108], and buckets, bottles, and steel samplers [16, 74, 90, 102–104, 109–116].

Conceptually, all MPs present in the environmental matrix can be sampled without any size limitation, thus avoiding the losses that may occur when sampling with the other methods mentioned.

It should also be noted that this type of sampling is faster and can help reduce the risk of contamination, primarily if a proper quality assurance/quality control (QA/QC) protocol is implemented and applied, as we will see in the appropriate section; considering that the short sample handling time and exposure time to the surrounding environment are reduced, potential contamination during sampling is minimized [11].

Nevertheless, there are some disadvantages concerning, for example, the sample's representativity relative to larger microplastics (>1 mm). The smaller the particles, the more numerous they are in the environment, so this sampling method is best suited for SMPs [11, 104]. Besides, several discrete samples can be collected to produce a composite averaged sample to increase the sample's representativity and reduce the variability within the matrix. However, it should be underlined that the choice of sampling method must always be made with respect to the size class under investigation.

As far as volume is concerned, the sample is considered representative when it provides a snapshot of the sampling site with findings that represent the actual concentration of microplastics in that area under investigation. The choice of the volume of water that should be sampled for representativity depends on the aqueous matrix to investigate and the size of the particles under scrutiny; as a matter of fact, the representative volume for the analysis of the largest microplastics is very different from the one for the analysis of SMPs [6, 11, 117, 118].

3.3 Quality Assurance/Quality Control

A strict QA/QC should be implemented and followed when analysing the abundance and distribution of microplastics in the natural waters [6, 11, 61]. It is of paramount importance to minimize at any step the potential contamination of plastic, which is ubiquitous in the environment.

Pretreatment procedures (e.g., extraction, purification, and filtration) and the preparation of all the tools and devices for samples should be decontaminated in plastic-free laboratories, e.g., plastic-free clean room in stainless steel, with controlled air circulation (air filtration system without polymers), and limited access [11, 61]. If a clean room with such features is not available, a stainless steel laminar flow hood should be set up in a space/laboratory that can be accessed only by those working on the microplastics investigation.

Special care should be taken in selecting operator safety equipment, e.g., lab coats, gloves, and protective shoes. Cotton lab coats and nitrile gloves are currently employed; regarding safety shoes, which may have plastic soles, it is good to cover them with cotton no-show socks, which can periodically be washed, like coats. A detergent free of fragrances, pigments, and additives is recommended for washing, as some of these substances may be common in microlitter. Work surfaces and hoods should be carefully wiped down with ethanol or a 70% ethanol/methanol solution (7:3).

Decontamination procedures should be performed on glassware, steelware, and sampling devices. Usually, for decontamination, alcohols, i.e., ethanol or a solution of ethanol/methanol, are employed. However, these and all the other reagents employed for the pretreatment should be checked thoroughly for the presence of microplastics. Ultrapure or double-distilled deionized water employed in the laboratory should also be periodically checked for microplastics. Other periodic tests are the reagent blanks, which should be run in triplicate and analysed as samples to check and minimize the potential microplastic contamination.

Blanks are of paramount importance for microplastic analysis (both for large-sized MPs and SMPs). When working in the field, the field blank should be collected, which consists of a decontaminated aluminium or glass box, to be exposed throughout the sampling session. At the end of sampling, the box should be closed and stored like the samples. Once in the laboratory, it will be rinsed with a known volume of ultrapure water (previously checked), usually equal to the volume of the sample that will be extracted and processed in the same manner as the sample (at least in triplicate) [6, 11, 56].

For field blanks, filters (e.g., glass fibre filters (GFF Whatman) or quartz filters) can also be employed, but these must first be checked for the presence of MPs [55, 56]. If samplers such as Niskin bottles are employed to sample MPs and SMPs in natural waters, it is necessary to collect a sampler blank, as these samplers are often made of plastic material and have various plastic seals. Besides field blanks, sampler blanks, and ultrapure and reagents blanks, procedural blanks should also be performed using a volume of ultrapure water equivalent to that of the water sample; this blank (performed at least in triplicate) should account for the potential contamination of the operator during the pretreatment procedure. Given the laborious checking of blanks, it could be advisable to operate in batches.

Once the pretreatment has been completed, whether the MPS have been collected in a previously decontaminated glass Petri dish or the SMPs have been filtered and the filter left to dry, they must remain in a plastic-free environment until the moment of analysis. As for SMPs, before the analysis, filters must be mounted on the stage of the micro-FTIR (Fourier-transform infrared) spectrometer for analysis under a decontaminated fume hood in the cleanroom or under a decontaminated stainless steel laminar flow hood and then carried carefully covered with aluminium foil to avoid any external contamination to the instrumental laboratory [6, 11, 51, 55, 56, 61].

Recovery tests are performed to assess the accuracy of the pretreatment method, employing certified reference standard materials (CRMs). However, due to the lack of CRMs for microplastics, commercially available plastic polymers with particles of known sizes and close to those of the particles under investigation can be employed. The recovery test can be performed by spiking with those plastic particles three further replicates of one of the collected samples under exam [6, 11, 55, 56]. Given that an experimental method's yield is unlikely to be 100%, it should be noted that a percentage method's yield is excellent when it is >90%, optimal between 80% and 90%, and very good between 75% and 80%. The method's yield higher than 100% may indicate contamination.

3.4 Pretreatment Methods

The nets and the cascade-filter samplers for microplastics can provide a size separation of the particles from the environmental matrices. MPs, especially the large ones (5 mm to 500 μm), can be sorted by removing them from the sieve or the filters [48, 71–74, 76,

77, 81, 87, 88, 95]. On the contrary, MPs in discrete samples collected by Niskin bottles, buckets, and other samplers [11, 16, 74, 83, 102, 104, 111, 112] should be extracted from the environmental matrices; however, interferents should be removed from the particles before the analysis by vibrational spectroscopy (FTIR and micro-FTIR). The interferents in natural waters can be the salts in the seawater and freshwater (hardness), and the organic matter of plant and animal origin.

Density separation is often employed for sediments but can be employed in retrieving MPs from freshwater [69, 119, 120]. This procedure is coupled with a step to remove the organic matter from the particles [14, 72, 88, 121–123]. Usually, hydrogen peroxide (H_2O_2) is employed; this oxidant agent is also employed during the digestion procedure since it is efficient in removing organic matter.

Different salts can be employed and, with respect to their densities, they can separate polymers smaller than this density. The following are used for this procedure: sodium chloride (NaCl) [124–126], sodium iodide (NaI) [127], sodium polytungstate ($Na_6(H_2W_{12}O_{40})$) [128], calcium chloride ($CaCl_2$) [129], lithium metatungstate (LMT) [130], potassium formate ($KHCO_2$) [131, 132], potassium fluoride (KF) [133], and zinc chloride ($ZnCl_2$) [134, 135]. However, this procedure seems insufficient for separating high-density polymers (e.g., fluorinated plastics). Therefore, along with the size class to be studied, it is crucial to consider which polymers to investigate carefully and, consequently, the best pretreatment for that purpose.

Acidic and alkaline digestions were employed for the pretreatment of the microplastics; the latter least employed in natural waters. Alkaline digestions (using sodium hydroxide (NaOH) and potassium hydroxide (KOH)) have been employed in digesting organisms to assess the microplastics' content [14]. However, it should be noted that alkaline digestion coupled with temperature above 40°C can lead to a degradation/denaturation of several polymers [14, 136–138].

On the other hand, strong acids, e.g., nitric acid, coupled with a temperature above 40°C (for more than 24 h), were employed in several studies to remove the organic matter from MPs in natural waters [14, 22, 94, 121, 139, 140]. A primary shortcoming of acidic digestion is due to the potential degradation of polymers. For instance, polyamides can completely disappear due to dissolution [11, 136, 138, 141, 142], while other polymers, e.g., polyethylene (PE), polypropylene (PP), and polyethylene terephthalate (PET), may show signs of degradation, such as yellowing, melting, and discoloration [136], especially when temperatures ranging from 50°C to 70°C are employed [143]. Besides, after dissolution, particles can agglutinate, resulting in a change in size, misidentification, and incorrect quantification [11, 136, 137, 144].

Oxidative digestion uses hydrogen peroxide (H_2O_2) and other strong oxidants such as Fenton's reagent. Different concentrations of H_2O_2 have been employed, even coupled with temperature (from room temperature to 100°C) [14, 145]. However, it should be noted that commercially available H_2O_2 is a 30–35% solution, which is then aliquoted and added to water samples, thus being diluted [6, 11]. As already mentioned for acidic digestions, temperature can adversely affect the pretreatment process, leading to the total disappearance or degradation of several polymers [136, 146]. Therefore, the result is suboptimal identification, which can lead to an underestimation of particle abundance. In addition, the combined action of oxidizing agents and elevated temperatures can change particle size or further fragmentation, leading to overestimation.

The Fenton reagent, a mixture of hydrogen peroxide (H_2O_2) and a solution containing Fe(II) ions, is in the National Oceanic and Atmospheric Administration's recommendation for removing organic matter from seawater and sediment samples, often coupled with temperature from 75°C up to boiling [14, 147]. However, it is worth highlighting again how high temperatures can lead to the disappearance of some polymers or their degradation and, consequently, to a non-optimal assessment of the abundance of microplastics in natural waters. Therefore, the temperature needs to be checked carefully [6, 11, 14, 56, 61, 136].

Enzymatic digestion may be a suitable alternative to employing strong oxidants, but it can be significantly time consuming, especially for samples containing different types of organic material [127, 148, 149]; for this reason, it may be coupled with additional digestion steps with strong oxidants (e.g., Fenton reagent). Besides, the enzymes employed need different digestion requirements (temperature and pH) at each step, resulting in an elaborate and tricky digestion procedure, which could also enhance the potential contamination. It is not often employed for natural waters, except stormwater ponds [150].

Since many of these pretreatments may not be efficient for removing interferents and especially for extracting SMPs, oleoextraction has proven to be an excellent, effective, and replicable alternative [11, 14, 93, 143, 151–156]. This pretreatment exploits the lipophilic nature of microplastics and their ability to transition from the aqueous to the oil phase. Different oils were employed for the oleoextraction: canola oil [151, 154], castor oil [157], olive oil [155], and sunflower oil [6, 11, 56]. Usually, H_2O_2 is employed with oleoextraction to remove organic matter from the particles [6, 11, 14, 56, 151, 157]. This procedure allows operating at room temperature, avoiding further denaturation, degradation, or even loss of the polymers during the pretreatment [11].

The main advantages of oleoextraction are (i) the efficient removal of all the interferents for micro-FTIR and micro-Raman analysis and (ii) the retrieval of polymers in a wide range of densities (from low-density polymers to high-density polymers, e.g., fluorinated polymers). Besides, oleoextraction allows the simultaneous collection of all the other microlitter components, including tyre wear and road particles (TWRPs), plastics additives, and natural and artificial fibres (e.g., rayon) [6, 11, 56]. The efficiency of the procedure (method's yield) was tested using commercially available plastic particles of known sizes, which were added to natural samples; the yield resulted optimal (<90%) [6, 11, 56, 157].

After the pretreatment, MPs, particularly SMPs, should be filtered. Filters with suitable mesh sizes and materials should be chosen; before the analysis, a blank filter should be checked for the presence of MPs. Besides, filters that allow working in transmission and with maximum IR energy should be preferred [11]. Vacuum pump filtration with a decontaminated glass filtering system is widely employed. A further step during filtration is performed to increase the removal of interferents further and to take away efficiently the oil employed in the extraction [11]; this enhances the optimal identification and a quantification free from potential under- or overestimates.

In addition to being significantly effective and reproducible, employing natural oils for oleoextraction is economically and environmentally sustainable. A further increase in sustainability would be the employment of green or low-impact solvents in the purification step.

3.5 Quantification via Microscopic Counting

Microscopic counting has been employed for bacteria, phytoplankton, pollen, spores, and MPs as well [6, 11, 158–168]. A significant advantage of microscopic counting is that there is no doubt how many organisms, cells, or particles are present within valid computable limits and degrees of chance.

Microscopic counting can be employed with any microscope slide or filter suitable for analysis using micro-FTIR and micro-Raman spectrometers (e.g., aluminium oxide filters, silicon oxide filters, etc.). Applying microscopic counting to MPs and SMPs that will be analysed with vibrational spectroscopy, i.e., micro-FTIR and micro-Raman spectroscopy, is an excellent advantage as quantification can be carried out simultaneously with identification.

The different approaches in choosing representative measurement areas for microscopic counting are shown in Fig. 3.6.

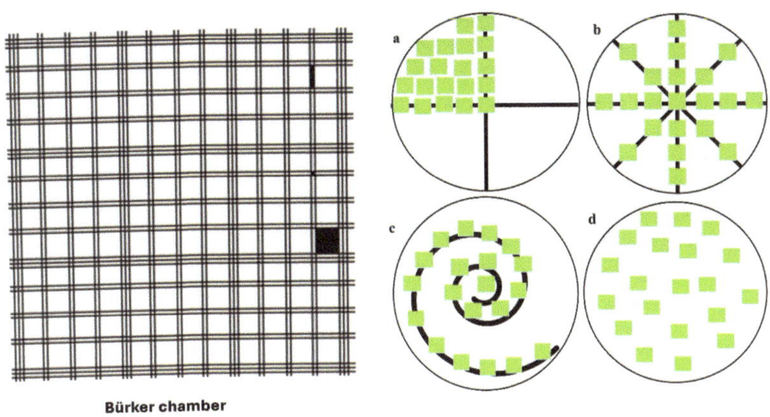

Figure 3.6 Different approaches for representative measurement areas on glass slides and filters; at least 20 count areas or count fields should be considered. The microscope counting can be performed in the Bürker chamber for slides and square silicon filters. On round filters (e.g., aluminium oxide, PTFE), four different approaches can be employed: (a) represents a quarter of the filters; (b) represents the cross section of the four axes; (c) represents a helical assembly; (d) represents a randomized assembly (modified from Refs. [11, 159, 162]).

When filters are employed as a support for counting, the measurement of complete filters can be very time consuming [162, 163]. However, analysed filter areas, i.e., counting areas or count fields, must represent the entire filter to avoid issues regarding representativeness and reproducibility. The Bürker chamber can be a model for microscopic counting on square silicon filters or glass slides, while four different approaches (Fig. 3.6a–d) can be employed for round filters, such as aluminium oxide filters. Since the loading of the filters cannot be known in advance, counting areas with different abundances should be considered to avoid issues regarding the accuracy of extrapolation of findings of microplastics, organisms, cells, or bacteria. Randomized counting areas or counting fields with no overlapping allow for accurate evaluation of differences in particle distribution [6, 11, 56].

The total counts from each filter must be multiplied by the appropriate microscope conversion (optical) factor (F) and volume or dilution factors to provide the absolute abundance.

3.6 Analytical Methods: FTIR and Micro-FTIR Spectroscopy

After collecting and pre-treating the sample in the laboratory, microplastics need to be quantified and unambiguously identified from the sample [11, 41, 61]. A visual inspection of potential plastic particles followed by a polymer composition analysis on a subset of particles is not adequate for the unequivocal characterization of microplastics. Therefore, a chemical analysis of the polymer composition of all quantified particles must be performed, using, also in combination with vibrational spectroscopy and thermoanalytical techniques [6, 11, 51, 55, 61, 62, 162, 169, 170].

Vibrational spectroscopy is widely used for the identification of microplastics, especially FTIR spectroscopy, since it is highly reliable in the identification of MPs in environmental samples [171, 172]. Microplastic analysis can be performed in transmission, reflectance, and attenuated total reflectance (ATR) both by FTIR and micro-FTIR (also known as μFTIR).

FTIR and ATR-FTIR spectrometers can be employed for the identification of large microplastics, i.e., > 200–300 μm (this also being the limit of detection (LOD) of the particle size of most of these spectrometers) [173]. Micro-FTIR spectrometers allow identification and quantification of MPs from the size of 10 μm upwards [150, 171, 174, 175]; in some studies, the LOD of micro-FTIR was lowered to 5 μm [6, 11, 51, 55, 56].

Other than the analysis by FTIR and ATR-FTIR, large-size MPs (5-1 mm) can be analysed by micro-FTIR (also ATR-micro-FTIR), employing BaF_2 discs and appropriate slides; maps and point-and-shoot analysis can be performed to retrieve the spectra of the particles. Also, MPs < 1 mm can be analysed in this manner, but the transfer is impractical as the size becomes smaller and smaller. Therefore, analysis is carried out using suitable supports where the MPs were collected (usually filters, after extraction and filtration).

Since analysing the whole filter is time consuming (as for in the quantification), the software can be used for automated particle analysis on various filter areas in the same manner as quantification, thus achieving simultaneous particle identification and quantification; this approach has been effectively tested for the analysis of SMPs in various environmental matrices [6, 11, 51, 55, 56, 61].

In the analysis of microplastics, reference libraries are fundamental; the spectra of the analysed particles are compared with those present in these libraries to be then identified. It is of paramount importance to employ various and different reference libraries with a high number of spectra within them, as different copolymers may be present; therefore, redundancy (intended both as the number of libraries and as the number of spectra present in the libraries) improves optimal identification. Besides, numerous assorted libraries with which to compare the collected spectra also allow other particles in the microlitter to be optimally identified, such as plastic additives and natural and artificial fibres [6, 11, 51, 55, 56, 61].

References

1. Dris, R., Gasperi, J., Saad, M., Mirande, C., Tassin, B. (2016). Synthetic fibers in atmospheric fallout: A source of microplastics in the environment? *Mar. Pollut. Bull.*, **104**(1–2), 290–293. https://doi.org/10.1016/j.marpolbul.2016.01.006

2. Dris, R., Gasperi, J., Mirande, C., Mandin, C., Guerrouache, M., Langlois, V., Tassin, B. (2017). A first overview of textile fibers, including microplastics, in indoor and outdoor environments. *Environ. Pollut.*, **221**, 453–458. https://doi.org/10.1016/j.envpol.2016.12.013
3. Habibi, N., Uddin, S., Fowler, S. W., Behbehani, M. (2022). Microplastics in the atmosphere: A review. *J. Environ. Expo. Assess.*, **1**(6), 10-20517. https://doi.org/10.20517/jeea.2021.07
4. O'Brien, S., Rauert, C., Ribeiro, F., Okoffo, E. D., Burrows, S. D., O'Brien, J. W., Wang, X., Wright, S. L., Thomas, K. V. (2023). There's something in the air: A review of sources, prevalence and behaviour of microplastics in the atmosphere. *Sci. Total Environ.*, **874**, 162193. https://doi.org/10.1016/j.scitotenv.2023.162193
5. Peries, S. D., Sewwandi, M., Sandanayake, S., Kwon, H. H., Vithanage, M. (2024). Airborne transboundary microplastics: A swirl around the globe. *Environ. Pollut.*, 124080. https://doi.org/10.1016/j.envpol.2024.124080
6. Rosso, B., Corami, F., Vezzaro, L., Biondi, S., Bravo, B., Barbante, C., Gambaro, A. (2022). Quantification and characterization of additives, plasticizers, and small microplastics (5–100 μm) in highway stormwater runoff. *J. Environ. Manage.*, **324**, 116348. https://doi.org/10.1016/j.jenvman.2022.116348
7. Singh, B. P. (2024). Understanding the dynamics and implications of airborne microplastics in atmosphere. *Air Qual. Atmos. Health*, 1–14. https://doi.org/10.1007/s11869-024-01593-4
8. Zhang, N., Zhang, C., Qin, Y., Wang, J., Ge, X., Li, H., Dai, Y., Aruffo, E. (2024). A review of atmospheric microplastics: Sources, characteristics, and detection method. *Curr. Pollut. Rep.*, 1–18. https://doi.org/10.1007/s40726-024-00316-8
9. Zhang, Y., Gao, T., Kang, S., Sillanpää, M. (2019). Importance of atmospheric transport for microplastics deposited in remote areas. *Environ. Pollut.*, **254**, 112953. https://doi.org/10.1016/j.envpol.2019.07.121
10. Bordós, G., Urbányi, B., Micsinai, A., Kriszt, B., Palotai, Z., Szabó, I., Hantosi, Z., Szoboszlay, S. (2019). Identification of microplastics in fish ponds and natural freshwater environments of the Carpathian basin, Europe. *Chemosphere*, **216**, 110–116. https://doi.org/10.1016/j.chemosphere.2018.10.110
11. Corami, F., Rosso, B., Morabito, E., Rensi, V., Gambaro, A., Barbante, C. (2021). Small microplastics (<100 μm), plasticizers and additives in seawater and sediments: Oleo-extraction, purification, quantification,

and polymer characterization using Micro-FTIR. *Sci. Total Environ.*, **797**, 148937. https://doi.org/10.1016/j.scitotenv.2021.148937

12. Corami, F., Rosso, B. (2023). Conveyance, bounty, and dangers of microplastics in nature. In: Khan, A., Wang, C., Asiri, A. M. (Eds.) *Microplastic Sources, Fate and Solution*. Springer Nature Singapore, Singapore. https://doi.org/10.1007/978-981-99-0695-6_6
13. Dar, A. A., Chen, Z., Sardar, M. F., An, C. (2024). Navigating the nexus: Coastal climate, microplastics, and the uncharted waters of coastal pollution. *Environ. Res.*, 118971. https://doi.org/10.1016/j.envres.2024.118971
14. Ermolin, M. S. (2024). Assessment of the microplastics content in natural waters and sediments: Sampling and sample preparation. *J. Anal. Chem.*, **79**, 500–519. https://doi.org/10.1134/S1061934824050058
15. Gregoris, E., Rosso, B., Roman, M., Corami, F. (2023). Microplastics in the freshwater and earthbound conditions: Prevalence, destinies, impacts, and supportable arrangements. In: Khan, A., Wang, C., Asiri, A. M. (Eds.) *Microplastic Sources, Fate and Solution*. Springer, Singapore. https://doi.org/10.1007/978-981-99-0695-6_2
16. La Daana, K. K., Gårdfeldt, K., Lyashevska, O., Hassellöv, M., Thompson, R. C., O'Connor, I. (2018). Microplastics in sub-surface waters of the Arctic Central Basin. *Mar. Pollut. Bull.*, **130**, 8–18. https://doi.org/10.1016/j.marpolbul.2018.03.011
17. Li, S., Wang, H., Feng, X., Zeng, Y., Shen, Y., Gu, Q. (2024). Microplastics in Chinese coastal waters: A mini-review of occurrence characteristics, sources and driving mechanisms. *Waste Manag. Res.*, 0734242X241248727. 10.1177/0734242X241248727
18. Santucci, L., Fernández-Severini, M. D., Rimondino, G. N., Colombo, C. V., Prieto, G., Forero-López, A. D., Carol, E. S. (2024). Assessment of meso- and microplastics distribution in coastal sediments and waters at the middle estuary of the Rio De La Plata, Argentina (SW Atlantic Ocean). *Sci. Total Environ.*, **914**, 170026. https://doi.org/10.1016/j.scitotenv.2024.170026
19. Soininen, T., Uurasjärvi, E., Hämäläinen, L., Huusari, N., Feodoroff, J., Moshnikoff, J., Niiranen, E., Feodoroff, P., Mustonen, T., Koistinen, A. (2024). Microplastics in Arctic waters of the Finnish Sámi area. *Sci. Total Environ.*, **940**, 173666. https://doi.org/10.1016/j.scitotenv.2024.173666
20. Yan, M., Wang, L., Dai, Y., Sun, H., Liu, C. (2021). Behavior of microplastics in inland waters: Aggregation, settlement, and transport. *Bull. Environ.*

Contam. Toxicol., **107**, 700–709. https://doi.org/10.1007/s00128-020-03087-2

21. Wang, X., Bolan, N., Tsang, D. C., Sarkar, B., Bradney, L., Li, Y. (2021). A review of microplastics aggregation in aquatic environment: Influence factors, analytical methods, and environmental implications. *J. Hazard. Mater.*, **402**, 123496. https://doi.org/10.1016/j.jhazmat.2020.123496
22. Prata, J. C., da Costa, J. P., Lopes, I., Andrady, A. L., Duarte, A. C., Rocha-Santos, T. (2021). A One Health perspective of the impacts of microplastics on animal, human and environmental health. *Sci. Total Environ.*, **777**, 146094. https://doi.org/10.1016/j.scitotenv.2021.146094
23. Akdemir, T., Gedik, K. (2023). Microplastic emission trends in Turkish primary and secondary municipal wastewater treatment plant effluents discharged into the Sea of Marmara and Black Sea. *Environ. Res.*, **231**, 116188. https://doi.org/10.1016/j.envres.2023.116188
24. Barchiesi, M., Chiavola, A., Di Marcantonio, C., Boni, M. R. (2021). Presence and fate of microplastics in the water sources: Focus on the role of wastewater and drinking water treatment plants. *J. Water Process Eng.*, **40**, 101787. https://doi.org/10.1016/j.jwpe.2020.101787
25. Edo, C., González-Pleiter, M., Leganés, F., Fernández-Piñas, F., Rosal, R. (2020). Fate of microplastics in wastewater treatment plants and their environmental dispersion with effluent and sludge. *Environ. Pollut.*, **259**, 113837. https://doi.org/10.1016/j.envpol.2019.113837
26. Monira, S., Roychand, R., Hai, F. I., Bhuiyan, M., Dhar, B. R., Pramanik, B. K. (2023). Nano and microplastics occurrence in wastewater treatment plants: A comprehensive understanding of microplastics fragmentation and their removal. *Chemosphere*, **334**, 139011. https://doi.org/10.1016/j.chemosphere.2023.139011
27. Oyewo, B., Tauringana, V., Tingbani, I. (2024). Microplastics in aquatic bodies: Assessing the role of governance mechanisms in industrial wastewater management. *J. Environ. Manage.*, **349**, 119563. https://doi.org/10.1016/j.jenvman.2023.119563
28. Ruffell, H., Pantos, O., Northcott, G., Gaw, S. (2023). Wastewater treatment plant effluents in New Zealand are a significant source of microplastics to the environment. *N. Z. J. Mar. Freshw. Res.*, **57**(3), 336–352. https://doi.org/10.1080/00288330.2021.1988647
29. Ta, A. T., Promchan, N. (2024). Microplastics in wastewater from developing countries: A comprehensive review and methodology suggestions. *TRAC Trend. Anal. Chem.*, 117537. https://doi.org/10.1016/j.trac.2024.117537

30. Wei, H., Wang, J., Pan, S., Liu, J., Ding, H., Smith, K., Yang, Z., Liu, P., Guo, X., Gao, S. (2023). Are wastewater treatment plants as the source of microplastics in surface water and soil? *J. Hazard. Mater.*, **459**, 132154. https://doi.org/10.1016/j.jhazmat.2023.132154

31. Zhuang, L., Zhang, Y., Xue, Y., Ren, Y., He, J., Sun, H. (2024). Influence of ongoing discharge from multiple wastewater treatment plants on microplastic patterns in small-scale receiving rivers. *Sci. Total Environ.*, **932**, 172880. https://doi.org/10.1016/j.scitotenv.2024.172880

32. Ziajahromi, S., Neale, P. A., Leusch, F. D. (2016). Wastewater treatment plant effluent as a source of microplastics: Review of the fate, chemical interactions and potential risks to aquatic organisms. *Water Sci. Technol.*, **74**(10), 2253–2269. https://doi.org/10.2166/wst.2016.414

33. Casella, C., Sol, D., Laca, A., Diaz, M. (2023). Microplastics in sewage sludge: A review. *Environ. Sci. Pollut. Res.*, **30**, 63382–63415. https://doi.org/10.1007/s11356-023-27151-6

34. Chand, R., Iordachescu, L., Bäckbom, F., Andreasson, A., Bertholds, C., Pollack, E., Malazadeh, M., Lorenz, C., Haaning Nielsen, A., Vollertsen, J. (2024). Treating wastewater for microplastics to a level on par with nearby marine waters. *Water Res.*, **256**, 121647. https://doi.org/10.1016/j.watres.2024.121647

35. Gao, Z., Chen, L., Cizdziel, J., Huang, Y. (2023). Research progress on microplastics in wastewater treatment plants: A holistic review. *J. Environ. Manage.*, **325**, 116411. https://doi.org/10.1016/j.jenvman.2022.116411

36. Kamani, H., Ghayebzadeh, M., Azari, A., Ganji, F. (2024). Characteristics of microplastics in a hospital wastewater treatment plant effluent and hazard risk assessment. *Environ. Process.*, **11**, 15. https://doi.org/10.1007/s40710-024-00694-7

37. Komorowska-Kaufman, M., Marciniak, W. (2024). Removal of microplastic particles during municipal wastewater treatment: A current review. *Desalin. Water Treat.*, 100006. https://doi.org/10.1016/j.dwt.2024.100006

38. Nguyen, P. D., Tran, Q. V., Le, T. T., Nguyen, Q. H., Kieu-Le, T. C., Strady, E. (2023). Evaluation of microplastic removal efficiency of wastewater-treatment plants in a developing country, Vietnam. *Environ. Technol. Inno.*, **29**, 102994. https://doi.org/10.1016/j.eti.2022.102994

39. Sun, J., Dai, X., Wang, Q., Van Loosdrecht, M. C., Ni, B. J. (2019). Microplastics in wastewater treatment plants: Detection, occurrence and removal. *Water Res.*, **152**, 21–37. https://doi.org/10.1016/j.watres.2018.12.050

40. Talvitie, J., Mikola, A., Setälä, O., Heinonen, M., Koistinen A. (2017). How well is microlitter purified from wastewater? A detailed study on the stepwise removal of microlitter in a tertiary level wastewater treatment plant. *Water Res.*, **109**, 164–172. https://doi.org/10.1016/j.watres.2016.11.046

41. ECHA (European Chemicals Agency) (2019). Annex XV restriction report proposal for a restriction. Report version number 1 (March 20th, 2019). Helsinki.

42. VV.AA. (2011). Marine Litter, Technical Recommendations for the Implementation of MSFD Requirements, MSFD GES Technical Subgroup on Marine Litter. Publications Office of the European Union. http://dx.doi.org/10.2788/92438

43. Allen, D., Allen, S., Abbasi, S., Baker, A., Bergmann, M., Brahney, J., Butler, T., Duce, R. A., Eckart, S., Evangeliou, N., Jickells, T., Kanakidou, M., Kershaw, P., Laj, P., Levermore, J., Li, D., Liss, P., Liu, K., Mahowald, N., Masque, P., Wright, S. (2022). Microplastics and nanoplastics in the marine-atmosphere environment. *Nat. Rev. Earth Environ.* **3**(6), 393-405. https://doi.org/10.1038/s43017-022-00292-x

44. Casagrande, N., Verones, F., Sobral, P., Martinho, G. (2024). Physical properties of microplastics affecting the aquatic biota: A review. *Environ. Adv.*, 100566. https://doi.org/10.1016/j.envadv.2024.100566

45. Frias, J. P., Nash, R. (2019). Microplastics: Finding a consensus on the definition. *Mar. Pollut. Bull.*, **138**, 145–147. https://doi.org/10.1016/j.marpolbul.2018.11.022

46. Ivleva, N. P. (2021). Chemical analysis of microplastics and nanoplastics: Challenges, advanced methods, and perspectives. *Chem. Rev.*, **121**(19), 11886–11936. https://doi.org/10.1021/acs.chemrev.1c00178

47. Rani-Borges, B., Moschini-Carlos, V., Pompêo, M. (2021). Microplastics and freshwater microalgae: What do we know so far? *Aquat. Ecol.*, **55**, 363–377 (2021). https://doi.org/10.1007/s10452-021-09834-9

48. Wu, F., Reding, L., Starkenburg, M., Leistenschneider, C., Primpke, S., Vianello, A., Zonneveld, K. A. F., Huserbråten, M. B. O., Versteeg, G. J. M., Gerdts, G. (2024). Spatial distribution of small microplastics in the Norwegian Coastal Current. *Sci. Tot. Environ.*, 173808. https://doi.org/10.1016/j.scitotenv.2024.173808

49. Alfonso, M. B., Lindsay, D. J., Arias, A. H., Nakano, H., Jandang, S., Isobe, A. (2023). Zooplankton as a suitable tool for microplastic research. *Sci. Tot. Environ.*, 167329. https://doi.org/10.1016/j.scitotenv.2023.167329

50. Chamanee, G., Sewwandi, M., Wijesekara, H., Vithanage, M. (2023). Global perspective on microplastics in landfill leachate: Occurrence, abundance, characteristics, and environmental impact. *Waste Manage.*, **171**, 10–25. https://doi.org/10.1016/j.wasman.2023.08.011
51. Corami, F., Rosso, B., Roman, M., Picone, M., Gambaro, A., Barbante, C. (2020a). Evidence of small microplastics (<100 μm) ingestion by Pacific oysters (Crassostrea gigas): A novel method of extraction, purification, and analysis using Micro-FTIR. *Mar. Pollut. Bull.*, **160**, 111606. https://doi.org/10.1016/j.marpolbul.2020.111606
52. Feng, S., Lu, H., Liu, Y. (2021). The occurrence of microplastics in farmland and grassland soils in the Qinghai-Tibet plateau: Different land use and mulching time in facility agriculture. *Environ. Pollut.*, **279**, 116939. https://doi.org/10.1016/j.envpol.2021.116939
53. Li, Q., Sun, H., Bai, Q., Li, P., Lai, Y., Yu, S., Liu, X., Yao, Z., Cai, Y., Liu, J. (2023). Spatial distribution of polystyrene nanoplastics and small microplastics in the Bohai Sea, China. *Sci. Tot. Environ.*, **881**, 163222. https://doi.org/10.1016/j.scitotenv.2023.163222
54. Luo, D., Wang, Z., Liao, Z., Chen, G., Ji, X., Sang, Y., Qu, L., Chen, Z., Wang, Z., Dahlgren, R. A., Zhang, M., Shang, X. (2024). Airborne microplastics in urban, rural and wildland environments on the Tibetan Plateau. *J. Hazard. Mater.*, **465**, 133177. https://doi.org/10.1016/j.jhazmat.2023.133177
55. Rosso, B., Corami, F., Barbante, C., Gambaro, A. (2023). Quantification and identification of airborne small microplastics (<100 μm) and other microlitter components in atmospheric aerosol via a novel elutriation and oleo-extraction method. *Environ. Pollut.*, **318**, 120889. https://doi.org/10.1016/j.envpol.2022.120889
56. Rosso, B., Scoto, F., Hallanger, I. G., Larose, C., Gallet, J. C., Spolaor, A., Bravo, B., Barbante C., Gambaro A., Corami, F. (2024). Characteristics and quantification of small microplastics (<100 μm) in seasonal Svalbard snow on glaciers and lands. *J. Hazard. Mater.*, **467**, 133723–133723. https://doi.org/10.1016/j.jhazmat.2024.133723
57. Smyth, K., Tan, S., Van Seters, T., Gasperi, J., Dris, R., Drake, J., Passeport, E. (2024). Small-size microplastics in urban stormwater runoff are efficiently trapped in a bioretention cell. *ACS ES&T Water*, **4**(6), 2522–2531. https://doi.org/10.1021/acsestwater.4c00037
58. Xu, Z., Sui, Q., Li, A., Sun, M., Zhang, L., Lyu, S., Zhao, W. (2020). How to detect small microplastics (20–100 μm) in freshwater, municipal wastewaters and landfill leachates? A trial from sampling to identification. *Sci. Tot. Environ.*, **733**, 139218. https://doi.org/10.1016/j.scitotenv.2020.139218

59. Zhao, S., Zettler, E. R., Bos, R. P., Lin, P., Amaral-Zettler, L. A., Mincer, T. J. (2022). Large quantities of small microplastics permeate the surface ocean to abyssal depths in the South Atlantic Gyre. *Glob. Chang. Biol.*, **28**(9), 2991–3006. https://doi.org/10.1111/gcb.16089

60. Zheng, Y., Su, Z., Liu, D., Huang, B., Mu, Q., Li, Y., Wen, D. (2024). Metagenomics reveals the influence of small microplastics on microbial communities in coastal sediments. *Sci. Tot. Environ.*, **914**, 169982. https://doi.org/10.1016/j.scitotenv.2024.169982

61. Corami, F., Rosso, B., Bravo, B., Gambaro, A., Barbante, C. (2020b). A novel method for purification, quantitative analysis and characterization of microplastic fibers using micro-FTIR. *Chemosphere*, **238**, 124564. https://doi.org/10.1016/j.chemosphere.2019.124564

62. Cordeiro, R. D. M., Cardoso, V. V., Carneiro, R. N., Almeida, C. C. M. (2024). Validation of an FT-IR microscopy method for the monitorization of microplastics in water for human consumption in Portugal: Lisbon case study. PRR: Preventing, Removing and Recycling. Pollutant Toxic Ions and Molecules. *Environ. Sci. Pollut. Res.* https://doi.org/10.1007/s11356-024-33966-8

63. De Frond, H., Cowger, W., Renick, V., Brander, S., Primpke, S., Sukumaran, S., Elkhatib, D., Barnett, S., Navas-Moreno, M., Rickabaugh, K., Vollanls, F., O'Donnel, B., Lusher, A., Lee, E., Lao, W., Amarpuri, G., Sarau, G., Christiansen, S. (2023). What determines accuracy of chemical identification when using microspectroscopy for the analysis of microplastics? *Chemosphere*, **313**, 137300. https://doi.org/10.1016/j.chemosphere.2022.137300

64. Hartmann, N. B., Hüffer, T., Thompson, R. C., Hassellöv, M., Verschoor, A., Daugaard, A. E., Rist, S., Karlsson, T., Brennholt, N., Cole, M., Herrling, M. P., Hess, M. C., Ivleva, N, P., Lusher, A. L., Wagner, M. (2019). Are we speaking the same language? Recommendations for a definition and categorization framework for plastic debris. *Environ. Sci. Technol.*, **53**(3), 1039–1047. https://doi.org/10.1021/acs.est.8b05297

65. Liza, A. A., Ashrafy, A., Islam, M. N., Billah, M., Arafat, S. T., Raham, M., Karim, R., Hasam, M., Promie, A. R., Rahman, S. M. (2024). Microplastic pollution: A review of techniques to identify microplastics and their threats to the aquatic ecosystem. *Environ. Monit. Assess*, **196**, 285. https://doi.org/10.1007/s10661-024-12441-4

66. Qiu, Q., Tan, Z., Wang, J., Peng, J., Li, M., Zhan, Z. (2016). Extraction, enumeration and identification methods for monitoring microplastics in the environment. *Estuar. Coast. Shelf S.*, **176**, 102–109. https://doi.org/10.1016/j.ecss.2016.04.012

67. Amenábar, M., Aguilera, M. A., Gallardo, C., Moore, C., De Vine, R., Lattin, G., Gamba, A., Luna-Acosta, A., Thiel, M. (2024). Spatial distribution of microplastics in a coastal upwelling region: Offshore dispersal from urban sources in the Humboldt Current System. *Environ. Pollut.*, **343**, 123157. https://doi.org/10.1016/j.envpol.2023.123157

68. Mani, T., Hauk, A., Walter, U., Burkhardt-Holm, P. (2015). Microplastics profile along the Rhine River. *Sci. Rep.*, **5**(1), 17988. https://doi.org/10.1038/srep17988

69. Hidalgo-Ruz, V., Gutow, L., Thompson, R. C., Thiel, M. (2012). Microplastics in the marine environment: A review of the methods used for identification and quantification. *Environ. Sci. Technol.*, **46**(6), 3060–3075. https://doi.org/10.1021/es2031505

70. Hung, C., Klasios, N., Zhu, X., Sedlak, M., Sutton, R., Rochman, C. M. (2021). Methods matter: Methods for sampling microplastic and other anthropogenic particles and their implications for monitoring and ecological risk assessment. *Integr. Environ. Assess. Manag.*, **17**(1), 282–291. https://doi.org/10.1002/ieam.4325

71. Jasmine, T. Y., Helm, P. A., Diamond, M. L. (2024). Source-specific categorization of microplastics in nearshore surface waters of the Great Lakes. *J. Great Lakes Res.*, **50**(1), 102256. https://doi.org/10.1016/j.jglr.2023.102256

72. Mai, L., Bao, L. J., Shi, L., Wong, C. S., Zeng, E. Y. (2018). A review of methods for measuring microplastics in aquatic environments. *Environ. Sci. Pollut. Res.*, **25**, 11319–11332. https://doi.org/10.1007/s11356-018-1692-0

73. Pasquier, G., Doyen, P., Kazour, M., Dehaut, A., Diop, M., Duflos, G., Amara, R. (2022). Manta net: The golden method for sampling surface water microplastics in aquatic environments. *Front. Environ. Sci.*, **10**, 811112. https://doi.org/10.3389/fenvs.2022.811112

74. Poli, V., Litti, L., Lavagnolo, M. C. (2024). Microplastic pollution in the North-east Atlantic Ocean surface water: How the sampling approach influences the extent of the issue. *Sci. Tot. Environ.*, 174561. https://doi.org/10.1016/j.scitotenv.2024.174561

75. Schönlau, C., Karlsson, T. M., Rotander, A., Nilsson, H., Engwall, M., van Bavel, B., Kärrman, A. (2020). Microplastics in sea-surface waters surrounding Sweden sampled by manta trawl and in-situ pump. *Mar. Pollut. Bull.*, **153**, 111019. https://doi.org/10.1016/j.marpolbul.2020.111019

76. Tamminga, M., Hengstmann, E., Fischer, E. K. (2018). Microplastic analysis in the South Funen Archipelago, Baltic Sea, implementing

manta trawling and bulk sampling. *Mar. Pollut. Bull.*, **128**, 601–608. https://doi.org/10.1016/j.marpolbul.2018.01.066

77. Uurasjärvi, E., Hartikainen, S., Setälä, O., Lehtiniemi, M., Koistinen, A. (2020). Microplastic concentrations, size distribution, and polymer types in the surface waters of a northern European lake. *Water Environ. Res.*, **92**(1), 149–156. https://doi.org/10.1002/wer.1229.
78. Dris, R., Gasperi, C. J., Rocher, A. V., Saad, B. M. (2015). Microplastic contamination in an urban area: A case study in Greater Paris. *Environ. Chem.*, **12**, 592–599, 10.1071/EN14167
79. Estahbanati, S., Fahrenfeld, N. L. (2016). Influence of wastewater treatment plant discharges on microplastic concentrations in surface water. *Chemosphere,* **162**, 277–284. 10.1016/j.chemosphere.2016.07.083
80. Lechner, A., Keckeis, H., Lumesberger-Loisl, F., Zens, B., Krusch, R., Tritthart, M., Glas, M., Schludermann, E. (2014). The Danube so colourful: A potpourri of plastic litter outnumbers fish larvae in Europe's second largest river. *Environ. Pollut.*, **188**, 177–181. 10.1016/j.envpol.2014.02.006
81. Lindeque, P. K., Cole, M., Coppock, R. L., Lewis, C. N., Miller, Z., Watts, A. J. R., Wilson-McNeal, A., Wright, S. L., Tamara, S. (2020). Are we underestimating microplastic abundance in the marine environment? A comparison of microplastic capture with nets of different mesh-size. *Environ. Pollut.*, **265**, 114721. https://doi.org/10.1016/j.envpol.2020.114721
82. Zhang, K., Gong, W., Lv, J., Xiong, X., Wu. C. (2015). Accumulation of floating microplastics behind the Three Gorges Dam. *Environ. Pollut.*, **204**, 117–123. 10.1016/j.envpol.2015.04.023
83. Barrows, A. P., Neumann, C. A., Berger, M. L., Shaw, S. D. (2017). Grab vs. neuston tow net: A microplastic sampling performance comparison and possible advances in the field. *Anal. Methods*, **9**(9), 1446–1453. https://doi.org/10.1039/C6AY02387H
84. Eriksen, M., Liboiron, M., Kiessling, T., Charron, L., Alling, A., Lebreton, L., Richards, H., Roth, B., Ory, N. C., Hidalgo-Ruz, V., Meerhof, E., Box, C., Cummins, A., Thiel, M. (2018). Microplastic sampling with the AVANI trawl compared to two neuston trawls in the Bay of Bengal and South Pacific. *Environ. Poll.*, **232**, 430–439. https://doi.org/10.1016/j.envpol.2017.09.058
85. Green, D. S., Kregting, L., Boots, B., Blockley, D. J., Brickle, P., Da Costa, M., Crowley, Q. (2018). A comparison of sampling methods for seawater microplastics and a first report of the microplastic litter in coastal

waters of Ascension and Falkland Islands. *Mar. Pollut. Bull.*, **137**, 695–701. https://doi.org/10.1016/j.marpolbul.2018.11.004

86. Nakano, H., Uchida, K., Aikawa, T., Hayashi, T., Arakawa, H. (2021). Reevaluation of microplastics identification based on Neuston net survey data. *Mar. Pollut. Bull.*, **171**, 112799. https://doi.org/10.1016/j.marpolbul.2021.112799
87. Nakano, H., Alfonso, M. B., Jandang, S., Phinchan, N., Chavanich, S., Viyakarn, V., Isobe, A. (2024). Influence of monsoon seasonality and tidal cycle on microplastics presence and distribution in the Upper Gulf of Thailand. *Sci. Tot. Environ.*, **920**, 170787. https://doi.org/10.1016/j.scitotenv.2024.170787
88. Prasad, G., Mohanty, S., Nayak, S. K., Bharat, G. K., Chakraborty, P. (2023). A scientific approach to the occurrence, isolation, and characterization of existing microplastic pollution in the marine environment: A review. *Water, Air, Soil Pollut.*, **234**(7), 480. https://doi.org/10.1007/s11270-023-06494-3
89. Tokai, T., Uchida, K., Kuroda, M., Isobe, A. (2021). Mesh selectivity of neuston nets for microplastics. *Mar. Pollut. Bull.*, **165**, 112111. https://doi.org/10.1016/j.marpolbul.2021.112111
90. Di Mauro, R., Kupchik, M. J., Benfield, M. C. (2017). Abundant plankton-sized microplastic particles in shelf waters of the northern Gulf of Mexico. *Environ. Pollut.*, **230**, 798–809. https://doi.org/10.1016/j.envpol.2017.07.030
91. Guo, P., Wang, Y., Moghaddamfard, P., Meng, W., Wu, S., Bao, Y. (2024). Artificial intelligence-empowered collection and characterization of microplastics: A review. *J. Hazard. Mater.*, 134405. https://doi.org/10.1016/j.jhazmat.2024.134405
92. Rist, S., Vianello, A., Winding, M. H. S., Nielsen, T. G., Almeda, R., Torres, R. R., Vollertsen, J. (2020). Quantification of plankton-sized microplastics in a productive coastal Arctic marine ecosystem. *Environ. Pollut.*, **266**, 115248. https://doi.org/10.1016/j.envpol.2020.115248
93. Stock, F., Kochleus, C., Bänsch-Baltruschat, B., Brennholt, N., Reifferscheid, G. (2019). Sampling techniques and preparation methods for microplastic analyses in the aquatic environment: A review. *TrAC Trend Anal. Chem.*, **113**, 84–92. https://doi.org/10.1016/j.trac.2019.01.014
94. Wu, M., Yang, C., Du, C., Liu, H. (2020). Microplastics in waters and soils: Occurrence, analytical methods and ecotoxicological effects. *Ecotoxicol. Environ. Saf.*, **202**, 110910. https://doi.org/10.1016/j.ecoenv.2020.110910

95. Lins-Silva, N., Marcolin, C. R., Kessler, F., Schwamborn, R. (2024). Influence of microplastics and biogenic particles on the size spectra of tropical estuarine and marine pelagic ecosystems. *Sci. Tot. Environ.*, **927**, 172244. https://doi.org/10.1016/j.scitotenv.2024.172244
96. Mendoza, L. M. R., Balcer, M. (2019). Microplastics in freshwater environments: A review of quantification assessment. *TrAC Trend. Anal. Chem.*, **113**, 402–408. https://doi.org/10.1016/j.trac.2018.10.020
97. Uurasjärvi, E., Pääkkönen, M., Setälä, O., Koistinen, A., Lehtiniemi, M. (2021). Microplastics accumulate to thin layers in the stratified Baltic Sea. *Environ. Pollut.*, **268**, 115700. https://doi.org/10.1016/j.envpol.2020.115700
98. Zhang, L., Liu, J., Xie, Y., Zhong, S., Yang, B., Lu, D., Zhong, Q. (2020). Distribution of microplastics in surface water and sediments of Qin River in Beibu Gulf, China. *Sci. Tot. Environ.*, **708**, 135176. https://doi.org/10.1016/j.scitotenv.2019.135176
99. Campanale C., Stock F., Massarelli C., Kochleus, C., Bagnuolo, G., Reifferscheid, G., Uricchio, V. F. (2020). Microplastics and their possible sources: The example of Ofanto river in Southeast Italy. *Environ. Pollut.*, **258**, 113284. https://doi.org/10.1016/j.envpol.2019.113284
100. Ryan, P. G., Suaria, G., Perold, V., Bornman, T. G., Aliani, S. (2020). Sampling microfibres at the sea surface: The effects of mesh size and water depth. *Environ. Pollut.*, **258**, 113413.
101. Liedermann, M., Gmeiner, P., Pessenlehner, S., Haimann, M., Hohenblum, P., Habersack, H. (2018). A methodology for measuring microplastic transport in large or medium rivers. *Water*, **10**(4), 414. https://doi.org/10.3390/w10040414
102. Razeghi, N., Hamidian, A. H., Wu, C., Zhang, Y., Yang., M. (2021). Microplastic sampling techniques in freshwaters and sediments: A review. *Environ. Chem. Lett.*, **19**, 4225–4252. https://doi.org/10.1007/s10311-021-01227-6
103. Campanale, C., Savino, I., Pojar, I., Massarelli, C., Uricchio, V. F. (2020). A practical overview of methodologies for sampling and analysis of microplastics in riverine environments. *Sustainability*, **12**(17), 6755. https://doi.org/10.3390/su12176755
104. Cutroneo, L., Reboa, A., Besio, G., Borgogno, F., Canesi, L., Canuto, S., Dara, M., Enrile, F., Forioso, I., Greco, G., Lenoble, V., Malatesta, A. Mounier, S., Petrillo, M., Rovetta, R., Stocchino, A., Tesan, J., Vagge, G., Capello, M. (2020). Microplastics in seawater: Sampling strategies, laboratory methodologies, and identification techniques applied to port environment. *Environ. Sci. Pollut. Res.*, **27**, 8938–8952. https://doi.org/10.1007/s11356-020-07783-8

105. Lenz, R., Labrenz, M. (2018). Small microplastic sampling in water: Development of an encapsulated filtration device. *Water,* **10**(8), 1055. https://doi.org/10.3390/w10081055

106. Liu, K., Courtene-Jones, W., Wang, X., Song, Z., Wei, N., Li, D. (2020). Elucidating the vertical transport of microplastics in the water column: A review of sampling methodologies and distributions. *Water Res.,* **186**, 116403. https://doi.org/10.1016/j.watres.2020.116403

107. Tamminga, M., Stoewer, S. C., Fischer, E. K. (2019). On the representativeness of pump water samples versus manta sampling in microplastic analysis. *Environ. Pollut.,* **254**, 112970. https://doi.org/10.1016/j.envpol.2019.112970

108. Farjami, H., Azizpour, J., Andi, S. (2024). Manifestation of internal waves in the Southern Caspian Sea using satellite imagery. *Reg. Stud. Mar. Sci.,* **69**, 103294. https://doi.org/10.1016/j.rsma.2023.103294

109. Antacli, J. C., Di Mauro, R., Rimondino, G. N., Alurralde, G., Schloss, I. R., González, G. A., Morales, S., Ottero, A., Vodopivez, C. (2024). Microplastic pollution in waters of the Antarctic coastal environment of Potter Cove (25 De Mayo Island/King George Island, South Shetlands). *Sci. Tot. Environ.,* 915, 170155. https://doi.org/10.1016/j.scitotenv.2024.170155

110. Barrows, A. P., Christiansen, K. S., Bode, E. T., Hoellein, T. J. (2018). A watershed-scale, citizen science approach to quantifying microplastic concentration in a mixed land-use river. *Water Res.,* **147**, 382–392. https://doi.org/10.1016/j.watres.2018.10.013

111. De-la-Torre, G. E., Santillán, L., Dioses-Salinas, D. C., Yenney, E., Toapanta, T., Okoffo, E. D., Kannan, G., Madadi, N., Dobaradaran, S. (2024). Assessing the current state of plastic pollution research in Antarctica: Knowledge gaps and recommendations. *Chemosphere,* 141870. https://doi.org/10.1016/j.chemosphere.2024.141870

112. Faull, L. E. M., Zaliznyak, T., Taylor, G. T. (2024). From the Caribbean to the Arctic, the most abundant microplastic particles in the ocean have escaped detection. *Mar. Pollut. Bull.,* **202**, 116338. https://doi.org/10.1016/j.marpolbul.2024.116338

113. Gunaalan, K., Almeda, R., Lorenz, C., Vianello, A., Iordachescu, L., Papacharalampos, K., Rohde Kiær, C. M., Vollertsen, J, Nielsen, T. G. (2023). Abundance and distribution of microplastics in surface waters of the Kattegat/Skagerrak (Denmark). *Environ. Pollut.,* **318**, 120853. https://doi.org/10.1016/j.envpol.2022.120853

114. Manullang, C. Y., Patria, M. P., Haryono, A., Anuar, S. T., Fadli, M., Susanto, R. D., Wei, Z. (2024). Vertical distribution of microplastic along the

main gate of Indonesian Throughflow pathways. *Mar. Pollut. Bull.*, **199**, 115954. https://doi.org/10.1016/j.marpolbul.2023.115954

115. McMullen, K., Calle, P., Alvarado-Cadena, O., Kowal, M. D., Espinoza, E., Domínguez, G. A., Tirape, A., Vargas, F. H., Grant, E., Hunt, B. P. V., Pakhomov, E., Alava, J. J. (2024). Ecotoxicological assessment of microplastics and cellulose particles in the Galápagos Islands and Galápagos penguin food web. *Environ. Toxicol. Chem.*, **43**(6), 1442–1457. https://doi.org/10.1002/etc.5874
116. Zhang, K., Shi, H., Peng, J., Wang, Y., Xiong, X., Wu, C., Lam, P. K. (2018). Microplastic pollution in China's inland water systems: A review of findings, methods, characteristics, effects, and management. *Sci. Tot. Environ.*, **630**, 1641–1653. https://doi.org/10.1016/j.scitotenv.2018.02.300
117. Koelmans, A. A., Nor, N. H. M., Hermsen, E., Kooi, M., Mintenig, S. M., De France, J. (2019). Microplastics in freshwaters and drinking water: Critical review and assessment of data quality. *Water Res.*, **155**, 410–422. https://doi.org/10.1016/j.watres.2019.02.054
118. Mintenig, S. M., Bäuerlein, P. S., Koelmans, A. A., Dekker, S. C., Van Wezel, A. P. (2018). Closing the gap between small and smaller: Towards a framework to analyze nano-and microplastics in aqueous environmental samples. *Environ. Sci. Nano*, **5**(7), 1640–1649. https://doi.org/10.1039/C8EN00186C
119. Kataoka, T., Nihei, Y., Kudou, K., Hinata, H. (2019). Assessment of the sources and inflow processes of microplastics in the river environments of Japan. *Environ. Pollut.*, **244**, 958–965. https://doi.org/10.1016/j.envpol.2018.10.111
120. Vaughan, R., Turner, S. D., Rose, N. L. (2017). Microplastics in the sediments of a UK urban lake. *Environ. Pollut.*, **229**, 10–18. https://doi.org/10.1016/j.envpol.2017.05.057
121. Razeghi, N., Hamidian, A. H., Mirzajani, A., Abbasi, S., Wu, C., Zhang, Y., Yang, M. (2022). Sample preparation methods for the analysis of microplastics in freshwater ecosystems: A review. *Environ. Chem. Lett.*, 1–27. https://doi.org/10.1007/s10311-021-01341-5
122. Tirkey, A., Upadhyay, L. S. B. (2021). Microplastics: An overview on separation, identification and characterization of microplastics. *Mar. Pollut. Bull.*, **170**, 112604. https://doi.org/10.1016/j.marpolbul.2021.112604
123. Zhang, Z., Wu, X., Liu, H., Huang, X., Chen, Q., Guo, X., Zhang, J. (2023). A systematic review of microplastics in the environment: Sampling, separation, characterization and coexistence mechanisms with pollutants. *Sci. Tot. Environ.*, **859**, 160151.

124. Briggs, E., de Moura, E. A., Furusawa, H. A., Cotrim, M. E., Oguzie, E. E., Lugao, A. B. (2019). Microplastics: A novel method for surface water sampling and sample extraction in Elechi Creek, Rivers State, Nigeria. In *Characterization of Minerals, Metals, and Materials* 2019 (pp. 269–281). Springer International Publishing.

125. Gray, A. D., Wertz, H., Leads, R. R., Weinstein, J. E. (2018). Microplastic in two South Carolina Estuaries: Occurrence, distribution, and composition. *Mar. Pollut. Bull.*, **128**, 223–233. https://doi.org/10.1016/j.marpolbul.2018.01.030

126. Yonkos, L. T., Friedel, E. A., Perez-Reyes, A. C., Ghosal, S., Arthur, C. D. (2014) Microplastics in four estuarine rivers in the Chesapeake Bay, USA. *Environ. Sci. Technol.*, **48**(24), 14195–14202. https://doi.org/10.1021/es5036317

127. Lusher, A., Buenaventura, N. T., Eidsvoll, D., Thrane, J. E., Økelsrud, A., Jartun, M. (2018). Freshwater microplastics in Norway: A first look at sediment, biota and historical plankton samples from Lake Mjøsa and Lake Femunden. Norwegian Insititute for Water Research. Available online http://hdl.handle.net/11250/2588713

128. Enders, K., Käppler, A., Biniasch, O., Feldens, P., Stollberg, N., Lange, X., Fischer, D., Eichorn, K. J., Pollenhe, F., Oberbeckmann, S., Labrenz, M. (2019). Tracing microplastics in aquatic environments based on sediment analogies. *Sci. Rep.*, **9**(1), 1–15. https://doi.org/10.1038/s41598-019-50508-2

129. Grbić, J., Helm, P., Athey, S., Rochman, C. M. (2020). Microplastics entering northwestern Lake Ontario are diverse and linked to urban sources. *Water Res.*, **174**, 115623. https://doi.org/10.1016/j.watres.2020.11562

130. Eo, S., Hong, S. H., Song, Y. K., Han, G. M., Shim, W. J. (2019). Spatiotemporal distribution and annual load of microplastics in the Nakdong River, South Korea. *Water Res.*, **160**, 228–237. https://doi.org/10.1016/j.watres.2019.05.053

131. Xiong, X., Wu, C., Elser, J. J., Mei, Z., Hao, Y. (2019). Occurrence and fate of microplastic debris in middle and lower reaches of the Yangtze River–from inland to the sea. *Sci. Total Environ.*, **659**, 66–73. https://doi.org/10.1016/j.scitotenv.2018.12.313

132. Zhang, K., Chen, X., Xiong, X., Ruan, Y., Zhou, H., Wu, C., Lam, P. K. (2019). The hydro-fluctuation belt of the three Gorges reservoir: Source or sink of microplastics in the water? *Environ. Pollut.*, **248**, 279–285. https://doi.org/10.1016/j.envpol.2019.02.043

133. Fan, Y., Zheng, K., Zhu, Z., Chen, G., Peng, X. (2019). Distribution, sedimentary record, and persistence of microplastics in the Pearl River catchment, China. *Environ. Pollut.*, **251**, 862–870. https://doi.org/10.1016/j.envpol.2019.05.056

134. Liu, F., Olesen, K. B., Borregaard, A. R., Vollertsen, J. (2019). Microplastics in urban and highway stormwater retention ponds. *Sci. Tot. Environ.*, **671**, 992–1000. https://doi.org/10.1016/j.scitotenv.2019.03.416

135. Zhao, S., Zhu, L., Wang, T., Li, D. (2014). Suspended microplastics in the surface water of the Yangtze Estuary System, China: First observations on occurrence, distribution. *Mar. Pollut. Bull.*, **86**(1–2), 562–568. https://doi.org/10.1016/j.marpolbul.2014.06.032

136. Al-Azzawi, M. S., Kefer, S., Weißer, J., Reichel, J., Schwaller, C., Glas, K., Knoop, O., Drewes, J. E. (2020). Validation of sample preparation methods for microplastic analysis in wastewater matrices—reproducibility and standardization. *Water*, **12**(9), 2445. https://doi.org/10.3390/w12092445

137. Hurley, R. R., Lusher, A. L., Olsen, M., Nizzetto, L. (2018). Validation of a method for extracting microplastics from complex, organic-rich, environmental matrices. *Environ. Sci. Tech.*, **52**(13), 7409–7417.

138. Pfeiffer, F., Fischer, E. K. (2020). Various digestion protocols within microplastic sample processing: Evaluating the resistance of different synthetic polymers and the efficiency of biogenic organic matter destruction. *Front. Environ. Sci.*, **8**, 572424. https://doi.org/10.3389/fenvs.2020.572424

139. Cole, M., Lindeque, P., Halsband, C., Galloway, T. S. (2011). Microplastics as contaminants in the marine environment: A review. *Mar. Pollut. Bull.*, **62**(12), 2588–2597. https://doi.org/10.1016/j.marpolbul.2011.09.025

140. Tuttle, E., Stubbins, A. (2023). An optimized acidic digestion for the isolation of microplastics from biota-rich samples and cellulose acetate matrices. *Environ. Pollut.*, **322**, 121198.

141. Dehaut, A., Cassone, A. L., Frère, L., Hermabessiere, L., Himber, C., Rinnert, E., Riviére, G., Lambert, C., Soudant, P., Huvet, A., Dufloss, G., Paul-Pont, I. (2016). Microplastics in seafood: Benchmark protocol for their extraction and characterization. *Environ. Pollut.*, **215**, 223–233.

142. Duan, J., Han, J., Zhou, H., Lau, Y. L., An, W., Wei, P., Cheung, S. G., Yang, Y., Tam, N. F. Y. (2020). Development of a digestion method for determining microplastic pollution in vegetal-rich clayey mangrove sediments. *Sci. Tot. Environ.*, **707**, 136030. https://doi.org/10.1016/j.scitotenv.2019.136030

143. Monteiro, S. S., da Costa, J. P. (2022). Methods for the extraction of microplastics in complex solid, water and biota samples. *Trends Environ. Anal. Chem.*, **33**, e00151. https://doi.org/10.1016/j.teac.2021.e00151

144. Avio, C. G., Gorbi, S., Regoli, F. (2015). Experimental development of a new protocol for extraction and characterization of microplastics in fish tissues: First observations in commercial species from Adriatic Sea. *Mar. Environ. Res.*, **111**, 18–26. https://doi.org/10.1016/j.marenvres.2015.06.014

145. Nava, V., Chandra, S., Aherne, J., Alfonso, M. B., Antao-Geraldes, A. M., Attermeyer, K., Bao, R., Bartrons, M., Berger, S. A., Biernaczyk, M., Bissen, R., Brookes, J. D., Brown, D., Canedo-Arguelles, M., Canle, M., Capelli, C., Carballeira, R., Cereijo, J. L., Chawchai, S., Christensen, S. T., Christoffersen, K. S., de Eyto, E., Delgado, J., Dornan, T. N., Doubek, J. P., Dusaucy, J., Erina, O., Ersoy, Z., Feuchtmayr, H., Frezzotti, M. L., Galafassi, S., Gateuille, D., Goncalves, V., Grossart, H. P., Hamilton, D. P., Harris, T. D., Kangur, K., Kankılıç, G. B., Kessler, R., Kiel, C., Krynak, E. M., Leiva-Presa, A., Lepori, F., Matias, M. G., Matsuzaki, S. S., McElarney, Y., Messyasz, B., Mitchell, M., Mlambo, M. C., Motitsoe, S. N., Nandini, S., Orlandi, V., Owens, C., Ozkundakci, D., Pinnow, S., Pociecha, A., Raposeiro, P. M., Room, E. I., Rotta, F., Salmaso, N., Sarma, S. S. S., Sartirana, D., Scordo, F., Sibomana, C., Siewert, D., Stepanowska, K., Tavsanoglu, U. N., Tereshina, M., Thompson, J., Tolotti, M., Valois, A., Verburg, P., Welsh, B., Wesolek, B., Weyhenmeyer, G. A., Wu, N., Zawisza, E., Zink, L., and Leoni, B. (2023). Plastic debris in lakes and reservoirs. *Nature*, **619**, 317. https://doi.org/10.1038/s41586-023-06168-4

146. Anderson, P. J., Warrack, S., Langen, V., Challis, J. K., Hanson, M. L., Rennie, M. D. (2017) Microplastic contamination in Lake Winnipeg, Canada. Environ. *Pollution* **225**, 223–231. https://doi.org/10.1016/j.envpol.2017.02.072

147. Rodrigues, M. O., Gonçalves, A. M. M., Gonçalves, F. J. M., Nogueira, H., Marques, J. C., Abrantes, N. (2018). Effectiveness of a methodology of microplastics isolation for environmental monitoring in freshwater systems. *Ecol. Indic.*, **89**, 488–495.

148. Löder, M. G., Imhof, H. K., Ladehoff, M., Löschel, L. A., Lorenz, C., Mintenig, S., Piehl, S., Primpke, S., Schrank, I., Laforsh, C., Gerdts, G. (2017). Enzymatic purification of microplastics in environmental samples. *Environ. Sci. Tech.*, **51**(24), 14283–14292.

149. Michida, Y., Chavanich, S., Chiba, S., Cordova, M. R., Cozsar Cabanas, A., Hagmann, P., Hinata, H., Isobe, A., Kershaw, P., Kozlovskii, N., Li, D., Lusher, A. L., Martì, E., Mason, S. A., Mu, J., Saito, H., Shim, W. J., Syakti, A.

D., Takada, H., Thompson, R., Tokai, T., Uchida, K., Vasilenko, K., Wang, J. (2019). Guidelines for harmonizing ocean surface microplastic monitoring methods. Version 1.1. Ministry of the Environment, Japan, p. 71. https://repository.oceanbestpractices.org/bitstream/handle/11329/983/20190514_First%20Version%20of%20Guideline%20v5.pdf?sequence=1

150. Olesen, K. B., Stephansen, D. A., van Alst, N., Vollertsen, J. (2019). Microplastics in a stormwater pond. *Water*, **11**(7), 1466. https://doi.org/10.3390/w11071466

151. Crichton, E. M., Noël, M., Gies, E. A., Ross, P. S. (2017). A novel, density-independent and FTIR-compatible approach for the rapid extraction of microplastics from aquatic sediments. *Anal. Methods*, **9**(9), 1419–1428. 10.1039/c6ay02733d

152. Debraj, D., Lavanya, M. (2023). Microplastics everywhere: A review on existing methods of extraction. *Sci. Tot. Environ.*, **893**, 164878. https://doi.org/10.1016/j.scitotenv.2023.164878

153. Dong, M., Luo, Z., Jiang, Q., Xing, X., Zhang, Q., Sun, Y. (2020). The rapid increases in microplastics in urban lake sediments. *Sci. Rep.*, **10**(1), 848. https://doi.org/10.1038/s41598-020-57933-8

154. Lechthaler, S., Hildebrandt, L., Stauch, G., Schüttrumpf, H. (2020). Canola oil extraction in conjunction with a plastic free separation unit optimises microplastics monitoring in water and sediment. *Anal. Methods*, **12**(42), 5128–5139. 10.1039/d0ay01574a

155. Lekše, N., Žgajnar Gotvajn, A., Zupančič, M., Bulc, T. G. (2024). Oil-based extraction as an efficient method for the quantification of microplastics in environmental samples. *Environ. Sci. Eur.*, **36**, 68. https://doi.org/10.1186/s12302-024-00898-6

156. Mani, T., Burkhardt-Holm, P. (2020). Seasonal microplastics variation in nival and pluvial stretches of the Rhine River – From the Swiss catchment towards the North Sea. *Sci. Tot. Environ.*, **707**, 135579. https://doi.org/10.1016/j.scitotenv.2019.135579

157. Mani, T., Frehland, S., Kalberer, A., Burkhardt-Holm, P. (2019). Using castor oil to separate microplastics from four different environmental matrices. *Anal. Methods*, **11**(13), 1788–1794. https://doi.org/10.1039/C8AY02559B

158. Andersen R. R. (2005) *Algal Culturing Technique*. Elsevier Press.

159. Brierley, B., Carvalho, L., Davies, S., Krokowski, J. (2007). Guidance on the quantitative analysis of phytoplankton in Freshwater Samples. NERC Open Research Archive. http://nora.nerc.ac.uk/id/eprint/5654/

160. Comtois, P., Alcazar, P., Neron, D. (1999). Pollen counts statistics and its relevance to precision. *Aerobiologia*, **15**, 19–28. https://doi.org/10.1023/A:1007501017470

161. Gough, H. L., Stahl, D. A. (2003). Optimization of direct cell counting in sediment. *J. Microbiol. Methods*, **52**, 39–46.

162. Huppertsberg, S., Knepper, T. P. (2018). Instrumental analysis of microplastics: Benefits and challenges. *Anal. Bioanal. Chem.*, **410**, 6343–6352. https://doi.org/10.1007/s00216-018-1210-8

163. Imhof, H. K., Laforsch, C., Wiesheu, A. C., Schmid, J., Anger P. M., Niessner, R., Ivlev, N. P. (2016). Pigments and plastic in limnetic ecosystems: A qualitative and quantitative study on microparticles of different size classes. *Water Res.*, **98**(1), 64–74.

164. Lisle, J. T., Hamilton, M. A., Willse, A. R., McFeters, G. A. (2004). Comparison of fluorescence microscopy and solid-phase cytometry methods for counting bacteria in water. *App. Environ. Microbiol.*, **70**(9), 5343–5348.

165. Mazziotti, C., Fiocca, A., Vadrucci, M. R. (2013). Phytoplankton in transitional waters: Sedimentation and counting methods. *TWB J.*, **2**, 90–99.

166. Muthukrishnan, T., Govender, A., Dobretsov, S., Abed R. M. M. (2017). Evaluating the reliability of counting bacteria using epifluorescence microscopy. *J. Mar. Sci. Eng.*, **5**, 4.

167. Oßmann, B. E., Sarau, G., Holtmannspötter, H., Pischetsrieder, M., Christiansen, S. H., Dicke, W. (2018). Small-sized microplastics and pigmented particles in bottled mineral water. *Water Res.*, **15**(141), 307–316.

168. Tong, H., Jiang, Q., Hu, X., Zhong, X. (2020). Occurrence and identification of microplastics in tap water from China. *Chemosphere*, **252**, 126493. https://doi.org/10.1016/j.chemosphere.2020.126493

169. Singh, B., Kumar, A. (2023). Advances in microplastics detection: A comprehensive review of methodologies and their effectiveness. *TrAC Trends in Analytical Chemistry*, 117440.

170. Wang, W., Wang, J. (2018). Investigation of microplastics in aquatic environments: An overview of the methods used, from field sampling to laboratory analysis. *TrAC Trend. Anal. Chem.*, **108**, 195–202. https://doi.org/10.1016/j.trac.2018.08.026

171. Shim, W. J., Hong, S. H., Eo, S. E. (2017). Identification methods in microplastic analysis: A review. *Anal. Meth.*, **9**(9), 1384–1391. https://doi.org/10.1039/C6AY02558G

172. Silva, A. B., Bastos, A. S., Justino, C. I., da Costa, J. P., Duarte, A. C., Rocha-Santos, T. A. (2018). Microplastics in the environment: Challenges in analytical chemistry—A review. *Anal. Chim. Acta*, **1017**, 1–19. https://doi.org/10.1016/j.aca.2018.02.043

173. Veerasingam, S., Ranjani, M., Venkatachalapathy, R., Bagaev, A., Mukhanov, V., Litvinyuk, D., Mugilarasan, M., Gurumoorthi, K., Guganathan, L., Aboobacker, V. M., Vethamony, P. (2020). Contributions of Fourier transform infrared spectroscopy in microplastic pollution research: A review. *Crit. Rev. Environ. Sci. Technol.*, **51**(22), 2681–2743. https://doi.org/10.1080/10643389.2020.1807450

174. Löder, M. G. J., Kuczera, M., Mintenig, S., Lorenz, C., Gerdts, G. (2015). Focal plane array detector-based micro-Fourier-transform infrared imaging for the analysis of microplastics in environmental samples. *Environ. Chem.*, **12**, 563–581. https://doi.org/10.1071/EN14205

175. Yang, J., Monnot, M., Sun, Y., Asia, L., Wong-Wah-Chung, P., Doumenq, P., Moulin, P. (2023). Microplastics in different water samples (seawater, freshwater, and wastewater): Methodology approach for characterization using micro-FTIR spectroscopy. *Water Res.*, **232**, 119711. https://doi.org/10.1016/j.watres.2023.119711

Chapter 4

Occurrence, Characteristics, and Removal of Microplastics in Wastewater Treatment Plants

Joaquín López-Castellanos, Sonia Olmos, Marta Doval-Miñarro, and Javier Bayo
Departamento de Ingeniería Química y Ambiental, Universidad Politécnica de Cartagena, Spain
javier.bayo@upct.es

This chapter presents a summary of the occurrence, characteristics, and removal of microplastics (MPs) in sewage treatment plants, considered sinks and sources of MPs to the neighbouring environment. It intends to highlight the importance of these systems in transporting MPs to the aquatic and terrestrial surrounding environment, previously studying the removal rates and the best available technologies described for their elimination. Despite being a barrier, wastewater treatment plants (WWTPs) have proved to be an important source of MPs in the environment, both from treated effluent and the application of sludges in agricultural soils. Through

Microplastic Pollution: Environmental Impact and Solutions
Edited by Julia Martín, Juan Luis Santos, Irene Aparicio, and Esteban Alonso

ISBN 978-981-5129-77-9 (Hardcover), 978-1-003-66006-4 (eBook)
www.jennystanford.com

DOI: 10.1201/9781003660064-4

literature data, the chapter discusses the abundance of MPs, the main methodologies for their identification and quantification, as well as the main polymers. It also discusses the removal rates for primary, secondary, and tertiary stages within the WWTPs, and the main technologies and operational parameters for their retention. Future perspectives regarding improvement in sewage treatment plants and research projects associated with them are also included; despite the high removal efficiencies achieved by the current treatment plants, the treated effluents still discharge millions of MPs into the receiving environment every single day.

4.1 Introduction

WWTPs act as a sink for MPs included in household wastewater, effluents from industry, or those carried away by stormwater run-off [1–4]. In the meantime, WWTPs act as a source [5], releasing into the aquatic environment a great amount of this type of micropollutant, especially in the form of fibres originating from laundry [6].

In this sense, the vast majority of MPs are not eliminated in the sewage wastewater treatment but relocated and stored into the sludge, although with a reduced size compared to the effluent due to mechanical erosion during the treatment process [7]. Because of the intensive use of these treated sludges or biosolids, with emphasis on land application in agriculture, MPs exist and accumulate in the environment, which can have major impacts on the surrounding ecosystem [8, 9]. Besides, MPs transferred from wastewater to sludge can be vectors for toxic substances such as antibiotics, heavy metals, or persistent organic compounds [3, 10]. It has been also reported that UV degradation enhances the uptake of Cu, Zn, and Pb by MPs, as well as the leaching of dissolved organic carbon [3].

The removal of MPs from wastewater, covering operation conditions and efficiencies, is still incipient. The lack of standardized procedural and analytical methods limits the accuracy assessment of these pollutants and data comparison between different studies [11]. Their abundance has been associated with several parameters such as chemical oxygen demand (COD) [12, 13], biochemical oxygen demand (BOD) [14], rainfall [4, 15], the formation process of biofilms on their surfaces [16], or even with various economic indicators [17]. The effect and impact of seasonality on the fate and transport, and the effectiveness in removing MPs by WWTPs, have also been studied [12, 14, 18], trying to develop a deeper understanding of the pathways, dynamic fate, and transport of MPs in WWTPs, which could guide the development of recommendations for plant operators, engineers, system designers, policymakers, and even educators, who can advise people on how to recycle plastics. An optimization of the process design of existing WWTP can be crucial for the prevention and control of MP pollution [19].

4.2 WWTPs as a Sink and a Source of MPs to the Environment

4.2.1 Introduction

WWTPs act as a sink for MPs contained in both domestic and industrial wastewater, but also as a source to the environment [20, 21] due to the continuous discharge of treated wastewater in large volumes, and MP analysis and monitoring in wastewater effluent and sludge samples is a growing field of study in recent years [22]. Sources of MPs in WWTPs are known; household washing machines are prone to releasing synthetic textile fibres [6, 23], MPs released from personal care products [24], paint residues [25], granules from the plastic industry and manufactured fibre [26], tyre dust [27], or MPs from plastic consumer goods [28] reach wastewater influent, thus being a route of entry of MPs into the aquatic environment. Conventional wastewater treatment processes based on primary and secondary activated sludge techniques can eliminate up to 99% of MPs, depending on different factors [29–31], although despite the high reduction capacity, conventional WWTPs are in turn an important source of MPs due to the large volume of effluents daily discharged. Although WWTPs have proved to be effective in removing MPs from wastewater, none of the treatment processes were designed for their removal.

4.2.2 Influence of the Treatment Process

4.2.2.1 Preliminary and primary treatments

After a preliminary treatment, where the removal of coarse solids, such as rags, sticks, floatable, and other large materials often found in raw wastewater, is carried out [32], primary treatment mainly focuses on physical processes to remove organic and inorganic suspended solids from wastewater, including coarse and fine screening, grit and grease chamber, and settlement process [33, 34]. These treatments reach, approximately, a 25–50% elimination of the incoming BOD, 50–70% of the total suspended solids, and 65% of oil and grease included in wastewater [32] and protect subsequent devices from damage or blockage.

Many authors have highlighted these stages as the most important for MP removal. It has been reported that 45% of MPs entering the WWTPs can be eliminated after grit and grease removal, and about 34% after primary sedimentation [29, 30], and Talvitie et al. [31] reported that textile fibres were mainly removed from wastewater during primary sedimentation.

Fine screens, typically with a mesh size less than 6 mm, allow the retaining of suspended and floating solids [33], collecting MPs no larger than 2 mm [7]. Zhang et al. [19] reported an MP removal rate of 60.3%, and Dris et al. [35] observed removal of fibres up to 80% during primary treatment. Gies et al. [36] reported the highest removal percentage of MPs after the primary treatment (91.6%), and Petroody et al. [37] indicated that more than 70% of plastic fibres were removed in a sewage plant in the south of the Caspian Sea after the primary settling tank, especially particles ≥500 μm. Tang et al. [1] and Liu et al. [7] observed a larger MP size in influent than that in effluent, and also in wastewater compared to MPs in sludge, mainly due to mechanical erosion and sedimentation into sludge.

In countries where primary sedimentation is absent in most WWTPs, the removal of suspended solids relies on biological processes and final clarification. In this sense, and because primary sedimentation plays an important role in removing fibres, there is still a large proportion of them in the effluent from secondary treatment [38].

4.2.2.2 Secondary treatment

In most developed countries, secondary wastewater treatment consists of a biological process to remove dissolved organic matter, nutrients, and [34]. The most widely used technologies are the classical activated sludge, anaerobic anoxic oxic (A2O) technology, the sequencing batch reactor (SBR), and the membrane bioreactor (MBR), which combines the activated sludge process with a membrane separation process [39, 40]. Zhang et al. [19] reported lower MP removal ability for SBR compared to A2O technology.

Secondary treatment reduces MPs through entrapment in solid flocs and ingestion by protozoa and metazoan [41, 42], and secondary sedimentation promotes their removal by transferring them into the sludge [33, 43]. In this sense, Jeong et al. [41] reported a reduced growth rate and fecundity in *Brachionus koreanus* after exposure to polystyrene microbeads, with significant size-dependent effects. As further discussed, a WWTP is a sink of MPs but also a source to the environment through agricultural biosolids application [9, 44].

The connection between microorganisms and MPs in WWTPs has been extensively studied, with a greater chance of adhesion to MPs with rough surfaces [45]. Extracellular polymeric substances (EPS), secreted by microorganisms once they adhere to a surface, make the MPs to aggregate and flocculate. Cunha et al. [46] reported a high bioflocculant activity of an EPS produced by *Cyanothece* sp. on polystyrene MPs, although, at the same time, these MPs induced a negative effect on the microalgal growth.

Overall, the secondary treatment reduces MPs between 1% and 52% with respect to the primary clarifier [33]. Murphy et al. [30] reported a further 20.1% removal. MPs may affect the biological conversion efficiency of inorganic nitrogen [12]. Cluzard et al. [47] reported the greatest recovery of ammonium in sediments with MPs, and Green et al. [48] indicated that carbonyl groups from plastic polymers may hydrate and easily sequester available nutrients. Wastewater influents with a high content of suspended solids were also related to a low MP burden because of a possible hetero aggregation [12].

4.2.2.3 Tertiary treatment

Some of the equipment and technologies developed for the tertiary treatment of wastewater have proved to be efficient in the removal of MPs. Rapid sand filtration (RSF) is a common strategy for the removal of different pollutants from water, due to its cost-effective, environmentally friendly, and easy-to-operate characteristics [49]. With this technology, MPs are effectively removed by 84–98%. The main drawbacks of this technology include the saturation processes, MPs smaller than the adsorbent, or reactor hydrodynamics, which can be alleviated by backwashing [50], together with the mechanical stress generated during the treatment, which can result in the release of smaller particles into the environment [51]. Also, selective removal of particulate MPs against microfibres has been reported [21]. For RSF, MPs can provide surfaces for microbial attachment and growth [31].

Flotation technology has been employed for more than a century in ore processing, and dissolved air flotation (DAF), using microbubbles between 30 and 100 μm, is the most used process to remove different pollutants from water, as colloids, fine and ultrafine particles, precipitates, or microorganisms, among others [52]. As a technology for MP removal, DAF is less sensitive to flow variations than typical sedimentation processes. Wang et al. [53] reported an optimal removal efficiency of MPs by conventional DAF between 32% and 38% at 0.4–0.5 MPa, with an increase of 13.6–33.7% when surface modifiers were used on microbubbles. Esfandiari and Mowla [54] used Al-based and Fe-based coagulants to remove polyethylene MPs from greywater in a DAF system, reporting a reduction up to 96.10% and 70.56%, respectively, and Swart et al. [55] observed a large reduction in MP count with the presence of NaCl salt that improved flotation performance.

The feasibility of removing MPs by dynamic membrane (DM) technology is gaining attraction, for its relatively low cost, low energy consumption, and easy cleaning, as it has been proven to be effective in MP removal in wastewater treatment [56]. A photocatalytic membrane reactor (PMR) is a combined system of photocatalytic treatment and membrane separation, which can enhance MP removal [57]. However, membrane fouling remains a major drawback to the application of membrane systems, including a membrane bioreactor (MBR) [58]; membrane technology has been used alongside traditional filtration systems that use membranes with pores of micrometre or nanometre size to retain small solids. It has been observed that the concentration of MPs decreased significantly after ultrafiltration (UF) and reverse osmosis (RO) [59]. Li et al. reported that the presence of MPs (<5 μm) caused higher fouling in their studied MBR system resulting from the pore-clogging of membranes by tiny MPs [59]. There is a need to investigate a more accurate simulation of the fouling behaviour of practical wastewater and provide a solid theoretical basis for fouling control strategies in membrane systems [60].

The use of constructed wetlands with horizontal subsurface flow has also been effective in MP removal, with up to 35.3% [19].

4.2.3 Retention of MPs in WWTP Sludge

Regardless of the amount and type of processes used in WWTPs, MPs are most intercepted in the sewage sludge, followed by different treatments, including sanitary landfilling, land utilization, burning, and preparation of building materials, among others [61]. Municipal solid waste landfills are an important source of MPs, and a huge bioreactor for plastic-degrading microorganisms [62]. Polymers such as polyethylene (PE), polypropylene (PP), polystyrene (PS), PET, and up to 20 different types have been detected [63], with a significant increase in landfill depth [62]. For this reason, research on the diffusion route of MPs in sludge should be considered a priority, and source control should be reinforced to prevent environmental risks. Besides, the wide range reported on the abundance of MPs strongly emphasizes the need for standardization and validation of analytical methods [64].

An examination of the role of sludge treatment processes indicates that the dewatering process along with the aerobic digestion process plays an important role in reducing the abundance of MPs in the final sludge, with an approximate decrease of 54% [65, 66]. Sludge stabilization may also affect the MP amount in the final sludge, although it depends on the stabilization technique applied to the sludge. Mahon et al. [67] reported small MP sizes after a full-scale anaerobic digestion, thermal drying, and lime stabilization process, which was attributed to a combination of elevated pH, temperature, and mechanical mixing. Moreover, no difference in MP concentration between thermophilic and mesophilic anaerobic digestion has also been reported [68]. Zhang et al. [10] reported that as MPs are just transferred from wastewater to sludge, they could bring problems to anaerobic digestion as they are vectors for toxic substances, as previously discussed. This is mainly due to the high hydrophobicity of MPs, which facilitates their adsorption onto organic surfaces of sewage sludge [65].

As a complex material, a plastic polymer can be degraded by microorganisms if it is previously exposed to abiotic factors, such as temperature, ultraviolet light, gases, or mechanical impacts, which initiate physical and chemical changes in its structure [69]. These changes unintentionally occurred in real life and can be monitored by different parameters, such as carbonyl index, vinyl index, or crystallinity, for instance in MPs recovered in marine sediments [70]. These processes facilitate biodegradation by reducing their high molecular weight and large polymer structure [65].

Furthermore, the escalating prevalence of MPs in sludge diminishes the effectiveness of anaerobic sludge digestion. This could be attributed to either the interruption of enzymatic processes or the harmful impacts of chemicals seeping from MPs on the microorganisms that facilitate sludge digestion [44].

4.2.4 Presence of MPs in the Surrounding Environment

WWTPs are identified as an important source of MPs in the aquatic environment, such as rivers and lakes, which eventually get into the marine environment [22]. Urban rivers serve as primary channels for the transportation of land-sourced MPs into the world's oceans, although the real-time dynamics of MP emissions into rivers during rainfall and run-off events remain poorly understood [15]. Kittipongvises et al. [18] demonstrated that rainfall influences the mobility and transport of MPs to the WWTP particularly during the wet season, while differences in sampling, extraction, and identification techniques must be considered, specific characteristics of the environment, morphology, surface, catchment area, wind patterns, and other factors are likely to play an important role in the distribution of MPs.

Because of their buoyancy and enduring nature, MPs have the potential to disperse extensively throughout the ocean via hydrodynamic processes and ocean currents [16], and due to their tiny size, they have been proven to be ingested by aquatic organisms, with still limited knowledge of their effects on these living beings. They are harmful to photosynthesizing algae, aquatic invertebrates, and fish [71], with detrimental effects on their body, i.e., decreasing their appetite and overall food consumption [72], neurological impairments, endocrine disruption, and increased susceptibility to cytotoxicity, among others [73]. Apart from the effects of direct MP ingestion, the combined effects with other environmental pollutants have also been identified [74]. The colour of MPs may affect the ingestion, as many organisms are visual predators and may feed on MPs that resemble their prey [75].

In current research, MPs are understood to impact the human body through two primary routes: inhalation and ingestion. MPs, owing to their small size and weight, can bypass nasal cilia and enter the trachea and lungs, potentially leading to chronic or acute inflammation [16, 35].

Different researchers have pointed out the role of WWTPs in this transportation [5, 76], with different estimations of the annual emission with the treated effluent or the sewage sludge. In coastal areas, MP pollution from the discharge of treated wastewater poses a significant environmental threat to marine life and ecosystems [8, 77, 78], and when tourism puts enormous pressure on WWTPs, some of them have adopted an innovative technology with variations in its size according to the season [14], and in some cases, a greater amount of MPs was reported in a WWTP during the dry season [18].

The presence of MPs in biosolids, used as a by-product of WWTPs in agriculture to enrich soil with nutrients, has also been studied [9]. The degree of effects on the ecosystems varies with concentration, size, type of polymer, and time exposition, although the level to which they should be limited or rejected for land applications due to the unavoidable presence of MPs is yet to be established [79]. Each analytical methodology has its limitations, emphasizing the need for standardized methods [9].

4.3 MP Analysis in WWTPS

The analysis of MPs in WWTPs is significantly hampered by the lack of a standardized methodology, which should include sampling techniques, digestion of organic matter, isolation, processing, and identification of MP in wastewater samples, similar to those guidelines published for MP analysis in solid samples [80, 81].

4.3.1 Sample Collection

There is a wide variety of techniques employed for sampling, processing, and quantifying MPs from WWTPs and, in many cases, the inconsistency of operation protocols is the main problem for comparisons among the available data. The number and amount of wastewater samples collected significantly vary among studies: Bayo et al. [12] collected 125 wastewater samples during 19 months in the southeast of a Spanish WWTP, and Petroody et al. [37], close to the Caspian Sea, and Wei et al. [13], in five WWTPs located in China, both during a week. Dris et al. [35] carried out three sampling campaigns on three consecutive days (8, 9, and 10 April 2014). Zhang et al. [82] collected wastewater from four WWTPs between March and June 2018 for three consecutive days, and Wei et al. [13] collected 28 samples from five WWTPs; both studies were carried out in China. Vardar et al. [83], Kasmuri et al. [84], and Ormaniec and Mikosz [85] collected wastewater samples only for one day, from a Turkish, a Malaysian, and a Polish WWTP, respectively.

Besides, three types of sample collection techniques have been described: (i) grab samples, or single samples taken at a specific time, and normally collected in plastic or glass jars, or steel buckets [12, 29, 30, 83, 86, 87], although with a limitation, as they only represent a snapshot in time [86]; (ii) composite samples, with equal volumes collected throughout the day (between 15 min and 24 h) and with or without autosampler [31, 37, 43]; and (iii) the volume reduction technique, where a pump and a filter system is used to directly feed steel screens with mesh sizes from about 10 to 500 μm [13, 29, 38, 85], or even a single sieve [88]. This last system can effectively increase the sampling volume, although the filters can get clogged, especially with raw wastewater [89].

4.3.2 Sample Pretreatment

Different pretreatment methods have been issued to reduce organic and inorganic matter from wastewater and sludge samples, in order not to interfere with MP analysis. Wet peroxidation, with reagents such as hydrogen peroxide (H_2O_2) [36, 37, 43, 83, 86, 90] or Fenton's reagent [38, 82, 85], is usually used, with more advantages than acid or alkaline at same reaction conditions [11]. Other practices described for wastewater samples include sodium hypochlorite ($NaClO$) [29], an enzymatic-oxidative protocol [91], an acid/alkaline digestion with sulphuric acid (H_2SO_4) and potassium hydroxide (KOH) [92], or a two-step digestion process with H_2O_2 and nitric acid (HNO_3), especially in organic-rich matrices, such as sludge samples [93].

However, the digestion of organic and inorganic matter in previous MP isolation has been also widely discussed. Van Cauwenberghe and Janssen [94] reported a detrimental effect on nylon fibres when HNO_3 is used for the extraction procedure, and Wagner et al. [42] indicated chemical destruction of plastic material with H_2O_2 and H_2SO_4, such as poly(methyl methacrylate) or polycarbonate.

4.3.3 Identification and Quantification Techniques

After the digestion process, most studies use vacuum filtration techniques to collect microparticles from wastewater samples for further analysis, together with density separation, including sodium chloride ($NaCl$), zinc chloride ($ZnCl_2$), sodium iodide (NaI), or zinc bromide ($ZnBr_2$) [11, 95–97]. Only a few studies have utilized centrifugation to process samples [34, 98, 99]. The most commonly reported identification and quantification methodologies are as follows:

4.3.3.1 Visual identification

A first approach to MP analysis can be naked eye sorting, although it can only separate plastics >2 mm or non-plastic matter [11]. It is more useful for their identification by microscopic counting or stereomicroscope. Samples with a relatively high concentration of large MPs can be easily identified by this methodology, although pure visual identification using only morphological criteria, colour, or size may lead to a misidentification, and should be complemented with a spectroscopic method [12,100]. It should be used as a visual discrimination step to yield an assessment of suspected MPs, also including other small items of non-plastic microdebris [36].

Visual identification is a preliminary step to observe treated particles, but it has drawbacks for fibres, particles smaller than 100 μm, or transparent microparticles [29]. In this sense, some authors have proposed specific characteristics to distinguish artificial or synthetic fibres from natural components, i.e., the form should not taper at either end, not rigidly straight form, no organic structures, uniformity of colour, and sheen once passed through the polarized light. Where ambiguity remains, the suspected polymer should be manipulated with a hot pin by which a melted form indicates a positive result [67].

4.3.3.2 Spectroscopy method

In these cases, samples are subjected to infrared radiation, for Fourier transform infrared spectroscopy (FTIR), or to the interaction with irradiated laser light, for Raman spectroscopy. For the first methodology, large particles can be analysed by attenuated total reflectance FTIR (ATR-FTIR) [5, 12, 21, 24, 30, 31, 36, 43, 61, 70] and small microparticles down to 20 μm can be analysed by microscopy coupled to FTIR (μFTIR) [38, 97, 101]. Sometimes, to facilitate particle counting, gridded Petri dishes are used [29].

Raman spectroscopy is a vibrational spectroscopy technique based on the inelastic scattering of light that provides information in the form of a vibrational spectrum, extensively used in MP analysis in WWTPs [7, 19, 102]. Microscopy coupled to Raman spectroscopy (μRaman) is the only method available for particles falling in the range of 1 to 20 μm [101] and has been deeply used for MP identification in WWTPs [13, 37]. Both spectroscopy methodologies share advantages, such as non-destructiveness, low sample amount requirement, possibility for high throughput screening and environmental friendliness [102], and clear advantage over chromatographic or calorimetric methods [103].

4.3.3.3 Chromatographic method

These include thermo-analytical methods, such as pyrolysis with gas chromatography and mass spectrometry (Py–GC-MS) and liquid chromatography (LC). Thermo-analytical methods have the advantage of minimal sample preparation with short analysis times, and they have been proposed for MP analysis in WWTPs in different studies carried out in Germany [91], South Africa [104], Republic of Korea [105], or China [106]. The study in South Africa includes a Py–GC-TOF (time of flight)-MS technique, a very appropriate method for the qualitative determination of the pyrolysis products, that requires a database of pyrograms created with the six most common MP standards, i.e., PE, PS, PP, PA, PET, and PVC [104].

4.3.3.4 Other methods

Staining agents have been widely used, as hydrophobic dyes are adsorbed onto MP surfaces, rendering fluorescence when they are irradiated with blue light [101]. It is a straightforward method that allows fast screening of MP at a low cost. However, dye molecules are easily desorbed from MPs because of just physical adsorption, and not all staining agents are suitable for all types of polymers [107].

Differential scanning calorimetry (DSC) is a fast and comparatively easy analytical tool, which has been used to detect and quantify semi-crystalline thermoplastics, although polymers without crystalline fraction, i.e., polystyrene (PS), polyvinylchloride (PVC), or polyurethane (PUR), are not quantifiable [2].

Kim et al. [23] reported the use of a laser diffraction particle size analyser to record the volume-based size distribution of microfibres, as well as fluid imaging microscopy (FlowCam) to quantify MP fibres in laundry wastewater.

4.3.4 Main Polymers Isolated in WWTPs

A wide variety of polymers have been observed in different studies related to MP in wastewater. While a vast majority of these polymers are low-density, they may lose buoyancy under the weight of fouling [58, 60]. According to Bitter and Lackner [2], polyethylene (PE) and polypropylene (PP) remain the predominant polymers detected in industrial wastewaters, with polyamide or nylon (PA) and polyethylene terephthalate (PET) also commonly identified. Liu et al. [7] reported that PA was the main plastic component in wastewater with 54.8%, and, according to Zhang et al. [10], PET, PE, PP, and polystyrene (PS) are the most presented MPs in wastewater. PE is generally detected as a transparent film, likely originating from its widespread use in packaging and plastic bags, as well as in personal care products [61]. Additionally, a significant proportion of identified fibres prove to be PET, likely stemming from the washing of clothing items [12, 61, 38].

Wei et al. [13] identified PP (54.6%), PS (29.7%), and PET (9.7%) in the influent of rural domestic wastewater, with fragments as the dominant shape (71.3%), followed by fibres (21.5%). Murphy et al. [30] reported polyester (PES) (28%), PP (12%), acrylic plastic (ACR) (12%), alkyd (ALK) (8%), PE (4%), PS (4%), and PET (4%) in the final effluent of domestic WWTPs.

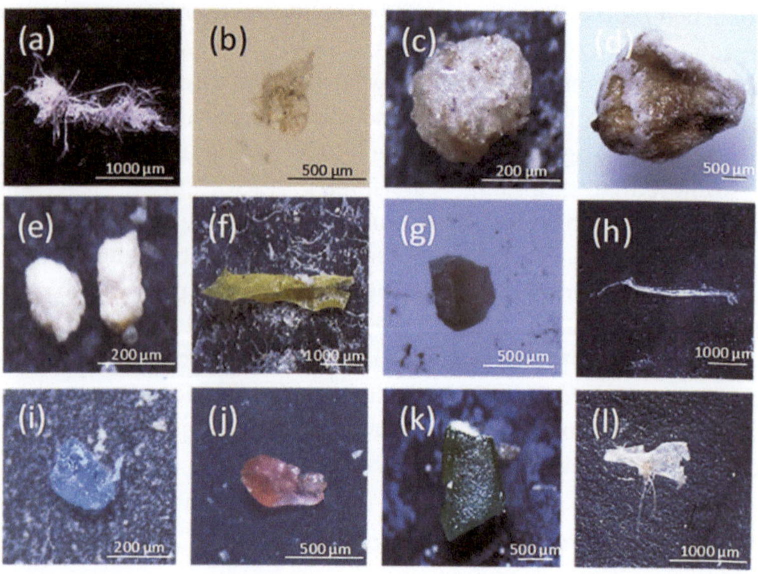

Figure 4.1 Microparticles (a–e) vs microplastics (f–l) in different stages of a WWTP identified by FTIR: (a) cellulose (effluent); (b) silica dioxide (biological); (c) soap (after roughing); (d) chipboard (biological); (e) calcium carbonate (effluent); (f) low-density polyethylene (LDPE) (after grinding); (g) ethylene propylenediene monomer (EPDM) rubber (biological); (h) polyethylene terephthalate (PET) (biological); (i) low-density polyethylene (LDPE) (after roughing); (j) polypropylene (PP) (biological); (k) polyethylene (PE) (effluent); (l) acrylate (ACRYL) (after roughing) [12].

Not all suspected MPs are always confirmed as polymers; indeed, a significant portion is identified as non-plastic particles upon closer examination (Fig. 4.1) [12]. Zhang et al. [19] reported that 49.8% of the suspected particles were identified as MPs and included PP, PE, ACR, PET, and polyvinyl chloride (PVC), the last one widely used as pipe material and buildings [61]. PES is the predominant type of fibre in the effluents of the sand chamber (57.0%), the primary sedimentation tank (70.0%) and the clarification unit (40.0%), followed by PA and ACR [37].

4.4 Discussion and Perspectives

There is no doubt that it is necessary to improve the management of plastics and reduce their use where feasible to minimize the amount of (micro)plastics released into the environment. There is a need to provide a comprehensive solution for the elimination of MPs by improving current technologies and processes or adjusting operational parameters (e.g. hydraulic residence time). Moreover, techniques need to be developed to recover MPs for subsequent reuse. For instance, conversion of plastics into fuel may be suitable for LDPE, HDPE, PP, PET, PVC, and PS, allowing energy recovery through a co-pyrolysis process; obtaining syngas as an alternative source of energy from various types of plastics (LDPE, HDPE, PS, PET, PP) by catalytic pyrolysis; use as a construction material in self-compacting concrete or clay composites. Although specific procedures to treat recovered MPs are not yet realistic, these techniques could be applied to MPs in the future.

Considering the current knowledge related to MPs in wastewater and WWTPs, it seems appropriate to address a few aspects of research and development for a better understanding and mitigation of MP pollution. Some of them relate to (a) harmonization of sampling methodologies and detection of MPs; (b) inclusion in future studies, treatment, and management of MPs smaller than 20 μm given their abundance and biotoxic potential; (c) enhancement of treatment processes aimed at reducing the discharge of MPs either through effluent or via sludge; (d) monitoring of MPs in sludge applied to soils, especially agricultural ones; and (e) development of technologies or measures in favour of the reduction at origin and control of the source of MPs in wastewater.

References

1. Tang, N., Liu, X., and Xing, W. (2020). Microplastics in wastewater treatment plants of Wuhan, Central China: Abundance, removal, and potential source in household wastewater, *Sci. Total Environ.*, **745**, 141026.
2. Bitter, H., and Lackner, S. (2020). First quantification of semi-crystalline microplastics in industrial wastewaters, *Chemosphere*, **258**, 127388.
3. Herath, A., Datta, D. K., Bonyadinejad, G., and Salehi, M. (2023). Partitioning of heavy metals in sediments and microplastics from stormwater runoff, *Chemosphere*, **332**, 138844.
4. Smyth, K., Drake, J., Li, Y., Rochman, C., Van Seters, T., and Passeport, E. (2021). Bioretention cells remove microplastics from urban stormwater, *Water Res.*, **191**, 116785.
5. Bayo, J., López-Castellanos, J., Olmos, S., and Rojo, D. (2023). A holistic approach on the impact of microplastic discharge from WWTPs to the neighboring environment in Southeast Spain, *Water Res.*, **244**, 120516.
6. Kelly, M. R., Lant, N. J., Kurr, M., and Burgess, J. G. (2019). Importance of water-volume on the release of microplastic fibers from laundry, *Environ. Sci. Technol.*, **53**, pp. 11735–11744.
7. Liu, X., Yuan, W., Di, M., Li, Z., and Wang, J. (2019). Transfer and fate of microplastics during the conventional activated sludge process in one wastewater treatment plant of China, *Chem. Eng. J.*, **362**, pp. 176–182.
8. Lu, Q., He, Z. L., and Stoffella, P. J. (2012). Land application of biosolids in the USA: A review, *Appl. Environ. Soil Sci.*, **2012**, pp. 1–11.
9. Christian, A. E., and Köper, I. (2023). Microplastics in biosolids: A review of ecological implications and methods for identification, enumeration, and characterization, *Sci. Total Environ.*, **864**, 161083.
10. Zhang, X., Chen, J., and Li, J. (2020). The removal of microplastics in the wastewater treatment process and their potential impact on anaerobic digestion due to pollutants association, *Chemosphere*, **251**, 126360.
11. Kang, P., Ji, B., Zhao, Y., and Wei, T. (2020). How can we trace microplastics in wastewater treatment plants: A review of the current knowledge on their analysis approaches, *Sci. Total Environ.*, **745**, 140943.
12. Bayo, J., Olmos, S., and López-Castellanos, J. (2020). Microplastics in an urban wastewater treatment plant: The influence of physicochemical parameters and environmental factors, *Chemosphere*, **238**, 124593.
13. Wei, S., Luo, H., Zou, J., Chen, J., Pan, X., Rousseau, D. P., and Li, J. (2020). Characteristics and removal of microplastics in rural domestic

wastewater treatment facilities of China, *Sci. Total Environ.*, **739**, 139935.

14. Bogdanova, M., Yotinov, I., and Topalova, Y. (2024). Comparison of the work of wastewater treatment plant "Ravda" in summer and winter influenced by the seasonal mass tourism industry and COVID-19, *Processes*, **12**, 192.
15. Imbulana, S., Tanaka, S., Moriya, A., and Oluwoye, I. (2024). Inter-event and intra-event dynamics of microplastic emissions in an urban river during rainfall episodes, *Environ. Res.*, **243**, 117882.
16. He, S., Jia, M., Xiang, Y., Song, B., Xiong, W., Cao, J., Peng, H., Yang, Y., Wang, W., Yang, Z., and Zeng, G. (2022). Biofilm on microplastics in aqueous environment: Physicochemical properties and environmental implications, *J. Hazard. Mat.*, **424**, 127286.
17. Miranda, M. N., Silva, A. M., and Pereira, M. F. R. (2020). Microplastics in the environment: A DPSIR analysis with focus on the responses, *Sci. Total Environ.*, **718**, 134968.
18. Kittipongvises, S., Phetrak, A., Hongprasith, N., and Lohwacharin, J. (2022). Unravelling capability of municipal wastewater treatment plant in Thailand for microplastics: Effects of seasonality on detection, fate and transport, *J. Environ. Manage.*, **302**, 113990.
19. Zhang, B., Wu, Q., Gao, S., Ruan, Y., Qi, G., Guo, K., and Zeng, J. (2023). Distribution and removal mechanism of microplastics in urban wastewater plants systems via different processes, *Environ. Pollut.*, **320**, 121076.
20. Gatidou, G., Arvaniti, O. S., and Stasinakis, A. S. (2019). Review on the occurrence and fate of microplastics in Sewage Treatment Plants, *J. Hazard. Mat.*, **367**, pp. 504–512.
21. Bayo, J., López-Castellanos, J., and Olmos, S. (2020). Membrane bioreactor and rapid sand filtration for the removal of microplastics in an urban wastewater treatment plant, *Mar. Pollut. Bull.*, **156**, 111211.
22. Okoffo, E. D., O'Brien, S., O'Brien, J. W., Tscharke, B. J., and Thomas, K. V. (2019). Wastewater treatment plants as a source of plastics in the environment: A review of occurrence, methods for identification, quantification and fate, *Environ. Sci.: Water Res. Technol.*, **5**, pp. 1908–1931.
23. Kim, S., Hyeon, Y., Rho, H., and Park, C. (2024). Ceramic membranes as a potential high-performance alternative to microplastic filters for household washing machines, *Sep. Purif. Technol.*, **344**, 127278.

24. Bayo, J., Martínez, A., Guillén, M., Olmos, S., Roca, M. J., and Alcolea, A. (2017). Microbeads in commercial facial cleansers: Threatening the environment, CLEAN–Soil Air Water, **45**, 1600683.
25. Gaylarde, C. C., Neto, J. A. B., and da Fonseca, E. M. (2021). Paint fragments as polluting microplastics: A brief review, *Mar. Pollut. Bull.*, **162**, 111847.
26. Kefer, S., Friedenauer, T., and Langowski, H. C. (2022). Characterisation of different manufactured plastic microparticles and their comparison to environmental microplastics, *Powder Technol.*, **412**, 117960.
27. Kole, P. J., Löhr, A. J., Van Belleghem, F. G., and Ragas, A. M. (2017). Wear and tear of tyres: A stealthy source of microplastics in the environment, *Int. J. Environ. Res. Public Health*, **14**, 1265.
28. Sipe, J. M., Bossa, N., Berger, W., von Windheim, N., Gall, K., and Wiesner, M. R. (2022). From bottle to microplastics: Can we estimate how our plastic products are breaking down? *Sci. Total Environ.*, **814**, 152460.
29. Carr, S. A., Liu, J., and Tesoro, A. G. (2016). Transport and fate of microplastic particles in wastewater treatment plants, *Water Res.*, **91**, pp. 174–182.
30. Murphy, F., Ewins, C., Carbonnier, F., and Quinn, B. (2016). Wastewater treatment works (WwTW) as a source of microplastics in the aquatic environment, *Environ. Sci. Technol.*, **50**, pp. 5800–5808.
31. Talvitie, J., Mikola, A., Setälä, O., Heinonen, M., and Koistinen, A. (2017). How well is microlitter purified from wastewater? A detailed study on the stepwise removal of microlitter in a tertiary level wastewater treatment plant, *Water Res.*, **109**, pp. 164–172.
32. Sonune, A., and Ghate, R. (2004). Developments in wastewater treatment methods, *Desalination*, **167**, pp. 55–63.
33. Iyare, P. U., Ouki, S. K., and Bond, T. (2020). Microplastics removal in wastewater treatment plants: A critical review, *Environ. Sci.: Water Res. Technol.*, **6**, pp. 2664–2675.
34. Ta, A. T., and Promcham, N. (2024). Microplastics in wastewater from developing countries: A comprehensive review and methodology suggestions, *TrAC, Trends Anal. Chem.*, **171**, 117537.
35. Dris, R., Gasperi, J., Rocher, V., Saad, M., Renault, N., and Tassin, B. (2015). Microplastic contamination in an urban area: A case study in Greater Paris, *Environ. Chem.*, **12**, pp. 592–599.
36. Gies, E. A., LeNoble, J. L., Noël, M., Etemadifar, A., Bishay, F., Hall, E. R., and Ross, P. S. (2018). Retention of microplastics in a major secondary wastewater treatment plant in Vancouver, Canada, *Mar. Pollut. Bull.*, **133**, pp. 553–561.

37. Petroody, S. S. A., Hashemi, S. H., and van Gestel, C. A. (2020). Factors affecting microplastic retention and emission by a wastewater treatment plant on the southern coast of Caspian Sea, *Chemosphere*, **261**, 128179.

38. Tadsuwan, K., and Babel, S. (2022). Unraveling microplastics removal in wastewater treatment plant: A comparative study of two wastewater treatment plants in Thailand, *Chemosphere*, **307**, 135733.

39. Al-Rekabi, W. S., Qiang, H., and Qiang, W. W. (2007). Improvements in wastewater treatment technology, *Pak. J. Nutr.*, **6**, pp. 104–110.

40. Rojas, M. R., Leung, C., Bonk, F., Zhu, Y., Edwards, L., Arnold, R. G., Sáez, A. E., and Klečka, G. (2013). Assessment of the effectiveness of secondary wastewater treatment technologies to remove trace chemicals of emerging concern, *Crit. Rev. Env. Sci. Tec.*, **43**, pp. 1281–1314.

41. Jeong, C. B., Won, E. J., Kang, H. M., Lee, M. C., Hwang, D. S., Hwang, U. K., Zhou, B., Souissi, S., Lee, S. J., and Lee, J. S. (2016). Microplastic size-dependent toxicity, oxidative stress induction, and p-JNK and p-p38 activation in the monogonont rotifer (*Brachionus koreanus*), *Environ. Sci. Technol.*, **50**, pp. 8849–8857.

42. Wagner, M., and Lambert, S. (2018). *Freshwater Microplastics: Emerging Environmental Contaminants?* 1st Ed. (Springer Nature, USA).

43. Koyuncuoğlu, P., and Erden, G. (2023). Microplastics in municipal wastewater treatment plants: A case study of Denizli/Turkey, *Front. Environ. Sci. Eng.*, **17**, 99.

44. Tang, K. H. D. (2023). Bioaugmentation of anaerobic wastewater treatment sludge digestion: A perspective on microplastics removal, *J. Clean. Prod.*, **387**, 135864.

45. Feng, L., He, L., Jiang, S., Chen, J., Zhou, C., Qian, Z. J., Hong, P., Sun, S., and Li, C. (2020). Investigating the composition and distribution of microplastics surface biofilms in coral areas. *Chemosphere*, **252**, 126565.

46. Cunha, C., Silva, L., Paulo, J., Faria, M., Nogueira, N., and Cordeiro, N. (2020). Microalgal based biopolymer for nano- and microplastic removal: A possible biosolution for wastewater treatment, *Environ. Pollut.*, **263**, 114385.

47. Cluzard, M., Kazmiruk, T. N., Kazmiruk, V. D., and Bendell, L. I. (2015). Intertidal concentrations of microplastics and their influence on ammonium cycling as related to the shellfish industry, *Arch. Environ. Contam. Toxicol.*, **69**, pp. 310–319.

48. Green, D. S., Boots, B., Sigwart, J., Jiang, S., and Rocha, C. (2016). Effects of conventional and biodegradable microplastics on a marine ecosystem engineer (*Arenicola marina*) and sediment nutrient cycling, *Environ. Pollut.*, **208**, pp. 426–434.

49. Yang, H., Tang, X., Bai, L., Yang, L., Ding, J., Chen, R., Du, X., li, G., and Liang, H. (2022). Synergistic effects of prokaryotes and oxidants in rapid sand filters treatment of groundwater versus surface water: Purification efficacy, stability and associated mechanisms, *Chemosphere*, **295**, 133804.

50. Chabi, K., Li, J., Ye, C., Kiki, C., Xiao, X., Li, X., Guo, L., Gad, M., Feng, M., and Yu, X. (2024). Rapid sand filtration for <10 μm-sized microplastic removal in tap water treatment: Efficiency and adsorption mechanisms, *Sci. Total Environ*, **912**, 169074.

51. Conesa, J. A., and Ortuño, N. (2022). Reuse of water contaminated by microplastics, the effectiveness of filtration processes: A review, *Energies*, **15**, 2432.

52. Rodrigues, R. T., and Rubio, J. (2007). DAF–dissolved air flotation: Potential applications in the mining and mineral processing industry, *Int. J. Min. Process.*, **82**, pp. 1–13.

53. Wang, Y., Li, Y. N., Tian, L., Ju, L., and Liu, Y. (2021). The removal efficiency and mechanism of microplastic enhancement by positive modification dissolved air flotation, *Water Environ. Res.*, **93**, pp. 693–702.

54. Esfandiari, A., and Mowla, D. (2021). Investigation of microplastic removal from greywater by coagulation and dissolved air flotation, *Process Saf. Environ. Prot.*, **151**, pp. 341–354.

55. Swart, B., Chew, Y. J., and Wenk, J. (2023). Simultaneous monitoring of flow patterns, and bubble, and plastics micro-particle characteristics in dissolved air flotation (DAF), *Chem. Eng. Res. Des.*, **197**, pp. 148–158.

56. Li, L., Xu, G., Yu, H., and Xing, J. (2018). Dynamic membrane for micro-particle removal in wastewater treatment: performance and influencing factors, *Sci. Total Environ.*, **627**, pp. 332–340.

57. Golmohammadi, M., Musavi, S. F., Habibi, M., Maleki, R., Golgoli, M., Zargar, M., Dumée, L. F., Baroutian, S., and Razmjou, A. (2023). Molecular mechanisms of microplastics degradation: A review, *Sep. Purif. Technol.*, **309**, 122906.

58. Esteki, S., Gharibian, S., Hazrati, H., and Taheri, E. (2024). Application of an electrochemical filter-press flowcell in an electrocoagulation-MBR system: Efficient membrane fouling mitigation, *J. Environ. Chem. Eng.*, **12**, 111769.

59. Li, L., Liu, D., Song, K., and Zhou, Y. (2020). Performance evaluation of MBR in treating microplastics polyvinylchloride contaminated polluted surface water, *Mar. Pollut. Bull.*, **150**, 110724.

60. Liu, X. Y., Chen, Y. B., Fu, J., Zhu, X., Lv, L. Y., Sun, L., Zhang, G. M., and Ren, Z. J. (2024). A review of combined fouling on high-pressure membranes in municipal wastewater reuse: Behaviors, mechanisms, and pretreatment mitigation strategies, *Chem. Eng. J.*, **485**, 150135.

61. Yuan, F., Zhao, H., Sun, H., Sun, Y., Zhao, J., and Xia, T. (2022). Investigation of microplastics in sludge from five wastewater treatment plants in Nanjing, China, *J. Environ. Manage.*, **301**, 113793.

62. Li, N., Han, Z., Guo, N., Zhou, Z., Liu, Y., and Tang, Q. (2022). Microplastics spatiotemporal distribution and plastic-degrading bacteria identification in the sanitary and non-sanitary municipal solid waste landfills, *J. Hazard. Mat.*, **438**, 129452.

63. Nayahi, N. T., Ou, B., Liu, Y., and Janjaroen, D. (2022). Municipal solid waste sanitary and open landfills: Contrasting sources of microplastics and its fate in their respective treatment systems, *J. Clean. Prod.*, **380**, 135095.

64. Müller, Y. K., Wernicke, T., Pittroff, M., Witzig, C. S., Storck, F. R., Klinger, J., and Zumbülte, N. (2020). Microplastic analysis—are we measuring the same? Results on the first global comparative study for microplastic analysis in a water sample, *Anal. Bioanal. Chem.*, **412**, pp. 555–560.

65. Hatinoğlu, M. D., and Sanin, F. D. (2021). Sewage sludge as a source of microplastics in the environment: A review of occurrence and fate during sludge treatment, *J. Environ. Manage.*, **295**, 113028.

66. Petroody, S. S. A., Hashemi, S. H., and van Gestel, C. A. (2021). Transport and accumulation of microplastics through wastewater treatment sludge processes, *Chemosphere*, **278**, 130471.

67. Mahon, A. M., O'Connell, B., Healy, M. G., O'Connor, I., Officer, R., Nash, R., and Morrison, L. (2017). Microplastics in sewage sludge: Effects of treatment, *Environ. Sci. Technol.*, **51**, pp. 810–818.

68. Hurley, R. R., Lusher, A. L., Olsen, M., and Nizzetto, L. (2018). Validation of a method for extracting microplastics from complex, organic-rich, environmental matrices, *Environ. Sci. Technol.*, **52**, pp. 7409–7417.

69. Liu, L., Xu, M., Ye, Y., and Zhang, B. (2022). On the degradation of (micro) plastics: Degradation methods, influencing factors, environmental impacts, *Sci. Total Environ.*, **806**, 151312.

70. Bayo, J., Rojo, D., and Olmos, S. (2022). Weathering indices of microplastics along marine and coastal sediments from the harbor of Cartagena (Spain) and its adjoining urban beach, *Mar. Pollut. Bull.*, **178**, 113647.

71. Harmon, S. M., Chen, Q., Ma, C., Ji, M., Yan, X., Ji, R., and Shi, H. (2024). The effects of microplastic pollution on aquatic organisms, Chapter 12 "Microplastic contamination in aquatic environments" (Elsevier Inc., pp. 355–379).
72. Banaee, M., Gholamhosseini, A., Sureda, A., Soltanian, S., Fereidouni, M. S., and Ibrahim, A. T. A. (2021). Effects of microplastic exposure on the blood biochemical parameters in the pond turtle (*Emys orbicularis*), *Environ. Sci. Pollut. Control Ser.*, **28**, pp. 9221–9234.
73. Shukla, S., Pei, Y., Li, W. G., and Pei, D. S. (2024). Toxicological research on nano and microplastics in environmental pollution: Current advances and future directions, *Aquat. Toxicol.*, **270**, 106894.
74. Nafea, T. H., Al-Maliki, A. J., and Al-Tameemi, I. M. (2024). Sources, fate, effects, and analysis of microplastic in wastewater treatment plants: A review, *Environ. Eng. Res.*, **29**, 230040.
75. Liu, P., Zhan, X., Wu, X., Li, J., Wang, H., and Gao, S. (2020). Effect of weathering on environmental behavior of microplastics: Properties, sorption and potential risks, *Chemosphere*, **242**, 125193.
76. Acarer Arat, S. (2024). A review of microplastics in wastewater treatment plants in Türkiye: Characteristics, removal efficiency, mitigation strategies for microplastic pollution and future perspective, *Water Sci. Technol.*, **89**, pp. 1771–1786.
77. Erkan, H. S., Emik, H. H., and Engin, G. O. (2024). Microplastics in advanced biological wastewater treatment plant of Kocaeli, Turkey: Point source of microplastics reaching Marmara Sea, *Int. J. Environ. Sci. Technol.*, **21**, pp. 1263–1284.
78. Ridall, A., Farrar, E., Dansby, M., and Ingels, J. (2023). Influence of wastewater treatment plants and water input sources on size, shape, and polymer distributions of microplastics in St. Andrew Bay, Florida, USA, *Mar. Polut. Bull.*, **187**, 114552.
79. Christian, A. E., and Köper, I. (2023). Microplastics in biosolids: A review of ecological implications and methods for identification, enumeration, and characterization, *Sci. Total Environ.*, **864**, 161083.
80. Galgani, F., Hanke, G., Werner, S., Oosterbaan, L., Nilsson, P., Fleet, D., Kinsey, S., Thompson, R. C., Vlachogianni, T., Scoullos, M., Viega, J. M., Palatinus, A., Matiddi, M., Maes, T., Korpinen, S., Budziak, A., Leslie, H., Gago, J., Liebezeit, G., and Franeker, J. V. (2013). Guidance on monitoring of marine litter in European Seas. 128. Publication Office of the European Union.
81. Masura, J., Baker, J. E., Foster, G. D., Arthur, C., and Herring, C. (2015). Laboratory methods for the analysis of microplastics in the marine

environment: Recommendations for quantifying synthetic particles in waters and sediments. Technical Memorandum NOAA technical memorandum NOS-OR&R; **48**.

82. Zhang, L., Liu, J., Xie, Y., Zhong, S., and Gao, P. (2021). Occurrence and removal of microplastics from wastewater treatment plants in a typical tourist city in China, *J. Clean. Prod.*, **291**, 125968.

83. Vardar, S., Onay, T. T., Demirel, B., and Kideys, A. E. (2021). Evaluation of microplastics removal efficiency at a wastewater treatment plant discharging to the Sea of Marmara, *Environ. Pollut.*, **289**, 117862.

84. Kasmuri, N., Rosli, M. S., Zaini, N., and Nayono, S. E. (2024). Reduction of microplastic in wastewater via electrocoagulation process, *IOP Conf. Series: Earth Environ. Sci.*, **1303**, 012020.

85. Ormaniec, P., and Mikosz, J. (2024). Circulation of microplastics in a municipal wastewater treatment plant with multiphase activated sludge, *Desalin. Water Treat.*, **317**, 100265.

86. Ziajahromi, S., Neale, P. A., Silveira, I. T., Chua, A., and Leusch, F. D. (2021). An audit of microplastic abundance throughout three Australian wastewater treatment plants, *Chemosphere*, **263**, 128294.

87. Hajji, S., Ben-Haddad, M., Abelouah, M. R., De-la-Torre, G. E., and Alla, A. A. (2023). Occurrence, characteristics, and removal of microplastics in wastewater treatment plants located on the Moroccan Atlantic: The case of Agadir metropolis, *Sci. Total Environ.*, **862**, 160815.

88. Gündoğdu, S., Çevik, C., Güzel, E., and Kilercioğlu, S. (2018). Microplastics in municipal wastewater treatment plants in Turkey: A comparison of the influent and secondary effluent concentrations, *Environ. Monit. Assess.*, **190**, pp. 1–10.

89. Ziajahromi, S., Neale, P. A., Rintoul, L., and Leusch, F. D. (2017). Wastewater treatment plants as a pathway for microplastics: Development of a new approach to sample wastewater-based microplastics, *Water Res.*, **112**, pp. 93–99.

90. Bao, R., Wang, Z., Qi, H., Mehmood, T., Cai, M., Zhang, Y., Yang, R., Peng, L., and Liu, F. (2022). Occurrence and distribution of microplastics in wastewater treatment plant in a tropical region of China, *J. Clean. Prod.*, **349**, 131454.

91. Roscher, L., Halbach, M., Nguyen, M. T., Hebeler, M., Luschtinetz, F., Scholz-Böttcher, B. M., Primpke, S., and Gerdts, G. (2022). Microplastics in two German wastewater treatment plants: Year-long effluent analysis with FTIR and Py-GC/MS, *Sci. Total Environ.*, **817**, 152619.

92. Lao, W., Dial, S., Salmon, M., and Wong, C. S. (2024). Development and validation of an acid/alkaline digestion method for efficient microplastic extraction from wastewater treatment plant effluents: Sulfuric acid concentration and contact time do matter, *Sci. Total Environ.*, **917**, 170528.

93. Zhang, Z., Liu, W., Gao, Q., Huang, F., Kang, Y., Pan, Y., Luo, G., Zhou, X., Fan, L., Wang, A., and Gao, S. H. (2023). Microplastics extraction from wastewater treatment plants: Two-step digestion pre-treatment and application, *Water Res.* **230**, 119569.

94. Van Cauwenberghe, L., and Janssen, C. R. (2014). Microplastics in bivalves cultured for human consumption, *Environ. Pollut.*, **193**, pp. 65–70.

95. Gui, J., Sun, Y., Wang, J., Chen, X., Zhang, S., and Wu, D. (2021). Microplastics in composting of rural domestic waste: Abundance, characteristics, and release from the surface of macroplastics, *Environ. Pollut.*, **274**, 116553.

96. Sun, Y., Ren, X., Rene, E. R., Wang, Z., Zhou, L., Zhang, Z., and Wang, Q. (2021). The degradation performance of different microplastics and their effect on microbial community during composting process, *Bioresour. Technol.*, **332**, 125133.

97. Edo, C., Fernández-Piñas, F., and Rosal, R. (2022). Microplastics identification and quantification in the composted organic fraction of municipal solid waste, *Sci. Total Environ.*, **813**, 151902.

98. Zhang, Y., Wang, H., Xu, J., Su, X., Lu, M., Wang, Z., and Zhang, Y. (2021). Occurrence and characteristics of microplastics in a wastewater treatment plant, *Bull. Environ. Contam. Toxicol.*, **107**, pp. 677–683.

99. Su, Y., Cheng, Z., Hou, Y., Lin, S., Gao, L., Wang, Z., Bao, R., and Peng, L. (2022). Biodegradable and conventional microplastics posed similar toxicity to marine algae Chlorella vulgaris, *Aquat. Toxicol.*, **244**, 106097.

100. Lenz, R., Enders, K., Stedmon, C. A., Mackenzie, D. M., and Nielsen, T. G. (2015). A critical assessment of visual identification of marine microplastic using Raman spectroscopy for analysis improvement, *Mar. Pollut. Bull.*, **100**, pp. 82–91.

101. Li, J., Liu, H., and Chen, J. P. (2018). Microplastics in freshwater systems: A review on occurrence, environmental effects, and methods for microplastics detection, *Water Res.*, **137**, pp. 362–374.

102. Araujo, C. F., Nolasco, M. M., Ribeiro, A. M., and Ribeiro-Claro, P. J. (2018). Identification of microplastics using Raman spectroscopy: Latest developments and future prospects, *Water Res.*, **142**, pp. 426–440.

103. Elert, A. M., Becker, R., Duemichen, E., Eisentraut, P., Falkenhagen, J., Sturm, H., and Braun, U. (2017). Comparison of different methods for MP detection: What can we learn from them, and why asking the right question before measurements matters? *Environ. Pollut.*, **231**, pp. 1256–1264.

104. Vilakati, B., Sivasankar, V., Nyoni, H., Mamba, B. B., Omine, K., and Msagati, T. A. (2021). The Py–GC-TOF-MS analysis and characterization of microplastics (MPs) in a wastewater treatment plant in Gauteng Province, South Africa, *Ecotoxicol. Environ. Saf.*, **222**, 112478.

105. Lee, J. H., Kim, M. J., Kim, C. S., Cheon, S. J., Choi, K. I., Kim, J., Jung, J., Yoon, J. K., Lee, S. H., and Jeong, D. H. (2023). Detection of microplastic traces in four different types of municipal wastewater treatment plants through FT-IR and TED-GC-MS, *Environ. Pollut.*, **333**, 122017.

106. Wu, Z., Dong, J., Wu, Y., Zhao, Y., Wang, H., Zhao, X., Zhang, B., Zhang, Y., and An, L. (2024). Mass-based fates of microplastics throughout wastewater treatment processes, *Chem. Eng. J.*, **487**, 150497.

107. Lv, L., Qu, J., Yu, Z., Chen, D., Zhou, C., Hong, P., Sun, S., and Li, C. (2019). A simple method for detecting and quantifying microplastics utilizing fluorescent dyes-Safranine T, fluorescein isophosphate, Nile red based on thermal expansion and contraction property, *Environ. Pollut.*, **255**, 113283.

Chapter 5

Distribution of Marine Debris in the Mangroves of a Semi-enclosed Lagoon System

Nur Hannah Abd Rahim,[a] Behara Satyanarayana,[a,b,c] Yusof Shuaib Ibrahim,[a,d] Christelle Not,[e] Izwandy Idris,[a,f] Jarina Mohd Jani,[a,g] Stefano Cannicci,[h] and Farid Dahdouh-Guebas[b,c,i]

[a]*Mangrove Research Unit (MARU), Institute of Oceanography and Environment (INOS), Universiti Malaysia Terengganu (UMT), Kuala Nerus 21300, Malaysia*
[b]*Systems Ecology and Resource Management Research Unit (SERM), Université Libre de Bruxelles-ULB, Brussels 1050, Belgium*
[c]*Mangrove Specialist Group (MSG), Species Survival Commission (SSC), International Union for the Conservation of Nature (IUCN), c/o Zoological Society of London, London, United Kingdom*
[d]*Microplastic Research Interest Group (MRIG), Faculty of Science and Marine Environment, Universiti Malaysia Terengganu (UMT), Kuala Nerus 21300, Malaysia*
[e]*Environmental Geochemistry & Oceanography Research Group, Department of Earth Sciences & Swire Institute for Marine Science, The University of Hong Kong, Hong Kong*
[f]*South China Sea Repository and Reference Centre, Institute of Oceanography and Environment (INOS), Universiti Malaysia Terengganu (UMT), Kuala Nerus 21300, Malaysia*

Microplastic Pollution: Environmental Impact and Solutions
Edited by Julia Martín, Juan Luis Santos, Irene Aparicio, and Esteban Alonso

ISBN 978-981-5129-77-9 (Hardcover), 978-1-003-66006-4 (eBook)
www.jennystanford.com

DOI: 10.1201/9781003660064-5

[g]*Biodiversity Conservation and Management Program, Faculty of Science and Marine Environment, Universiti Malaysia Terengganu (UMT), Kuala Nerus 21300, Malaysia*
[h]*Department of Biology, University of Florence, via Madonna del Piano 6-50019, Sesto Fiorentino, Italy*
[i]*Ecology & Biodiversity Research Unit, Department of Biology, Vrije Universiteit Brussel-VUB, Brussels 1050, Belgium*
satyam2149@gmail.com

In recent years, marine debris has become a serious environmental concern, particularly for coastal ecosystems such as mangroves. This chapter focuses on Setiu Wetlands (Malaysia), a semi-enclosed lagoon system with some luxuriant mangrove cover vulnerable to plastic entrapment. The assessment was conducted during low tide in dry and wet seasonal periods at three sites, from the upstream (Pulau Layat), midstream (Kampung Pengkalan Gelap), and downstream (Pulau Sutung). Three 25 m long parallel transects, at a distance of 5, 10, and 25 m from the water edge to the back mangrove, were established for each site. All marine debris within 1 m × 1 m quadrats along each transect (25 plots/transect) were observed, collected, and identified. Plastic bags were the most dominant type of marine debris, followed by water bottles and Styrofoam blocks. Plastic bags and water bottles are largely associated with domestic usage (including the chances for a small fraction through tourism activities), while large-sized Styrofoam blocks belong to fishing activities. Notable differences, although not significant, were observed in the findings. Higher accumulation of marine debris in the dry season be linked to reduced water exchange, while its highest entrapment around 10 m inside the forest was due to mixed (*Avicennia-Rhizophora*) mangrove species with dense pneumatophores and prop root systems. Overall, the abundance of marine debris in Setiu Wetlands was rather low compared to other published reports from mangrove areas in Asia. Enhanced plastic waste management practices are still crucial for mitigating marine debris pollution and conserving the mangroves in Setiu Wetlands.

5.1 Introduction

Marine debris—defined by the United Nations Environment Programme in 2009 as any solid material, whether persistent, manufactured, or processed, that is discarded, disposed of, or abandoned in the marine and coastal environment—poses a significant threat to the ecosystem. This debris originates from various anthropogenic activities such as fishing, food packaging, recreational activities, and medicines and ends up in landfills or natural habitats [1]. Plastics can enter the marine environment directly through improper waste management or irresponsible littering. The upstream discharge/stormwater run-off also brings plastics into the rivers, which eventually lead to end up in the sea and/or get trapped by mangroves due to their aboveground structural complexity [2, 3]. Hence, mangrove forests are identified as the primary repository for plastics among coastal wetlands [4, 5].

Mangrove ecosystems are well documented for their vast array of ecological services such as erosion control, shoreline protection, breeding/nursery ground for finfish and shellfish species, and carbon sequestration [6, 7]. They have been identified as exceptional carbon sinks, capable of absorbing carbon 100 times more than terrestrial forests, and the world's largest carbon reservoir [8]. Moreover, mangroves play a pivotal role in supporting the local communities and their livelihood through bivalve and oyster collection, boat fishing, and handicraft making (using *Nypa* leaves) [9]. However, the expansion of aquaculture and agricultural sites, settlements, fuelwood harvesting, and pollution are still threatening the extent of mangroves in many places. The range of pollution varies from wastewater, heavy metals as well as marine debris or plastics.

Accumulation of marine debris in mangroves can show adverse impacts on vegetation health [10], by entangling the trunk and roots, contaminating the fauna, and affecting the growth of the seedlings [11]. Although several studies have identified marine debris accumulation in the mangroves in Asia [12–15], there remains a gap in knowledge for semi-enclosed water/lagoon systems [10]. A study by Suyadi and Corry [10] found that semi-enclosed mangroves exhibit a higher abundance of marine debris compared to open bays. Hence, this study investigates the accumulation and distribution of marine debris in the mangrove of Setiu Wetlands—a semi-enclosed lagoon system on the east coast of Peninsular Malaysia.

Setiu Wetlands is a unique place where nine different ecosystems—including mangroves, seagrass beds, peat swamp, beach forest, lagoon, river, and estuary—are interconnected and influence each other. Nowadays, plastic pollution is rapidly increasing due to discarded fishing nets, Styrofoam buoys and containers, fragmented plastic barrels, and old/abandoned aquaculture cages in the vicinity. Previous studies have recorded microplastics in the water, sediment, and marine organisms at Setiu Wetlands [16–18], yet no research has specifically addressed the source of microplastics: marine debris. Furthermore, the shifted (Setiu) river mouth brought considerable changes to the hydrodynamics and affected water circulation in the semi-enclosed lagoon [19].

The present study aims to enhance our comprehension of marine debris accumulation in an ecosystem facing the dual challenges of irresponsible littering and altered water circulation. The findings provide valuable insights for marine debris management as well as effective conservation of the mangroves at Setiu Wetlands, which acquired the status of a State Park recently.

5.2 Materials and Methods

5.2.1 Study Site

Setiu Wetlands in State Terengganu is connected to the South China Sea on the east coast of Peninsular Malaysia (Fig. 5.1). Besides the daily flood and ebb conditions, the hydrodynamics in the lagoon are strongly influenced by seasonal changes, characterized by distinct wet and dry periods throughout the year [20].

5.2.2 Sample Collection and Analysis

Marine debris samples were collected from three 25 m long transects at the upstream (Pulau Layat), midstream (Kampung Pengkalan Gelap), and downstream (Pulau Sutung) locations (Fig. 5.1). The transects are established parallelly at 5, 10, and 25 m distances from the water edge to the back mangrove (Fig. 5.2). Each transect was subdivided into 25 (1 m × 1 m) plots, and all anthropogenic marine litter made of plastics and sized >5 mm was collected. The sampling covered both dry (April 2021) and wet seasonal periods (January 2022) in a year. All samples were brought to the laboratory for washing and air drying. The identified samples were then separated into different types before weighing (using a weighing balance, GX-20K Precision Balance, United States). Data were recorded as mass density (g/m^2) and quantity density (item/m^2).

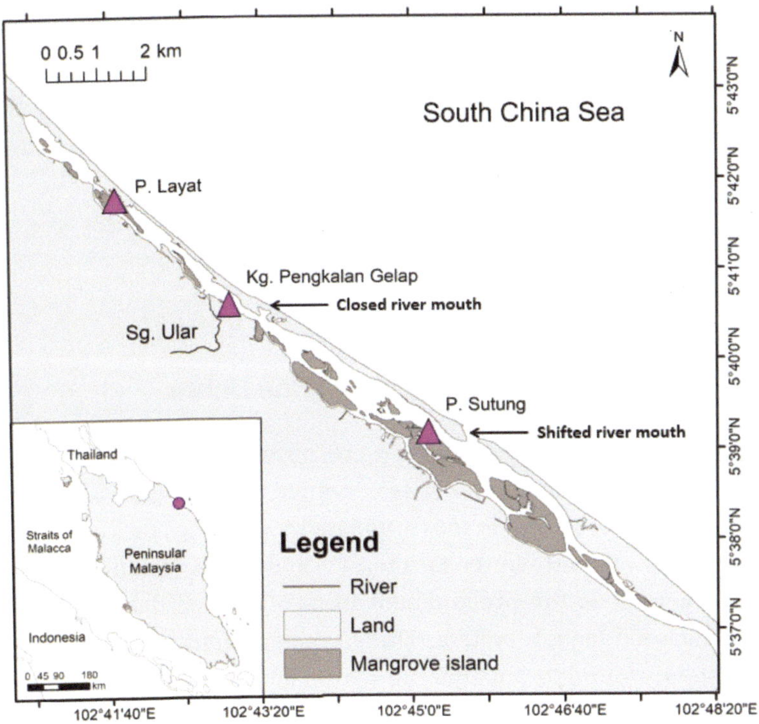

Figure 5.1 Study area showing upstream (Pulau Layat), midstream (Kampung Pengkalan Gelap), and downstream (Pulau Sutung) sampling sites for marine debris observations at the Setiu Wetlands.

The type of marine debris encountered was given in a table (Microsoft Word), while its abundance was represented by boxplots (IBM SPSS v.28, Academic license, UMT). A general linear model (ANOVA) was employed to analyse the differences in the datasets. Statistical significance was determined at $p < 0.05$ for all comparisons.

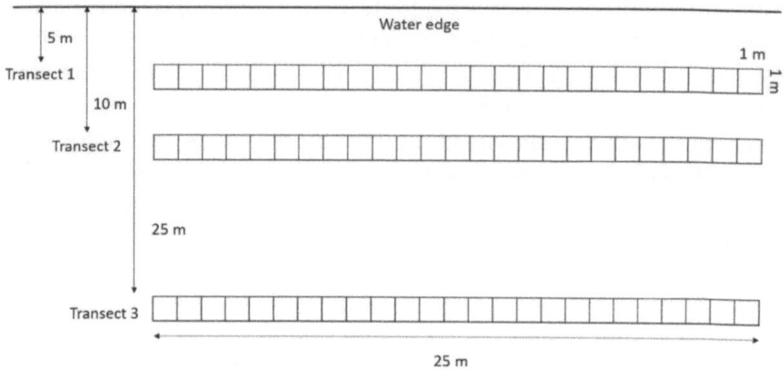

Figure 5.2 Sampling design showing the three 25 m long parallel transects (with 1 m × 1 m sub-plots) at 5, 10, and 25 m distances from the water edge to the back mangrove for anthropogenic marine debris collection at Setiu Wetlands.

5.3 Results

5.3.1 Type and Abundance of Marine Debris

Various types of marine debris were observed within the Setiu Wetlands, including plastic bags, bottles, Styrofoam cups and plates, large Styrofoam pieces (originated from buoys or ice boxes), fishing nets, cigarette butts, face masks, and ropes. Notably, plastic bags emerged as the predominant form of marine debris in this semi-enclosed lagoon system. These types of marine debris were consistently found throughout the year in both dry and wet seasons (Table 5.1).

The mass and item density exhibited considerable variations, ranging from 0.7 to 0.296 kg/m^2 and 0 to 2 items/m^2, respectively. Marine debris could be found in 14 out of 18 transects. Overall, the abundance of marine debris in Setiu Wetlands was relatively low compared to other mangrove sites in Malaysia and elsewhere, as reported by previous studies (Table 5.2).

Table 5.1 Number of items for each type of marine debris collected from different sites during dry and wet seasons

Type	Number of Items						Total Number of Items
	Dry Season			Wet Season			
	PL	KPG	PS	PL	KPG	PS	
Plastic bag	2	1	6	1	0	1	11
Bottle	4	0	1	1	2	0	8
Cup and plate	2	0	0	2	2	0	6
Styrofoam box/block	3	0	2	1	1	0	7
Others	0	3	2	0	0	3	8
Total marine debris	11	4	11	5	5	4	

PL = Pulau Layat, KPG = Kampung Pengkalan Gelap, PS = Pulau Sutung

Table 5.2 Summary of marine debris findings in Peninsular Malaysia and Indonesia

Site, Country	Item per m^2	Weight (kg) per m^2	Reference
Pantai Indah Kapuk, Jakarta, Indonesia	20–533	—	[12]
Tunda Island, Indonesia	6	—	[14]
Penang Island, Malaysia	2.15–73.12	—	[13]
Nelayan Beach, Negeri Sembilan	—	0.1064	[21]
Sungai Lurus, Johor	—	66.15	[22]
Present study	0.12	0–3	[16]

5.3.2 Distribution of Marine Debris across Seasons, Sites and Transects

Marine debris accumulation was higher during the dry period than the wet period, but no significant difference was revealed in two-way ANOVA (p = 0.99). Despite the lack of statistical significance between seasons, and across sampling sites (two-way ANOVA, p = 0.52) and transects (two-way ANOVA, p = 0.25), some noteworthy differences were observed in the findings. For the dry season, Kampung Pengkalan Gelap in the midstream location showed the lowest abundance in contrast to Pulau Layat and Pulau Sutung in the upstream and downstream locations (Fig. 5.3). The pattern of marine debris abundance in the wet season was different with almost similar counts at the three study sites.

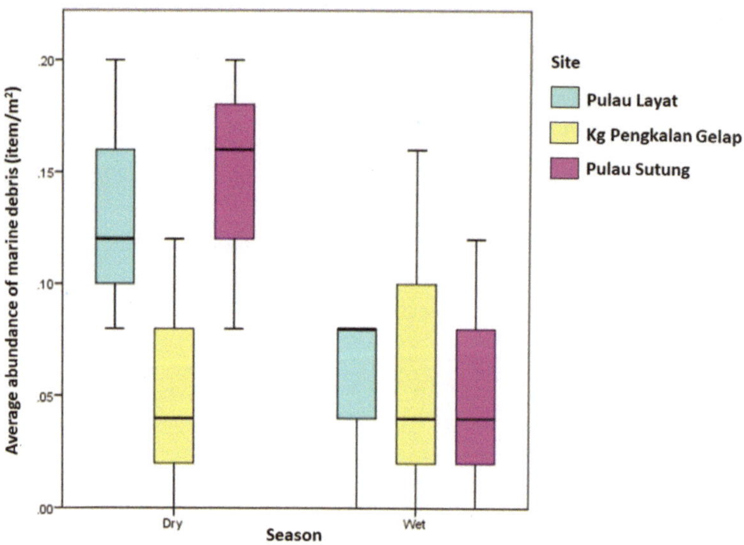

Figure 5.3 Summary of marine debris findings in Peninsular Malaysia and Indonesia.

The abundance of marine debris was higher at a distance of 10 m from the water edge during both dry and wet seasons. Interestingly, the marine debris at 25 m showed little variation between the seasons (Fig. 5.4).

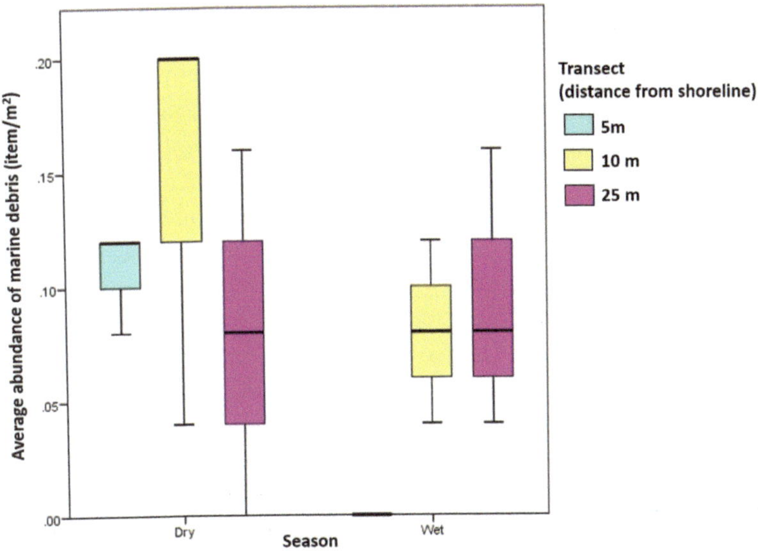

Figure 5.4 Average abundance of marine debris across transects at 5 m, 10 m, and 25 m distances from the shoreline during the dry and wet seasons.

5.4 Discussion

This study reports the distribution of marine debris in the mangroves of Setiu Wetlands for the first time and establishes a critical baseline for future research and management initiatives. Plastic bags emerged as the most prevalent type of marine debris in the mangroves, a trend consistent with other researchers elsewhere [13]. Plastic bags, with their flexible and lightweight characteristics compared to other items, are easily trapped by the mangroves [23, 24]. These entangled plastic bags, caught by pneumatophores and prop roots, are less likely to be washed away by tides [25]. The plastic bags in Setiu must have originated from land sources such as nearby settlements and direct littering. The same reasons are also valid for other items such as bottles, Styrofoam cups and plates, cigarette butts, and face masks found in the vicinity. In this context, tourism activities such as camping and recreation by local and non-local visitors on the beach are responsible. The items of large Styrofoam pieces (likely from buoys or ice boxes), fishing nets, and ropes belong to fishery-related activities, as Setiu Wetlands is one of the fishery hubs for Malaysia [26, 27].

The average mass and density of marine debris in Seitu Wetlands were found to be lower than in other mangrove areas in Peninsular Malaysia and Indonesia [12, 14, 28]. Additionally, there were transects with no trace of plastics, especially the transect located 5 m from the water edge across all three sites during the wet season. The absence of marine debris along these transects was likely due to reduced roots for debris entrapment [4] and improved water circulation during this period. This variability underscores the pristine nature of conditions at Setiu, despite localized pollution pressures. However, seasonal and spatial variations in marine debris abundance highlight the dynamic environmental interactions within the mangrove ecosystem.

The higher abundance of marine debris during the dry season was possibly due to reduced water circulation and increased entrapment within mangrove structures [19]. In addition, oil palm plantations in the northern part of Setiu Wetlands affected regular freshwater discharge into the lagoon [29]. This was compounded by the shifted river mouth located further south, thus increasing the residence time of pollutants, including marine debris. Meanwhile, a few researchers elsewhere also found higher marine debris during the wet season due to greater surface run-off [30]. This suggests varying impacts at different biogeographical systems, particularly in terms of pollutant trapping. Open areas such as beaches may show a higher abundance of marine debris during the wet season [31, 32], compared to semi-enclosed systems such as Setiu Wetlands.

The lower abundance and distribution of marine debris at Kampung Pengkalan Gelap, despite its location closer to the ecotourism boardwalk, which is expected to show the highest amount of marine debris [31, 33], could be linked to the availability of State Park personnel and regular monitoring. In addition, the discharge of Sungai Ular nearby (Fig. 5.1) might be facilitating the transport of debris to the north (during high tide) or south (during low tide) within this semi-enclosed lagoon system [29]. The other two sites—Pulau Layat and Pulau Sutung—are still in the planning phase for their gazettement as part of the State Park.

The transects revealed a trend of higher debris accumulation around 10 m from the water edge due to mixed mangrove species, especially *Avicennia* and *Rhizophora* spp. with dense pneumatophores and prop root systems [3, 25]. On the other hand, dense patches of *Rhizophora* trees at a distance of 25 m inside the forest showed lesser entrapment of marine debris. Other factors that may influence the accumulation of marine debris in mangroves, including wind speed and direction [34], were not investigated in this present chapter.

In conclusion, this chapter underscores the importance of understanding marine debris dynamics in semi-enclosed lagoon systems such as Setiu Wetlands. Future research should explore comprehensive factors influencing marine debris distribution, including hydrological processes and anthropogenic influences. Such insights are crucial for developing targeted conservation strategies and enhancing the resilience of mangrove ecosystems against marine pollution.

Acknowledgements

This study was jointly supported by the FRGS research grant (FRGS/1/2022/STG03/UMT/02/4) (Vote #59694) and the Mangrove Research Unit (MARU) (Vote #63906). The authors gratefully acknowledge the financial support provided by the Ministry of Higher Education (MOHE), Malaysia, and MARU. We thank the Setiu Wetlands State Park Management Council for granting permission to conduct this study. We also extend our appreciation to all who contributed to the research facilities and sample collection, including MRIG members, UMT staff and students, and local fishermen.

References

1. Geyer, R., J. R. Jambeck, and K. L. Law, Production, use, and fate of all plastics ever made. *Science Advances*, 2017. **3**(7): e1700782.
2. Vorsatz, L. D., P. Pattrick, and F. Porri, Quantifying the in situ 3-dimensional structural complexity of mangrove tree root systems: Biotic and abiotic implications at the microhabitat scale. *Ecological Indicators*, 2021. **121**: 107154.

3. Luo, Y. Y., et al., Landward zones of mangroves are sinks for both land and water borne anthropogenic debris. *Science of The Total Environment*, 2022. **818**: 151809.

4. Luo, Y. Y., C. Not, and S. Cannicci, Mangroves as unique but understudied traps for anthropogenic marine debris: A review of present information and the way forward. *Environmental Pollution*, 2020: 116291.

5. Ouyang, X., et al., Fate and effects of macro- and microplastics in coastal wetlands. *Environmental Science & Technology*, 2022. **56**(4): 2386–2397.

6. Macreadie, P. I., et al., Blue carbon as a natural climate solution. *Nature Reviews Earth & Environment*, 2021. **2**(12): 826–839.

7. Lee, S. Y., et al., Ecological role and services of tropical mangrove ecosystems: A reassessment. *Global Ecology and Biogeography*, 2014. **23**(7): 726–743.

8. Donato, D. C., et al., Mangroves among the most carbon-rich forests in the tropics. *Nature Geoscience*, 2011. **4**(5): 293.

9. Satyanarayana, B., et al., Mangroves fueling livelihoods: A socio-economic stakeholder analysis of the charcoal and pole production systems in the world's longest managed mangrove forest. *Frontiers in Ecology and Evolution*, 2021. **9**. https://doi.org/10.3389/fevo.2021.621721.

10. Suyadi and C. Y. Manullang, Distribution of plastic debris pollution and it is implications on mangrove vegetation. *Marine Pollution Bulletin*, 2020. **160**: 111642.

11. Kesavan, S., et al., Anthropogenic pressure on mangrove ecosystems: Quantification and source identification of surficial and trapped debris. *Science of The Total Environment*, 2021. **794**: 148677.

12. Hastuti, A. R., F. Yulianda, and Y. Wardiatno, Distribusi spasial sampah laut di ekosistem mangrove Pantai Indah Kapuk, Jakarta. *Bonorowo Wetlands*, 2014. **4**(2): 94–107.

13. Chee, S. Y., et al., Anthropogenic marine debris accumulation in mangroves on Penang island, Malaysia. *Journal of Sustainability Science and Management*, 2020. **15**(6): 36–60.

14. Maharani, A., N. P. Purba, and I. Faizal. Occurrence of beach debris in Tunda Island, Banten, Indonesia. In *E3S Web of Conferences*. 2018. EDP Sciences.

15. Smith, S. D. A., Marine debris: A proximate threat to marine sustainability in Bootless Bay, Papua New Guinea. *Marine Pollution Bulletin*, 2012. **64**(9): 1880–1883.

16. Abd Rahim, N. H., et al., Commercially important mangrove crabs are more susceptible to microplastic contamination than other brachyuran species. *Science of The Total Environment*, 2023. **903**: 166271.

17. Hamzah, S. R., et al., Ingestion of microplastics by the estuarine polychaete, *Namalycastis* sp. in the Setiu Wetlands, Malaysia. *Marine Pollution Bulletin*, 2021. **170**: 112617.

18. Ibrahim, Y. S., et al., Spatiotemporal microplastic occurrence study of Setiu Wetland, South China Sea. *Science of The Total Environment*, 2021. **788**: 147809.

19. Zainol, Z., M. F. Akhir, and Z. Zainol, Pollutant transport and residence time of a shallow and narrow coastal lagoon estimated using a numerical model. *Marine Pollution Bulletin*, 2021. **164**: 112011.

20. Zainol, Z., et al., Mangrove propagule dispersal in a shallow and narrow coastal lagoon: A simulation-based assessment of the Setiu Wetlands, Malaysia. *Forests*, 2022. **13**(9): 1525.

21. Yi, C. J. and N. Kannan, Solid waste transportation through ocean currents: Marine debris sightings and their waste quantification at Port Dickson Beaches, Peninsular Malaysia. *EnvironmentAsia*, 2016. **9**(2).

22. Kadir, A. A., A. Hasni, and N. A. Sarani, Marine debris composition in Batupahat, Johor: A comparison between Sungai Lurus and Minyakbeku beaches. *ARPN Journal of Engineering and Applied Sciences*, 2015. **10**(15).

23. Ivar do Sul, J. A., et al., Plastic debris retention and exportation by a mangrove forest patch. *Marine Pollution Bulletin*, 2014. **78**(1): 252–257.

24. Cappa, P., et al., Impact of mangrove forest structure and landscape on macroplastics capture. *Marine Pollution Bulletin*, 2023. **194**: 115434.

25. Martin, C., H. Almahasheer, and C. M. Duarte, Mangrove forests as traps for marine litter. *Environmental Pollution*, 2019. **247**: 499–508.

26. Mohd Salim, J., M. Faridah, and S. Rohani, Setiu: More than a Wetland. In: *Setiu Wetlands: Species, Ecosystem and Livelihoods*, 1st edn, pp. 87–100. Penerbit Universiti Malaysia Terengganu, 2015.

27. Faridah, M., et al., *Setiu Wetlands: Species, Ecosystems and Livelihoods*, 1st edn. Penerbit Universiti Malaysia Terengganu, 2015.

28. Fauziah, S. H., et al., Marine debris in Malaysia: A review on the pollution intensity and mitigating measures. *Marine Pollution Bulletin*, 2021. **167**: 112258.

29. Suratman, S., et al., Reassessment of physico-chemical water quality in Setiu Wetland, Malaysia. *Sains Malaysiana*, 2014. **43**(8): 1127–1131.

30. Garcés-Ordóñez, O., et al., Seasonal variation in plastic litter pollution in mangroves from two remote tropical estuaries of the Colombian Pacific. *Marine Pollution Bulletin*, 2023. **193**: 115210.

31. Vorsatz, L. D., et al., Anthropogenic debris pollution in peri-urban mangroves of South China: Spatial, seasonal, and environmental drivers in Hong Kong. *Marine Pollution Bulletin*, 2023. **195**: 115495.

32. Cheung, P. K., L. T. O. Cheung, and L. Fok, Seasonal variation in the abundance of marine plastic debris in the estuary of a subtropical macro-scale drainage basin in South China. *Science of The Total Environment*, 2016. **562**: 658–665.

33. Plantado, L. C., et al., Characterization and distribution of macroplastics along the coastal municipalities of Partido district, Camarines Sur, Philippines. *Regional Studies in Marine Science*, 2024. **74**: 103520.

34. Lechthaler, S., et al., The way of macroplastic through the environment. *Environments*, 2020. **7**(10): 73.

Chapter 6

Microplastic Pollution in Mountain Environments

Paolo Pastorino,[a] **Marino Prearo,**[a] **and Damià Barceló**[b]

[a]*Istituto Zooprofilattico Sperimentale del Piemonte, Liguria e Valle d'Aosta, 10154 Torino, Italy*

[b]*Chemistry and Physics Department, University of Almeria, Ctra Sacramento s/n, 04120 Almería, Spain*

paolo.pastorino@izsplv.it

6.1 Introduction

Plastic pollution has emerged as one of the most urgent environmental challenges of our era, with microplastics (MPs) contributing substantially to the problem. MPs, plastic particles smaller than 5 mm in diameter, have emerged as a pervasive pollutant, infiltrating even the most remote corners of the planet [1]. These particles, along with even smaller nanoplastics (1 nm and 100 nm) [2], originate from a myriad of sources, both terrestrial and atmospheric, and are transported over vast distances, ultimately accumulating in diverse environmental compartments [3]. Mountain

Microplastic Pollution: Environmental Impact and Solutions
Edited by Julia Martín, Juan Luis Santos, Irene Aparicio, and Esteban Alonso

ISBN 978-981-5129-77-9 (Hardcover), 978-1-003-66006-4 (eBook)
www.jennystanford.com

DOI: 10.1201/9781003660064-6

ecosystems, often perceived as pristine and isolated, are increasingly recognized as vulnerable recipients of microplastic pollution [4]. The unique topography, climate, and hydrological processes of these regions create complex pathways for microplastic transport and fate, posing potential risks to fragile flora and fauna. This chapter delves into the current understanding of microplastic sources, transport mechanisms, and fate in mountain environments. By synthesizing recent research findings, we aim to provide a comprehensive overview of the pathways by which these ubiquitous pollutants reach mountain environments, highlighting the key factors influencing their distribution and persistence. Furthermore, we identify critical knowledge gaps and propose future research directions to advance our understanding of microplastic dynamics in these valuable and vulnerable ecosystems.

6.2 Sources of Microplastics in Mountain Environments

Human activities and atmospheric conditions, which vary seasonally, are the main sources of MPs in mountain ecosystems. The primary mechanisms for MP deposition include atmospheric transport from urban sources and precipitation events such as rain and snow [5]. Snow plays a key role by capturing pollutants and particulate matter from the atmosphere, which are then deposited on the ground through wind, storms, and rain [6]. Due to their low density, MPs can be lifted into the upper atmosphere by wind currents and subsequently deposited in higher altitude areas through snowfall or rainfall, posing environmental risks [7]. MPs can reach remote ecosystems and spread into both terrestrial and aquatic environments through atmospheric transport and deposition [5]. They have been discovered in isolated regions such as the polar Arctic, the deep-sea environment, sea surfaces, and glaciers [8]. MPs can remain trapped in glaciers indefinitely, and as glaciers melt, these particles are released into freshwater and marine systems [4, 9, 10]. For example, the increased melting of glaciers during the pre-monsoon period in the high mountains of Asia leads to higher MP concentrations downstream, as MPs are mainly contained in snow-covered glaciers [11–13]. In addition to atmospheric deposition, MPs can be introduced into remote environments, mountainous areas, and high-mountain lakes through the fragmentation of plastic items [14]. In urban areas, MPs are aerosolized pollutants and can be measured through atmospheric sampling from specific sources and remote locations. The widespread presence of MPs in various remote ecosystems indicates their ability to travel from major industrial areas or other human sources over medium to long distances from lower altitudes. Research on MP pollution in extreme environments has revealed contamination in places once considered pristine, such as Mount Everest and the Vatnajökull Ice Cap [15, 16]. Synthetic fibres, such as polyester and polyamides from tourist garments, are commonly found in remote regions. Activities such as trekking, hiking, and the use of mountain climber clothing and equipment contribute to MP shedding in mountainous lakes and glacial areas such as Mount Everest, Antisana Glacier, and Forni Glacier [10, 15, 17, 18]. Understanding the origins of MPs in mountain environments is crucial for developing effective mitigation strategies.

6.3 Transport Mechanisms

Once MPs enter the mountain environment, a complex interplay of hydrological, atmospheric, and potentially biological processes governs their transport and distribution [19, 20] (Fig. 6.1).

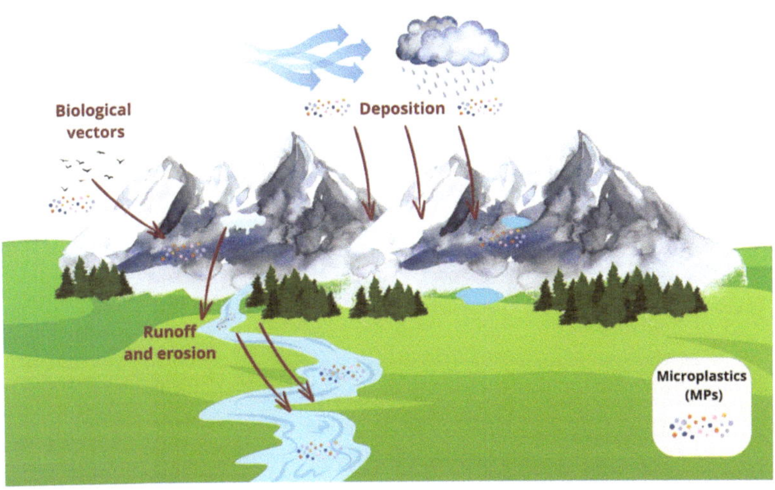

Figure 6.1 Mechanisms of transport and distribution of microplastics in mountain environments.

6.3.1 Hydrological Processes

- **Run-off and erosion:** Rainfall and snowmelt generate surface run-off, which can mobilize and transport MPs across the landscape. The efficiency of this transport depends on factors such as slope, vegetation cover, and soil properties. Steep slopes and sparse vegetation can enhance run-off velocity and erosion, leading to increased microplastic transport [21].
- **Influence of topography:** The rugged topography of mountain regions, characterized by steep slopes, valleys, and interconnected waterbodies, creates a complex hydrological network that influences microplastic movement. For instance, MPs can accumulate in depressions and along riverbanks, acting as temporary sinks before being remobilized during high-flow events [22].
- **Cryosphere:** The cryosphere, encompassing snow, glaciers, and permafrost, plays a crucial role in microplastic transport and storage in mountain environments. MPs can be scavenged from the atmosphere by snowfall and subsequently incorporated into the snowpack. As snow melts, these particles are released, potentially transporting them over long distances [23].

6.3.2 Atmospheric Transport

- **Wind patterns:** Prevailing wind patterns and mountain meteorology significantly influence the atmospheric transport and deposition of MPs. Mountain ranges can act as barriers, altering wind flow and leading to localized deposition patterns. Upslope winds can transport MPs from lower elevations, while downslope winds may carry them to valley bottoms or adjacent lowlands [5].
- **Precipitation events:** Rainfall, snowfall, and fog can effectively scavenge MPs from the atmosphere, depositing them onto mountain surfaces. The size and type of precipitation, as well as the physicochemical properties of the MPs, influence the efficiency of this scavenging process. For example, smaller MPs may be more effectively transported within raindrops, while larger particles might settle out more quickly [24, 25].

6.3.3 Biological Vectors

- **Role of animals:** While less studied than hydrological and atmospheric transport, biological vectors could potentially contribute to dispersal of MPs in mountain environments. Birds, insects, and other animals might ingest MPs directly or indirectly through their prey, potentially transporting them to higher elevations or across geographical barriers [26–28]. However, more research is needed to quantify the significance of this transport pathway.

6.4 Fate of Microplastics

The ultimate fate of MPs in mountain environments is determined by a complex interplay of factors, including their physicochemical properties, environmental conditions, and interactions with biotic and abiotic components [29]. Understanding these processes is crucial for assessing the long-term risks posed by these pollutants.

6.4.1 Mountains

- **Infiltration and retention:** Once deposited on the soil surface, MPs can infiltrate the soil profile through various mechanisms, including gravitational forces, leaching by rainwater, and bioturbation by soil organisms. The extent of infiltration depends on factors such as soil texture, porosity, and the size and shape of the MPs. For example, smaller particles are more likely to be transported deeper into the soil profile, while larger particles may be retained in surface layers [30, 31].
- **Vertical transport:** The vertical transport of MPs within the soil profile can be influenced by factors such as freeze–thaw cycles, root growth, and animal activity. These processes can either enhance or hinder the downward movement of MPs, potentially leading to their accumulation in specific soil horizons [32].
- **Impacts on soil biota:** The presence of MPs in mountain soils can potentially impact soil organisms, although research in

this area is still in its early stages. MPs may be ingested by soil invertebrates, potentially affecting their feeding behaviour, growth, and reproduction. Additionally, MPs can alter soil properties, such as porosity and water retention, which could indirectly impact soil communities [33].

6.4.2 Freshwater Systems

- **Lakes and reservoirs:** Mountain lakes and reservoirs act as sinks for MPs transported from surrounding catchments. These waterbodies can accumulate significant quantities of MPs, particularly in sediments, where they can persist for extended periods. The distribution and abundance of MPs in lakes and reservoirs are influenced by factors such as water flow patterns, sedimentation rates, and the proximity to pollution sources [4, 10].
- **Rivers and streams:** Rivers and streams play a crucial role in transporting MPs through mountain landscapes [34]. The flow dynamics of these systems, characterized by rapids, waterfalls, and meandering channels, can influence the transport, deposition, and fragmentation of MPs [35]. Additionally, riparian vegetation and in-stream structures can trap MPs [36].

6.4.3 Cryosphere

- **Long-term storage:** Glaciers and ice caps can act as long-term storage reservoirs for MPs, trapping them within their icy matrices for decades or even centuries. As these frozen environments melt due to climate change, they release stored MPs into downstream ecosystems, potentially posing a significant threat to water quality and aquatic life [37].
- **Release mechanisms:** The release of MPs from the cryosphere is influenced by factors such as melting rates, glacier dynamics, and the physicochemical properties of the particles. For example, smaller MPs may be more readily released during melt events, while larger particles might be transported within glacial meltwater over longer distances [6, 37].

6.5 Occurrence and Characteristics of Microplastics in Mountain Ecosystems

6.5.1 Occurrence

In northern Tibet, MP concentrations were reported as follows: Lake Siling Co had 563 ± 1219 × 10^6 particles/km^2, Lake Geren Co had 42 ± 47 × 10^6 particles/km^2, Lake Wuru Co had 117 ± 126 × 10^6 particles/km^2, and Lake Mujiu Co had 17 ± 20 × 10^6 particles/km^2 [38]. In the remote Lake Hovsgol in Mongolia, MP levels ranged from 997 to 44,435 particles/km^2, attributed to littering from plastic bottles, bags, and fishing gear near the lake [39]. This rise in MP pollution in both the Siling Co Basin lakes and Lake Hovsgol is primarily due to inadequate waste management [38, 39]. Lake Hovsgol exhibited an average MP abundance of 20,264 particles/km^2 [39]. In the French Pyrenees, atmospheric transport from rural areas resulted in an average daily MP deposition of 365 ± 69 $m^{-2}d^{-1}$ in field samples [5]. In the Swat River, MP concentrations averaged 305.79 ± 289.66 MPs m^{-3} in surface water, 588.29 ± 253.95 MPs kg^{-1} in sediment, and 12.54 ± 8.02 MPs per fish sample [40]. In remote parts of Europe, over 50% of the detected MPs measured between 30 and 50 μm, with very few larger than 2600 μm [5, 18]. In the snow samples from Lake Dimon (Eastern Alps), the average MP concentration was 0.11 ± 0.19 particles/L, indicating a relatively low level of pollution and minimal human impact [41]. Similar values were recorded in Balma Lakes (Western Alps) where no MPs were found in water and zooplankton samples, whereas the mean MPs in sediment samples were 1.33 ± 0.67 items/m^3 and 1.75 ± 0.62 items/m^3 in Lower and Upper Lake Balma, respectively [10]. The mean MPs in the tadpoles of *Rana temporaria* were 0.33 ± 0.58 items/individual and 0.66 ± 0.58 items/individual in Lower and Upper Lake Balma, respectively [10]. The mean number of MPs items found in the gastrointestinal tract (GIT) of brook trout (*Salvelinus fontinalis*) was considerably higher in specimens from the Lower Lake (0.45 items/g GIT) than in those from the Upper Lake (0.20 items/g GIT) [10]. In Lake Sassolo, the highest MP concentration was found in snow (sediment) samples from a cliff at 14 m depth (60 MPs/kg). Meanwhile, in the main cavity, MP concentrations at 8 and 33 m were 40 and 0 MPs/

kg, respectively. Despite the MP concentration being zero at 33 m, it contained the highest number of fibres (both natural and synthetic) at 340 fibres/kg, suggesting that fibre concentration increases with depth and that fibres are more persistent in the environment. Additionally, 18% of MPs (2.6 MPs/L) were found in the water column, and 6% (2.6 MPs/L) were found in the lake outlet sample. The average accumulation rate of MPs in the snow was 33 MPs/kg and 514 fibres/kg [42]. On Antisana Glacier, surface snow had nearly 88 MP particles per sample, while ice core samples had 52 MPs per sample [18]. On Mount Everest, MPs in snow samples ranged from 3 to 119 MPs/L, averaging 30 ± 11 MPs/L, whereas stream samples ranged from 0 to 2 MPs/L, with an average of 1 ± 0.3 MPs/L [15]. In contrast, the terrestrial environment of the Forni glacier had a slightly higher MP concentration, with 74.4 ± 28.3 particles/kg [17]. Among the eight stream samples from Mount Everest, only three contained MPs. As the altitude increased, MP concentration in snow samples decreased (Everest Base Camp, 79 MPs/L; Camp I, 13 MPs/L; Camp II, 11 MPs/L; Lobuche, 14 MPs/L; South Col, 3 MPs/L), with the highest concentration at the Everest Base Camp and the lowest at the South Col [15]. Similarly, only four microplastic fragments and two plastic fibres were identified in the Vatnajökull ice cap [16].

6.5.2 Characteristics: Size, Shape, and Chemical Types

MPs exhibit variation in size, shape, colour, and polymer composition based on their origin, degradation level, and time spent in the environment [43]. The characteristics of atmospheric MPs differ significantly due to diverse sources such as urban, suburban, and rural areas [5, 44]. Particle size and shape are crucial factors influencing MP interactions with biota [45]. The fragmentation of plastic polymers by UV radiation is a major contributor to MP presence in lake regions [46]. For instance, sediment samples from Crater Lake contained MPs in the range of 8–15 μm [46], while Dimon Lake had a polyethylene terephthalate (PET) fragment sized 75–220 μm [41]. In the Tibetan Plateau glaciers, MPs less than 100 μm were common, with some as small as 10 μm and others up to 500 μm [47]. MPs found in Mount Everest snow and stream samples measured 36–3800 μm in length and 18–2000 μm in diameter [15].

Vatnajökull ice cap had fragments ranging from 30 to 3000 μm and fibres from 1300 to 3000 μm [16]. At the Pyrenee site, fibre lengths were predominantly ≤300 μm, with most MP fragments being ≤50 μm [5].

Polyethylene (PE), a widely used polymer in single-use plastics such as bottles and bags, was the most prevalent plastic polymer in Lake Sassolo, comprising 76.9% (2 MPs/L) in the water column and 53.3% (0.6 MPs/L) at the lake outlet [42]. Polypropylene (PP), commonly used in foam packaging, was the second most abundant polymer [42]. Lakes on the Northern Tibetan Plateau predominantly contained PE, PP, polystyrene (PS), PET, and polyvinyl chloride (PVC). Notably, PVC and PET were only found in Lake Geren Co, while PS was unique to Lake Siling Co [38]. MP presence has also been documented in snow compartments; for example, Dimon Lake's snow samples contained MPs at very low concentrations, with no macroplastics or mesoplastics detected [41]. MPs in Mount Everest's snow and stream samples varied in length, with significant differences in polymer concentrations [15]. In Swat River, fibres (80%–90%) derived from fishing nets and household waste were predominant, with large MPs (~300 μm) prevalent due to human activities [40]. High-altitude ecosystems show anthropogenic contributions to MP pollution, especially near tourist areas [39, 48]. Qinghai Lake, a large inland lake in China, accumulates MPs due to tourism and its lack of outflow [48].

MP shapes commonly found in various ecosystems include fibres, films, fragments, and spheres. These shapes were observed in Lake Crater, Lake Sassolo, Lake Hovsgol, Antisana ice cap, River Swat, and the Tibetan Plateau [18, 39, 40, 42, 46, 47]. Common colours of MPs are blue, red, transparent, and black, with less frequent colours being cream, yellow, and violet [4, 10, 17, 18, 39, 41].

The Forni Glacier contains 65.2% fibres and 34.8% fragments in various colours [17]. The Antisana ice cap has MPs sized 60–2500 μm, with fibres accounting for 71% of the snow sample and 66% of the ice core [18]. Most fibres in Tibetan Plateau glaciers are black, similar to those in urban aerial depositions [47]. At the Pyrenees site, the longest plastic fibre reached 3000 μm, with 749 fragments, 44 fibres, and 73 films/m^2 [5]. Tibetan lakes show various mechanical weathering textures on plastic polymers, such as fractures and pits, likely from transport erosion [38].

The predominance of plastic fibres in the Swat River, Mount Everest, Tibetan Plateau, and Antisana ice cap suggests pollution primarily from anthropogenic activities and atmospheric deposition rather than industrial or municipal waste [15, 18, 40, 47]. Pellet-shaped MPs are nearly absent in Lake Hovsgol samples (0.006%, four pellets). The absence of microbeads in the River Swat, Siling Co, and Tibetan glaciers indicates the lack of cosmetic or industrial waste [38, 39, 40, 47, 49, 50]. FTIR and Raman spectroscopic analyses have identified various MP polymers in high-altitude ecosystems, including PET, PTFE, PVC, PU, and ABS [5, 15, 16, 17, 38, 39, 46, 47]. EDS analysis of MPs from the Alakananda River showed various elements on their surface, facilitating their role as carriers for heavy metals and toxic elements [51].

6.6 Ecological Implications

The presence of MPs in mountain ecosystems raises concerns about their potential impacts on the delicate balance of these environments [10] (Fig. 6.2). While research is ongoing, and a comprehensive understanding of the long-term ecological consequences is still developing, several key areas of concern have emerged [52].

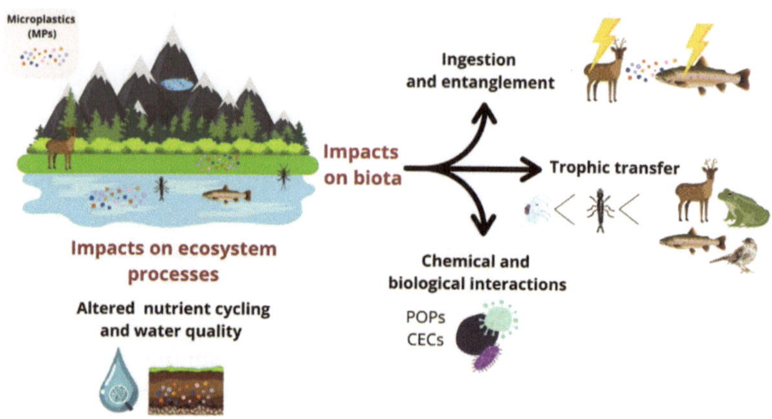

Figure 6.2 Potential impacts of microplastics in mountain environments.

6.6.1 Impacts on Biota

- **Ingestion and entanglement:** MPs can be ingested by a wide range of organisms in mountain ecosystems, from microscopic zooplankton to larger animals such as fish and birds [4]. Ingestion can lead to physical harm, such as internal abrasion or blockage of the digestive tract [53]. Additionally, MPs can create a false sense of satiation, potentially leading to reduced feeding and energy depletion [54].
- **Trophic transfer:** A significant concern is a potential for MPs to bioaccumulate and transfer through food webs [55]. As smaller organisms ingest MPs, these particles can accumulate in their tissues and be passed to predators, potentially leading to higher concentrations at higher trophic levels [56]. This process, known as biomagnification, raises concerns about the potential for MPs to impact top predators and the overall health of mountain food webs [57].
- **Chemical interactions:** MPs can act as vectors for other pollutants, such as persistent organic pollutants (POPs), contaminants of emerging concern (CECs) and trace elements, which can adsorb to their surfaces [58]. When ingested, these contaminated MPs can release these harmful chemicals into the organisms' digestive systems, potentially leading to toxic effects [59].

6.6.2 Ecosystem Processes

- **Nutrient cycling:** MPs can interfere with nutrient cycling in mountain ecosystems [60]. For example, they can alter the decomposition rates of organic matter, potentially affecting the availability of essential nutrients for plants and other organisms. Additionally, MPs can disrupt the activity of soil microorganisms involved in nutrient cycling, potentially impacting soil fertility and overall ecosystem functioning [61].
- **Water quality:** The accumulation of MPs in mountain lakes, rivers, and groundwater raises concerns about potential impacts on water quality. While the direct effects of MPs on human health are still being investigated, their presence in

drinking water sources is a growing concern [62]. Additionally, MPs can alter the physical and chemical properties of waterbodies, potentially impacting aquatic life and overall ecosystems [35].

6.7 Future Directions and Conclusion

The study of MP pollution in mountain environments is a rapidly evolving field, and significant knowledge gaps remain. Future research should focus on the following:

- **Standardized monitoring:** Developing standardized methods for sampling, extracting, and identifying MPs in different mountain environments (soil, water, snow, ice) is crucial for generating comparable data and assessing the true extent of the problem.
- **Source apportionment:** Conducting more detailed source apportionment studies using advanced analytical techniques can help pinpoint the major sources of MPs in specific mountain regions, enabling targeted mitigation efforts.
- **Harmonization efforts:** Conduct harmonization through inter-laboratory studies and the use of reference materials on both local and global scales, mirroring efforts made for regulated and emerging pollutants in previous years.
- **Long-term monitoring:** Establishing long-term monitoring programmes in key mountain ecosystems is essential for understanding the temporal trends of microplastic pollution, assessing the effectiveness of mitigation measures, and detecting potential ecological shifts.
- **Ecotoxicological studies:** Conducting more comprehensive ecotoxicological studies using environmentally relevant concentrations of MPs and representative mountain species is crucial for understanding the long-term impacts of these pollutants on the health and functioning of mountain ecosystems.
- **Green analytical chemistry and green ecotoxicology:** Integrate the principles of green analytical chemistry and green ecotoxicology into MP research and monitoring practices.

- **International database:** Create an international database to collect and share data on the occurrence of MPs.
- **Comparison with other threats:** Compare the impact of MP pollution with other environmental threats, such as climate change.
- **Human health implications:** Despite the current lack of evidence that MPs threaten human health, continue to monitor and gather data on their occurrence in biological fluids, particularly in megacities.
- **Community engagement:** Fostering collaboration among local communities, scientists, policymakers, and tourism operators is crucial for developing and implementing effective mitigation strategies tailored to specific mountain regions.
- **Public awareness campaigns:** Launching targeted public awareness campaigns can educate the public about the issue of MP pollution in mountain environments, promote responsible consumer choices, and encourage individual actions to reduce plastic consumption and waste.
- **Citizen science initiatives:** Engaging the public in citizen science initiatives, such as MP monitoring programs, can help collect valuable data, raise awareness, and foster a sense of stewardship for mountain ecosystems.

In conclusion, MP pollution poses a significant and growing threat to the ecological integrity of mountain environments worldwide. Moving forward, it is imperative that we prioritize research efforts, promote international collaboration, and engage policymakers and stakeholders in finding sustainable solutions to address this global environmental challenge. By adopting a comprehensive and collaborative approach that addresses both the sources and impacts of MP pollution, we can strive to preserve the ecological integrity and societal value of mountain environments for generations to come.

References

1. Xi, B., Wang, B., Chen, M., Lee, X., Zhang, X., Wang, S., Yu, Z., and Wu, P. (2022). Environmental behaviors and degradation methods of microplastics in different environmental media. *Chemosphere*, **299**, 134354.

2. Shen, M., Zhang, Y., Zhu, Y., Song, B., Zeng, G., Hu, D., Wen, X., and Ren, X. (2019). Recent advances in toxicological research of nanoplastics in the environment: A review. *Environmental Pollution*, **252**, 511–521.
3. Pastorino, P., and Barceló, D. (2023). Microplastics and their environmental effects. *Environmental Toxicology and Pharmacology*, **104**, 104324.
4. Pastorino, P., Prearo, M., Pizzul, E., Elia, A. C., Renzi, M., Ginebreda, A., and Barceló, D. (2022). High-mountain lakes as indicators of microplastic pollution: Current and future perspectives. *Water Emerging Contaminants & Nanoplastics*, **13**(1).
5. Allen, S., Allen, D., Phoenix, V. R., Le Roux, G., Durántez Jiménez, P., Simonneau, A., Binet, S., and Galop, D. (2019). Atmospheric transport and deposition of microplastics in a remote mountain catchment. *Nature Geoscience*, **12**(5), 339–344.
6. Zhang, Y., Gao, T., Kang, S., Shi, H., Mai, L., Allen, D., and Allen, S. (2022). Current status and future perspectives of microplastic pollution in typical cryospheric regions. *Earth-Science Reviews*, **226**, 103924.
7. Padha, S., Kumar, R., Dhar, A., and Sharma, P. (2022). Microplastic pollution in mountain terrains and foothills: A review on source, extraction, and distribution of microplastics in remote areas. *Environmental Research*, **207**, 112232.
8. Mishra, A. K., Singh, J., and Mishra, P. P. (2021). Microplastics in polar regions: An early warning to the world's pristine ecosystem. *Science of the Total Environment*, **784**, 147149.
9. Cabrera, M., Moulatlet, G. M., Valencia, B. G., Maisincho, L., Rodríguez-Barroso, R., Albendín, G., Sakali, A., Lucas-Solis, O., Conicelli, B., and Capparelli, M. V. (2022). Microplastics in a tropical Andean Glacier: A transportation process across the Amazon basin? *Science of the Total Environment*, **805**, 150334.
10. Pastorino, P., Anselmi, S., Esposito, G., Bertoli, M., Pizzul, E., Barceló, D., Elia, A. C., Dondo, A., Prearo, M., and Renzi, M. (2023). Microplastics in biotic and abiotic compartments of high-mountain lakes from Alps. *Ecological Indicators*, **150**, 110215.
11. Byers, A. (2005). Contemporary human impacts on Alpine ecosystems in the Sagarmatha (Mt. Everest) National Park, Khumbu, Nepal. *Annals of the Association of American Geographers*, **95**(1), 112–140.
12. Nicholson, K., Hayes, E., Neumann, K., Dowling, C., and Sharma, S. (2016). Drinking water quality in the Sagarmatha National Park, Nepal. *Journal of Geoscience and Environment Protection*, **4**(4), 43–53.

13. Wood, L. R., Neumann, K., Nicholson, K. N., Bird, B. W., Dowling, C. B., and Sharma, S. (2020). Melting Himalayan glaciers threaten domestic water resources in the Mount Everest Region, Nepal. *Frontiers in Earth Science*, **8**, 128.

14. Coe, J. M., and Rogers, D. (Eds.) (2012). *Marine Debris: Sources, Impacts, and Solutions*. Springer Science & Business Media.

15. Napper, I. E., Davies, B. F., Clifford, H., Elvin, S., Koldewey, H. J., Mayewski, P. A., Miner, K. R., Potocki, M., Elmore, A. C., Gajurel, A. P., and Thompson, R. C. (2020). Reaching new heights in plastic pollution: Preliminary findings of microplastics on Mount Everest. *One Earth*, **3**(5), 621–630.

16. Stefánsson, H., Peternell, M., Konrad-Schmolke, M., Hannesdóttir, H., Ásbjörnsson, E. J., and Sturkell, E. (2021). Microplastics in glaciers: First results from the Vatnajökull ice cap. *Sustainability*, **13**(8), 4183.

17. Ambrosini, R., Azzoni, R. S., Pittino, F., Diolaiuti, G., Franzetti, A., and Parolini, M. (2019). First evidence of microplastic contamination in the supraglacial debris of an alpine glacier. *Environmental Pollution*, **253**, 297–301.

18. Cabrera, M., Valencia, B. G., Lucas-Solis, O., Calero, J. L., Maisincho, L., Conicelli, B., Massaine Moulatlet, G., and Capparelli, M. V. (2020). A new method for microplastic sampling and isolation in mountain glaciers: A case study of one Antisana glacier, Ecuadorian Andes. *Case Studies in Chemical and Environmental Engineering*, **2**, 100051.

19. Su, L., Xiong, X., Zhang, Y., Wu, C., Xu, X., Sun, C., and Shi, H. (2022). Global transportation of plastics and microplastics: A critical review of pathways and influences. *Science of The Total Environment*, **831**, 154884.

20. Margenat, H., Guasch, H., Le Roux, G., Vila-Gispert, A., Cornejo, D., Chouache, L., Martì, E., Sonke, J. E., El Zrelli, R. B., Laffont, L., and Hansson, S. V. (2024). Distinct dynamics in mountain watersheds: Exploring mercury and microplastic pollution—Unraveling the influence of atmospheric deposition, human activities, and hydrology. *Environmental Research*, **242**, 117760.

21. Han, N., Zhao, Q., Ao, H., Hu, H., and Wu, C. (2022). Horizontal transport of macro-and microplastics on soil surface by rainfall induced surface runoff as affected by vegetations. *Science of the Total Environment*, **831**, 154989.

22. Rolf, M., Laermanns, H., Kienzler, L., Pohl, C., Möller, J. N., Laforsch, C., Löder, M. G. J., and Bogner, C. (2022). Flooding frequency and floodplain topography determine abundance of microplastics in an alluvial Rhine soil. *Science of the Total Environment*, **836**, 155141.

23. Zhang, Y., Gao, T., Kang, S., Allen, D., Wang, Z., Luo, X., Yang, L., Chen, J., Hu, Z., Chen, P., Du, W., and Allen, S. (2023). Cryosphere as a temporal sink and source of microplastics in the Arctic region. *Geoscience Frontiers,* **14**(4), 101566.
24. Xia, W., Rao, Q., Deng, X., Chen, J., and Xie, P. (2020). Rainfall is a significant environmental factor of microplastic pollution in inland waters. *Science of the Total Environment,* **732**, 139065.
25. Rosso, B., Scoto, F., Hallanger, I. G., Larose, C., Gallet, J. C., Spolaor, A., Bravo, B., Barbante, C., Gambaro, A., and Corami, F. (2024). Characteristics and quantification of small microplastics (<100 μm) in seasonal svalbard snow on glaciers and lands. *Journal of Hazardous Materials,* **467**, 133723.
26. de Souza, S. S., Freitas, Í. N., de Oliveira Gonçalves, S., da Luz, T. M., da Costa Araújo, A. P., Rajagopal, R., Balasubramani, G., Rahman, Md.M., and Malafaia, G. (2022). Toxicity induced via ingestion of naturally-aged polystyrene microplastics by a small-sized terrestrial bird and its potential role as vectors for the dispersion of these pollutants. *Journal of Hazardous Materials,* **434**, 128814.
27. Hoang, T. C., and Mitten, S. (2022). Microplastic accumulation in the gastrointestinal tracts of nestling and adult migratory birds. *Science of The Total Environment,* **838**, 155827.
28. Simakova, A., Varenitsina, A., Babkina, I., Andreeva, Y., Bagirov, R., Yartsev, V., and Frank, Y. (2022). Ontogenetic transfer of microplastics in bloodsucking mosquitoes *Aedes aegypti* L. (Diptera: Culicidae) is a potential pathway for particle distribution in the environment. *Water,* **14**(12), 1852.
29. Haque, F., and Fan, C. (2023). Fate and impacts of microplastics in the environment: Hydrosphere, pedosphere, and atmosphere. *Environments,* **10**(5), 70.
30. Zhao, S., Zhang, Z., Chen, L., Cui, Q., Cui, Y., Song, D., and Fang, L. (2022). Review on migration, transformation and ecological impacts of microplastics in soil. *Applied Soil Ecology,* **176**, 104486.
31. Yu, H., Zhang, W., Zheng, L., Li, T., Hai, C., Wang, Y., and Lyu, T. (2024). A review of the migration mechanisms of microplastics in terrestrial environments. *Environmental Engineering Research,* **29**(5).
32. Koutnik, V. S., Borthakur, A., Leonard, J., Alkidim, S., Koydemir, H. C., Tseng, D., Ozcan, A., Ravi, S., and Mohanty, S. K. (2022). Mobility of polypropylene microplastics in stormwater biofilters under freeze-thaw cycles. *Journal of Hazardous Materials Letters,* **3**, 100048.

33. Wang, J., Liu, X., Li, Y., Powell, T., Wang, X., Wang, G., and Zhang, P. (2019). Microplastics as contaminants in the soil environment: A mini-review. *Science of the Total Environment*, **691**, 848–857.

34. Talukdar, A., Bhattacharya, S., Bandyopadhyay, A., and Dey, A. (2023). Microplastic pollution in the Himalayas: Occurrence, distribution, accumulation and environmental impacts. *Science of the Total Environment*, **874**, 162495.

35. Bertoli, M., Lesa, D., Pastorino, P., Mele, A., Anselmi, S., Barceló, D., Prearo, M., Renzi, M., and Pizzul, E. (2023). Microplastic patterns in riverine waters and leaf litter: Leaf bag technique to investigate the microplastic accumulation trends in lotic ecosystems. *Journal of Contaminant Hydrology*, **259**, 104253.

36. Cesarini, G., and Scalici, M. (2022). Riparian vegetation as a trap for plastic litter. *Environmental Pollution*, **292**, 118410.

37. Ásmundsdóttir, Á. M., and Schulz, B. (2020). Effects of microplastics in the cryosphere. *Handbook of Microplastics in the Environment*, 1–46.

38. Zhang, K., Su, J., Xiong, X., Wu, X., Wu, C., and Liu, J. (2016). Microplastic pollution of lakeshore sediments from remote lakes in Tibet plateau, China. *Environmental Pollution*, **219**, 450–455.

39. Free, C. M., Jensen, O. P., Mason, S. A., Eriksen, M., Williamson, N. J., and Boldgiv, B. (2014). High-levels of microplastic pollution in a large, remote, mountain lake. *Marine Pollution Bulletin*, **85**(1), 156–163.

40. Bilal, M., Qadir, A., Yaqub, A., Irfan, M., Aslam, M., and Ullah, R. (2021). Microplastics Monitoring in Environmental Matrices of an Alpine River originating from Hindu Kush Mountain Range. *Research Square*. DOI: 10.21203/rs.3.rs-481545/v1.

41. Pastorino, P., Pizzul, E., Bertoli, M., Anselmi, S., Kušće, M., Menconi, V., Prearo, M., and Renzi, M. (2021). First insights into plastic and microplastic occurrence in biotic and abiotic compartments, and snow from a high-mountain lake (Carnic Alps). *Chemosphere*, **265**, 129121.

42. Negrete Velasco, A. D. J., Rard, L., Blois, W., Lebrun, D., Lebrun, F., Pothe, F., and Stoll, S. (2020). Microplastic and fibre contamination in a remote mountain lake in Switzerland. *Water*, **12**(9), 2410.

43. Rodríguez-Seijo, A., and Pereira, R. (2017). Morphological and physical characterization of microplastics. In *Comprehensive Analytical Chemistry* (Vol. 75, pp. 49–66). Elsevier.

44. Cai, L., Wang, J., Peng, J., Tan, Z., Zhan, Z., Tan, X., and Chen, Q. (2017). Characteristic of microplastics in the atmospheric fallout from Dongguan city, China: Preliminary research and first evidence. *Environmental Science and Pollution Research*, **24**, 24928–24935.

45. Scherer, C., Weber, A., Lambert, S., and Wagner, M. (2018). *Interactions of Microplastics with Freshwater Biota* (pp. 153–180). Springer International Publishing.
46. Çomaklı, E., Bingöl, M. S., and Bilgili, A. (2020). Assessment of microplastic pollution in a crater lake at high altitude: A case study in an urban crater lake in Erzurum, Turkey. *Water, Air, & Soil Pollution*, **231**, 1–6.
47. Zhang, Y., Gao, T., Kang, S., Allen, S., Luo, X., and Allen, D. (2021). Microplastics in glaciers of the Tibetan Plateau: Evidence for the long-range transport of microplastics. *Science of the Total Environment*, **758**, 143634.
48. Xiong, X., Zhang, K., Chen, X., Shi, H., Luo, Z., and Wu, C. (2018). Sources and distribution of microplastics in China's largest inland lake – Qinghai Lake. *Environmental Pollution*, **235**, 899–906.
49. Fendall, L. S., and Sewell, M. A. (2009). Contributing to marine pollution by washing your face: Microplastics in facial cleansers. *Marine Pollution Bulletin*, **58**(8), 1225–1228.
50. Lambert, S., Scherer, C., and Wagner, M. (2017). Ecotoxicity testing of microplastics: Considering the heterogeneity of physicochemical properties. *Integrated Environmental Assessment and Management*, **13**(3), 470–475.
51. Chauhan, J. S., Semwal, D., Nainwal, M., Badola, N., and Thapliyal, P. (2021). Investigation of microplastic pollution in river Alaknanda stretch of Uttarakhand. Environment. *Development and Sustainability*, 1–15.
52. Pastorino, P., Elia, A. C., Pizzul, E., Bertoli, M., Renzi, M., and Prearo, M. (2024). The old and the new on threats to high-mountain lakes in the Alps: A comprehensive examination with future research directions. *Ecological Indicators*, **160**, 111812.
53. Ziani, K., Ioniţă-Mîndrican, C. B., Mititelu, M., Neacşu, S. M., Negrei, C., Moroşan, E., Drăgănescu, D., and Preda, O. T. (2023). Microplastics—A real global threat for environment and food safety: a state of the art review. *Nutrients*, **15**(3), 617.
54. Chen, Q., Lackmann, C., Wang, W., Seiler, T. B., Hollert, H., and Shi, H. (2020). Microplastics lead to hyperactive swimming behaviour in adult zebrafish. *Aquatic Toxicology*, **224**, 105521.
55. McIlwraith, H. K., Kim, J., Helm, P., Bhavsar, S. P., Metzger, J. S., and Rochman, C. M. (2021). Evidence of microplastic translocation in wild-caught fish and implications for microplastic accumulation dynamics in food webs. *Environmental Science & Technology*, **55**(18), 12372–12382.

56. Walkinshaw, C., Lindeque, P. K., Thompson, R., Tolhurst, T., and Cole, M. (2020). Microplastics and seafood: lower trophic organisms at highest risk of contamination. *Ecotoxicology and Environmental Safety*, **190**, 110066.

57. Miller, M. E., Hamann, M., and Kroon, F. J. (2020). Bioaccumulation and biomagnification of microplastics in marine organisms: A review and meta-analysis of current data. *PloS One*, **15**(10), e0240792.

58. Tumwesigye, E., Nnadozie, C. F., Akamagwuna, F. C., Noundou, X. S., Nyakairu, G. W., and Odume, O. N. (2023). Microplastics as vectors of chemical contaminants and biological agents in freshwater ecosystems: Current knowledge status and future perspectives. *Environmental Pollution*, **330**, 121829.

59. Ribeiro, F., O'Brien, J. W., Galloway, T., and Thomas, K. V. (2019). Accumulation and fate of nano-and micro-plastics and associated contaminants in organisms. *TrAC Trends in Analytical Chemistry*, **111**, 139–147.

60. Shen, M., Song, B., Zhou, C., Almatrafi, E., Hu, T., Zeng, G., and Zhang, Y. (2022). Recent advances in impacts of microplastics on nitrogen cycling in the environment: A review. *Science of the Total Environment*, **815**, 152740.

61. Zhang, X., Li, Y., Ouyang, D., Lei, J., Tan, Q., Xie, L., Li, Z., Liu, T., Xiao, Y., Farooq, T. H., Wu, X., Chen, L., and Yan, W. (2021). Systematical review of interactions between microplastics and microorganisms in the soil environment. *Journal of Hazardous Materials*, **418**, 126288.

62. Koelmans, A. A., Nor, N. H. M., Hermsen, E., Kooi, M., Mintenig, S. M., and De France, J. (2019). Microplastics in freshwaters and drinking water: Critical review and assessment of data quality. *Water Research*, **155**, 410–422.

Chapter 7

Microplastic Pollution in Antarctic Environments

Gabriel Enrique De-la-Torre,[a] Mohamed Ben-Haddad,[b] and Luis Santillán[a]
[a]*Grupo de Investigación de Biodiversidad, Medio Ambiente y Sociedad, Universidad San Ignacio de Loyola, Lima, Perú*
[b]*Laboratory of Aquatic Systems, Marine and Continental Environments (AQUAMAR), Faculty of Sciences, Ibn Zohr University, Agadir 80000, Morocco*
gabriel.delatorre@usil.pe

7.1 Introduction

Antarctica, the southernmost continent on Earth, is a unique and critical region for scientific research due to its extreme environment, significant influence on global climate systems, and unique fauna (Fig. 7.1) [1]. This vast, icy expanse, covering an area of approximately 14 million square kilometres, holds around 70% of the world's fresh water in its massive ice sheets [2]. Despite its harsh conditions, Antarctica plays a vital role in Earth's climate regulation, ocean circulation, and global sea level. The continent's remote and pristine

Microplastic Pollution: Environmental Impact and Solutions
Edited by Julia Martín, Juan Luis Santos, Irene Aparicio, and Esteban Alonso

ISBN 978-981-5129-77-9 (Hardcover), 978-1-003-66006-4 (eBook)
www.jennystanford.com

DOI: 10.1201/9781003660064-7

ecosystems provide an exceptional scenario for climate change, glaciology, and marine biology research, offering insights into both historical climate patterns and future environmental shifts.

Figure 7.1 Antarctic megafauna: (a) Black-browed albatross (*Thalassarche melanophris*) (photograph by Luis Santillán). (b) Southern elephant seal (*Mirounga leonina*) (photograph by Luis Santillán). (c) Leopard seal (*Hydrurga leptonyx*) (photograph by Luis Santillán). (d) Weddell seal (*Leptonychotes weddellii*) (photograph by Gabriel E. De-la-Torre). (e) Chinstrap penguin (*Pygoscelis antarcticus*) (photograph by Gabriel E. De-la-Torre). (f) Gentoo penguin (*Pygoscelis papua*) (photograph by Gabriel E. De-la-Torre).

Microplastic pollution has emerged as a pervasive environmental issue, infiltrating even the most remote and pristine regions of the world, including Antarctica [3]. Despite its isolation and sparse human presence, recent studies have revealed the presence of microplastics in Antarctic waters, ice, and biota, indicating the far-reaching impact of global pollution [4–6]. Microplastics are defined as particles primarily made of synthetic polymers in the size range of 1 μm to 5 mm [7, 8]. These particles originate from various sources, including the degradation of larger plastic debris, synthetic fibres from clothing, and microbeads from personal care products, among other sources [9–11]. The introduction of microplastics into the Antarctic ecosystem poses significant threats to its unique wildlife and fragile environment, as these particles can be ingested by marine organisms, potentially leading to physical and chemical harm. Furthermore, microplastics can act as vectors for persistent organic pollutants, exacerbating the contamination of the food web [12, 13].

In light of the recent evidence regarding the contamination of Antarctica and the Southern Ocean and its implications for conservation efforts, this chapter will address the current state of microplastic pollution in Antarctic biota and environmental matrices, as well as their potential effects, ecotoxicity, and future research perspectives.

7.2 Microplastics in Antarctic Biota

Microplastic pollution has infiltrated the Antarctic biota, with studies finding these tiny particles in krill, fish, and seabirds. These pollutants can disrupt feeding habits and introduce toxins into the food web, posing a significant threat to the health and survival of these organisms in their pristine environment [14].

7.2.1 Invertebrates

Antarctic invertebrates are the base of the marine food web. Fish, penguins, seals, and whales all depend on these abundant invertebrates for survival. These organisms also contribute to the overall health of the ecosystem in other ways, such as breaking down organic matter and nutrient cycling [15]. Studies have recently reported the presence of microplastics in macroinvertebrates of multiple taxa. Researchers have found that bivalves from Livingston Island were contaminated with microplastics [16]. Out of 69 bivalves collected, 45.6% had at least one item of microdebris. The average number of items per individual was 0.71 ± 0.89, and the number of items per gram of soft tissue was 1.49 ± 2.35. The most common types of microdebris found were fibres, primarily cellulose, followed by synthetic polymers such as polyethylene terephthalate and nylon. In the case of krill and salps, Wilkie Johnston et al. [17] reported higher concentrations in krill (2.13 ± 0.26 microplastics per individual) than in salps (1.38 ± 0.42 microplastics per individual), primarily composed of fibres. Last, Sfriso et al. [18] found that 83% of 12 analysed macrobenthic species contained microplastics, with contamination levels ranging from 0.01 to 3.29 items per mg. The microplastics identified were predominantly nylon (86%) and polyethylene (5%), with particle sizes mostly between 50 and 100 μm. Filter feeders and grazers showed higher contamination levels than omnivores and predators, suggesting no significant bioaccumulation through the food web. The study highlighted local pollution sources, particularly from the nearby Mario Zucchelli Base, as significant contributors to the microplastic presence in the benthic community.

Research confirms the presence of microplastics in marine invertebrates across Antarctica. This troubling discovery extends the reach of plastic pollution to even the most remote corners of the Earth. Filter feeders and organisms at the base of the food web seem particularly susceptible to ingesting microplastics. The long-term effects of microplastics on Antarctic invertebrates are still being investigated. However, the potential for harm is significant. Microplastics could obstruct digestive systems, reduce nutrient uptake, or even leach harmful chemicals [19]. Furthermore, the presence of microplastics in these organisms raises concerns about potential biomagnification throughout the Antarctic food chain.

7.2.2 Fish

The consumption of microplastics from contaminated prey is expected at higher trophic levels. In the case of fish, several studies have reported microplastics in their gastrointestinal tract. Mancuso et al. [20] investigated the occurrence of microplastics in the gastrointestinal tracts of emerald rockcod (*Trematomus bernacchii*) caught near the Mario Zucchelli Station (MZS) in the Ross Sea. The study involved analysing 78 fish specimens collected during the 2021–2022 Italian Antarctic expedition and comparing the results with those obtained in 1998. The findings showed a significant presence of anthropogenic microparticles, predominantly polyester, with indigo dye being the most common industrial dye found. The study reports a higher frequency of occurrence (43.2%) and abundance (0.9 items/specimen) of AMs in 2021–2022 compared to 1998. The results highlight the increased impact of AMs over the last 25 years, likely due to the continued use of polyester-based technical clothing and insufficient efficiency of the wastewater treatment plant at MZS. Zhu et al. [21] comprehensively studied the presence of microplastics in four species of Antarctic fish (*Chionodraco rastrospinosus, Electrona carlsbergi, Notolepis coatsi, Trematomus eulepidotus*) collected from the South Shetland Islands region. The study found that over 90% of the fish contained detectable microplastics, with an average abundance ranging from 0.72 to 1.7 items per individual. The predominant shapes of microplastics were fibres, accounting for 60%–82% of total items across different species. The microplastics were primarily composed of polypropylene, polyamide, and polyethylene, which collectively made up 63%–86% of the total microplastics. Other studies reported that 89% of *Harpagifer antarcticus*, 73% of *Harpagifer bispinis*, and 31.3% of *Chionodraco rastrospinosus* caught contained microfibres primarily composed of polyethylene terephthalate and acrylics in their gastrointestinal tracts [22, 23].

Studies have confirmed the presence of microplastics in Antarctic fish, highlighting the vast reach of plastic pollution. This discovery raises significant concerns for the health of Antarctic ecosystems. Fish of various feeding habits have been found to ingest microplastics, suggesting widespread contamination throughout the food web.

7.2.3 Mammals

Studies on Antarctic mammals are fairly limited. The study by Garcia-Garin et al. [24] analysed 42 scats of male fur seals collected in late summer to assess microplastic contamination in the Bransfield Strait's pelagic food web. However, no microplastics were detected in the analysed scats. The particles initially suspected to be microplastics were identified as silicate minerals and chitin through Fourier-transform infrared (FTIR) spectroscopy. Conversely, a recent study found that 100% of scat samples from three different species (crabeater, leopard, and Weddell seals) contained anthropogenic debris, primarily fibres, followed by fragments and filaments [25]. The predominant materials identified included plastic polymers (polystyrene, polyester, polyethylene terephthalate), pigments (reactive blue 238, indigo blue, copper phthalocyanine), and carbon black. The findings highlight significant contamination in the Antarctic ecosystem, particularly at higher trophic levels, emphasizing the need for effective waste management and pollution mitigation strategies. On the other hand, studies on whales are significantly lacking.

7.2.4 Birds

Penguins are some of the most studied birds in Antarctica. Kim et al. [6] investigated the presence and characteristics of microplastics ingested by gentoo penguin chicks on King George Island. The research involved collecting stomach and upper intestine samples from 14 deceased penguin chicks and analysing these for microplastic content using FTIR spectroscopy. The results showed a significant presence of microplastics, with an average of 27 microplastics per individual. Most of these microplastics were fragments smaller than 300 μm, primarily made of polyethylene and polypropylene. The study also found that the local marine environment near the breeding colonies contained a high concentration of plastic debris, suggesting that the penguins' parents might be foraging in polluted areas, leading to the ingestion of microplastics by the chicks. Another study on gentoo penguins found that 20% of the penguin scats collected contained microplastics, with microfibres being the most abundant type, followed by fragments and films [26]. The average number of microplastics was 0.23 per scat and were composed of polyester, with other polymers such as rayon, polyacrylonitrile, polypropylene, polyethylene, and polyacrylate. Similarly, the occurrence of microplastics in gentoo, chinstrap, and Adélie penguins has been reported in the range of 15% to 29% across various breeding colonies in the Antarctic Peninsula and Scotia Sea [27]. Conversely, in 41 gizzards of emperor penguin chicks, 85 putative particles were sorted, none were identified as microplastics, and most were traced back to contamination during sample processing [28]. This suggests that microplastic concentrations in the local food web of the Weddell Sea and Dronning Maud Land are currently negligible, contrasting with studies on other Antarctic penguins.

Other than penguins, skuas have also been investigated. Perold et al. [29] investigated the use of regurgitated skua pellets to monitor marine plastic pollution. The study involved collecting skua pellets from Inaccessible Island, which contained the remains of four seabird taxa: Fregetta storm petrels, white-faced storm petrels, broad-billed prions, and great shearwaters. The findings indicated that skua pellets reliably reflect the plastic loads in seabird prey, although they might underrepresent smaller items. The study identified hard fragments as the most common plastic type ingested, with polyethylene being the predominant polymer. Other studies have also reported the presence of plastic debris in skuas from King George Island and the Antarctic Peninsula [30, 31].

7.3 Microplastics in Abiotic Compartments

Microplastic pollution has emerged as a critical environmental issue globally, affecting even the most remote and pristine regions such as Antarctica. This compilation of studies provides a comprehensive overview of the presence, distribution, and impact of microplastics in the abiotic compartments of Antarctica (seawater, sediment, air, soil, snow, and ice), a group of ecosystems that interact between them in a unique environment (Fig. 7.2). The research spans multiple locations, methodologies, and characterization techniques (i.e., ATR-FTIR, μFTIR, Raman, and μRaman), offering relevant insights into the extent and sources of microplastic contamination in this valuable area of the Earth [32].

Figure 7.2 Graphic representation of the Antarctic environmental compartments where microplastics have been reported. Created using Canva.

7.3.1 Seawater

Microplastics have been detected in seawater across various regions of Antarctica, indicating the pervasive nature of plastic pollution even in these remote areas. A comprehensive study by Suaria et al. [33] during the 2016/17 Antarctic Circumnavigation Expedition revealed significantly low concentrations of plastics in polar waters south of the Subtropical Front (STF), with only five microplastics detected. This suggests that the STF acts as a barrier to southward plastic transport. However, Isobe et al. [34] highlighted the presence of microplastics smaller than 5 mm in the Southern Ocean, estimating around 100,000 pieces per km^2 near Antarctica. Similarly, Cincinelli et al. [35] found microplastics in subsurface water samples from the Ross Sea, with concentrations ranging from 0.0032 to 1.18 particles per m^3. These studies underscore the significant presence of microplastics in Antarctic waters.

Likewise, Kim et al. [36] investigated the behaviour of microplastics from a wastewater treatment plant in Marian Cove, revealing that local flow conditions and release timing significantly affect microplastic accumulation in seawater. Zhang et al. [37] found higher microplastic concentrations in subsurface waters compared to surface waters, with the Ross Sea exhibiting the highest levels. Their findings suggest both local sources and southward transport from mid- and low-latitude oceans contribute to microplastic pollution in Antarctica. Additionally, Leistenschneider et al. [38] reported high concentrations of small microplastics (11–500 μm) in the southern Weddell Sea, emphasizing the need to include smaller particles in pollution assessments. Hence, paint fragments were examined to discern vessel-induced contamination, with 45.5% of recovered microplastic attributed to vessel sources. Micro-X-ray fluorescence spectroscopy (μXRF) helped identify differences in elemental composition, highlighting challenges in solely relying on visual and FTIR spectroscopy for contamination source determination [38].

Otherwise, Pakhomova et al. [39] conducted a comparative investigation of microplastic variability in subsurface waters from the Arctic to Antarctica, using a consistent sampling and analytical approach across five cruises. A total of 121 subsurface water samples (100 μm to 5 mm) were collected using an underway pump-through system, covering regions including the East-Siberian Sea, Atlantic

Ocean, and Antarctic Peninsula. The overall average microplastic abundance was 0.7 ± 0.6 items/m^3, with equal average abundances of fragments and fibres (0.34 items/m^3). While no significant regional differences in microplastic abundance were found, variations were noted in size, morphology, polymer types, and weight concentrations. The Central Atlantic and Barents Sea had higher microplastic weight concentrations (7–7.5 μg/m^3) compared to the North Atlantic and Siberian Arctic (0.6 μg/m^3). The study suggests terrestrial sources dominate microplastic in the Northern Hemisphere, whereas offshore industries are key sources in the Southern Hemisphere. Additionally, fibre pollution was higher in the Northern Hemisphere, indicating fibres can be transported long distances by air and water, while fragments tend to remain near their source [39].

7.3.2 Sediments

Sediments in various Antarctic regions have been found to contain significant amounts of microplastics. Munari et al. [40] conducted the first survey on microplastics in sediments from Terra Nova Bay, identifying plastic debris in all 31 samples, with fibres being the most common type. The study highlighted a decreasing concentration of plastic debris with increasing distance from the Mario Zucchelli Base. Similarly, Reed et al. [41] examined marine sediments near the Rothera Research Station and found the highest concentrations near the sewage treatment plant outfall, suggesting local pollution sources.

Cunningham et al. [42] reported microplastic pollution in 93% of deep-sea sediment cores from Antarctica and the Southern Ocean, with higher abundances than expected. Their study highlighted the correlation between microplastic fragments and clay percentage, indicating similar dispersion behaviour. Alurralde et al. [43] analysed sediment trap samples from Potter Cove and identified anthropogenic microfibres, revealing temporal patterns of microplastic pollution in Antarctic coastal ecosystems. In addition, Perfetti-Bolaño et al. [44] investigated microplastic contamination in soils and intertidal sediments at Fildes Bay, finding fragments predominantly in soils and fibres in intertidal sediments. Physicochemical analysis showed larger particle sizes in intertidal sediments and higher organic matter content in soils. Fragments were composed of phenoxy resin, resembling coatings of facilities, while fibres were identified as polyethylene terephthalate.

7.3.3 Air

Microplastic pollution in Antarctic air has been less studied but provides crucial insights into the atmospheric transport of these particles. Cunningham et al. [45] investigated air samples from the Weddell Sea, identifying 47 microplastic categories in air samples, predominantly fibres, that overlapped between air and other media, suggesting atmospheric transport from sources such as southern South America. This study highlighted the role of air masses in dispersing microplastics over long distances, contributing to the contamination of remote regions such as Antarctica. Indirectly, Aves et al. [4] documented microplastics in Antarctic snow, collected from the Ross Island region, with backward air mass trajectories indicating potential long-range transportation up to 6000 km. These findings underscore the significant role of atmospheric processes in distributing microplastics to and within Antarctica, emphasizing the need for further research on airborne microplastics in polar regions.

7.3.4 Soil

Soil contamination by microplastics in Antarctica has been rarely documented. However, the limited studies revealed significant pollution levels and diverse sources. Perfetti-Bolaño et al. [44] found abundant plastic fragments in soils around Fildes Bay, with higher counts at certain sites and associations with phenoxy resin coatings from local facilities. These findings highlight the impact of human activities on soil contamination. Similarly, González-Pleiter et al. [46] reported microplastics in a freshwater stream on Byers Peninsula, indicating terrestrial sources and local human activities as significant contributors to soil pollution. These studies demonstrate the pervasive nature of microplastic pollution in Antarctic soils, driven by local human presence and activities.

7.3.5 Snow and Ice

Microplastics have been detected in Antarctic snow and ice, providing evidence of contamination in these compartments. Aves et al. [4]

presented the first evidence of microplastics in Antarctic snow, with an average concentration of 29 particles per litre, predominantly fibres, and polyethylene terephthalate. A similar study investigated microplastic contamination in East Antarctic sea ice [5]. The analysis of one ice core revealed 96 microplastic particles averaging 11.71 particles per litre. The most common polymers were polyethylene, polypropylene, and polyamide, consistent with marine microplastic studies. Sea-ice microplastic concentrations correlated positively with chlorophyll a, indicating potential incorporation by living biomass. These findings suggest sea ice could serve as a reservoir for microplastic debris in the Southern Ocean, potentially impacting local food webs and biogeochemistry.

In the same way, González-Pleiter et al. [47] documented plastic pollution on Collins Glacier, highlighting local activities and wind transport as sources. Materić et al. [48] investigated nanoplastics in polar ice from Greenland and Antarctica using Thermal Desorption–Proton Transfer Reaction–Mass Spectrometry (TD-PTR-MS). Nanoplastics of various types, including PE, PP, PET, PS, PVC, and tyre wear nanoparticles, were detected in both Greenland firn core and Antarctic sea ice. Nanoplastic concentrations averaged 13.2 ng/mL in Greenland firn samples and 52.3 ng/mL in Antarctic sea ice. The study shed light on the presence of nanoplastics in remote polar regions, likely transported through complex processes involving emission from land and sea surfaces, as well as atmospheric and marine circulation.

7.4 Potential Effects of Microplastics in Antarctic Environments

The presence of microplastics in Antarctica evidences the pervasiveness of plastic pollution. Studies have confirmed microplastics in various Antarctic organisms, including krill, a keystone species at the base of the marine food web, as well as fish, mammals, seabirds, and even benthic invertebrates. Microplastics likely originate from a multitude of sources, including disintegrating plastic debris from research stations, tourism activities, and even atmospheric transport from distant continents and urban areas [49, 50].

The consequences of microplastic ingestion for Antarctic wildlife are concerning. Filter feeders such as krill are particularly vulnerable due to their non-selective feeding behaviour [51]. This can lead to a feeling of fullness, reducing their appetite for essential nutrients. In some cases, microplastics can physically block digestive systems, leading to starvation [52]. Research suggests that microplastics may also leach harmful chemicals or harbour contaminants already present in the environment [53, 54], further compromising the health of these organisms. The threat extends beyond individual organisms. Microplastics have the potential to disrupt entire food webs. When organisms lower on the food chain ingest microplastics, these pollutants can biomagnify as they are consumed by larger predators [55]. This biomagnification can lead to a build-up of harmful chemicals in top predators such as penguins, seals, and whales, potentially impacting their health and reproductive success. However, the long-term ecological implications of microplastic contamination in Antarctica are still being explored. Regardless, the potential for harm is undeniable.

Ecotoxicology is a scientific discipline that investigates the adverse effects of toxic chemicals on biological organisms within an ecological context. Ecotoxicological studies play an indispensable role in environmental science by unveiling the intricate web of effects caused by contaminants. Multiple studies have conducted ecotoxicological studies with microplastics [56], including several exposing Antarctic species. Dawson et al. [57] investigated how Antarctic krill (*Euphausia superba*) contributed to the fragmentation of microplastics into nanoplastics through their digestive processes. The results showed that krill could fragment microplastics (31.5 μm) into significantly smaller pieces (<1 μm in diameter), demonstrating that the digestive system of krill plays a crucial role in breaking down microplastics into nanoplastics. This process, termed "digestive fragmentation," suggests that krill and potentially other marine organisms can influence the degradation of microplastics in the marine environment. Later investigations on Antarctic krill showed through the exposure of gravid female krill and their embryos to nanoplastics that the presence of food during maternal exposure to nanoplastics led to higher levels of essential fatty acids, such as polyunsaturated fatty acids (PUFAs), eicosapentaenoic acid (EPA), and docosahexaenoic acid (DHA) [58]. This indicates that the provision of a food source can mitigate some of the negative effects of nanoplastic exposure on embryonic development. Furthermore, the total lipid content in embryos was not significantly affected by nanoplastic exposure. Manno et al. [59] showed that short-term exposure (48 h) to polystyrene nanoplastics and ocean acidification significantly increased mortality rates in pteropods, especially when both stressors were combined. The nanoplastics adhered to and penetrated the shells, causing physical damage, while the acidified conditions exacerbated shell dissolution. The combined stress of ocean acidification and nanoplastics was more detrimental than either stressor alone, highlighting the need to consider multiple environmental stressors when assessing the impact on marine ecosystems. Last, Oliveira de Miranda et al. [60] explored the effects of polyethylene fragments and polyacrylonitrile fibres on soil properties and microbial activity in the Fildes Peninsula, King George Island, Antarctica. The research conducted a 22-day pot incubation experiment under natural field conditions, exposing the soil to different doses of polyethylene fragments and polyacrylonitrile fibres. The results showed that polyethylene fragments mainly decreased soil bulk density and hydraulic conductivity without significantly affecting soil porosity. In contrast, polyacrylonitrile fibres significantly altered soil porosity, macroporosity, and bulk density, especially at higher doses. Both microplastic types increased microbial activity, as indicated by fluorescein diacetate hydrolysis, and decreased the cation exchange capacity. The highest doses of polyacrylonitrile fibres increased soil nitrogen, while the highest doses of polyethylene fragments increased total organic carbon. These findings underscore the potential for microplastics to alter soil's physical and chemical properties and microbial functions even in remote environments such as Antarctica.

Microplastics have penetrated what appears to be a pristine ecosystem, posing a significant threat both ecologically and ecotoxicologically. Filter feeders such as krill, which play a vital role at the base of the food web, are particularly at risk of ingesting microplastics. This ingestion could lead to reduced feeding efficiency and compromised health. The possibility of biomagnification through the food chain raises further concerns, as harmful chemicals associated with microplastics could accumulate in top predators such as penguins and whales. Ecotoxicological studies are urgently required to comprehensively understand the impact of microplastic contamination on Antarctic wildlife. These studies should examine not only the direct effects on individual organisms but also the wider ecological consequences, such as disruptions to food web dynamics and ecosystem stability.

7.5 Perspectives

Continued research is needed to fully understand the scope of microplastic contamination and its impact on Antarctic wildlife. This knowledge will be vital in developing effective strategies to protect this unique and irreplaceable ecosystem. The next phase should delve deeper into investigating the types and sources of microplastics prevalent in different regions. This will help identify the most significant contributors to the problem and inform targeted mitigation strategies. In terms of its impact, research has documented microplastics in various Antarctic organisms. However, a more nuanced understanding of their impact is needed. Ecotoxicological studies should examine the physiological effects on individuals, including potential harm to digestive systems, nutrient uptake, and overall health. Furthermore, research needs to explore how microplastics disrupt population dynamics and potentially alter food web interactions.

The small number of studies, as well as their extension in terms of sampling efforts, might be limited by the difficulties of conducting fieldwork and surveys under extreme weather conditions. For this reason, quantifying and characterizing microplastics in such a remote and challenging environment present unique difficulties. Research efforts should focus on refining existing techniques and developing new methodologies to accurately measure microplastic contamination in Antarctic ecosystems.

Combating this issue requires a multi-pronged approach. Stricter regulations on plastic waste disposal in research stations and on tourist vessels are essential. Developing sustainable alternatives to single-use plastics and promoting responsible waste management practices are crucial steps in reducing the influx of microplastics into Antarctica. It should also be noted that microplastic pollution in Antarctica is not an isolated issue. International collaboration is vital to address sources beyond the continent itself and develop global solutions. Sharing data, expertise, and best practices will be crucial in tackling this widespread environmental challenge.

By addressing these critical perspectives, future research can illuminate the full scope of microplastic contamination in Antarctica. This deeper understanding will allow policymakers, scientists, and the global community to implement effective mitigation strategies, safeguard the future of this unique ecosystem, and ensure the pristine condition of Antarctica.

7.6 Conclusion

The threat of microplastic pollution casts a significant shadow over Antarctica's pristine ecosystems. The detection of microplastics in Antarctic biota, ranging from foundational krill to top predators, raises serious concerns regarding potential ecological disruption and the biomagnification of harmful chemicals. Although research has highlighted the extent and possible consequences of this contamination, critical gaps persist. Future efforts must deepen our understanding of microplastic impacts on individual organisms and food webs, while also addressing the unique challenges posed by Antarctica's harsh environment. International collaboration and a commitment to innovative research are essential to tackling this multifaceted issue. Such efforts are necessary to ensure that Antarctica's stunning landscapes and unparalleled biodiversity remain undisturbed by microplastic pollution.

Acknowledgements

The authors are thankful to the Peruvian Antarctic Program—Ministry of Foreign Affairs, Peru, for providing logistical support in the framework of the Peruvian Expeditions to Antarctica XXVII (2019–2020) and XXIV (2023) in which Universidad San Ignacio de Loyola participated. The authors are also thankful to the members of the Antarctic Operations Company—Peruvian Army and the BAP Carrasco crew—Peruvian Navy for providing logistical support. The authors are also thankful to Universidad San Ignacio de Loyola for providing the Antarctic expeditions with financial support.

References

1. Chaturvedi, S., 2009. Antarctica, in: Kitchin, R., Thrift, N. (Eds.), *International Encyclopedia of Human Geography*. Elsevier, pp. 133–139. https://doi.org/10.1016/B978-008044910-4.00247-9
2. Pan, X.L., Li, B.F., Watanabe, Y.W., 2022. Intense ocean freshening from melting glacier around the Antarctica during early twenty-first century. *Sci. Rep.* **12**, 383. https://doi.org/10.1038/s41598-021-04231-6
3. Waller, C.L., Griffiths, H.J., Waluda, C.M., Thorpe, S.E., Loaiza, I., Moreno, B., Pacherres, C.O., Hughes, K.A., 2017. Microplastics in the Antarctic marine system: An emerging area of research. *Sci. Total Environ.* **598**, 220–227. https://doi.org/10.1016/j.scitotenv.2017.03.283
4. Aves, A.R., Revell, L.E., Gaw, S., Ruffell, H., Schuddeboom, A., Wotherspoon, N.E., Larue, M., McDonald, A.J., 2022. First evidence of microplastics in Antarctic snow. *Cryosphere* **16**, 2127–2145. https://doi.org/10.5194/tc-16-2127-2022
5. Kelly, A., Lannuzel, D., Rodemann, T., Meiners, K.M., Auman, H.J., 2020. Microplastic contamination in east Antarctic sea ice. *Mar. Pollut. Bull.* **154**, 111130. https://doi.org/10.1016/j.marpolbul.2020.111130
6. Kim, Y., Kim, H., Jeong, M.S., Kim, D., Kim, J., Jung, J., Seo, H.M., Han, H.J., Lee, W.S., Choi, C.Y., 2023. Microplastics in gastrointestinal tracts of gentoo penguin (*Pygoscelis papua*) chicks on King George Island, Antarctica. *Sci. Rep.* **13**, 13016. https://doi.org/10.1038/s41598-023-39844-6
7. Frias, J.P.G.L., Nash, R., 2019. Microplastics: Finding a consensus on the definition. *Mar. Pollut. Bull.* **138**, 145–147. https://doi.org/10.1016/j.marpolbul.2018.11.022
8. Hartmann, N.B., Hüffer, T., Thompson, R.C., Hassellöv, M., Verschoor, A., Daugaard, A.E., Rist, S., Karlsson, T., Brennholt, N., Cole, M., Herrling, M.P., Hess, M.C., Ivleva, N.P., Lusher, A.L., Wagner, M., 2019. Are we speaking the same language? Recommendations for a definition and categorization framework for plastic debris. *Environ. Sci. Technol.* **53**, 1039–1047. https://doi.org/10.1021/ACS.EST.8B05297/ASSET/IMAGES/ACS.EST.8B05297.SOCIAL.JPEG_V03
9. De-la-Torre, G.E., Forero López, A.D., Dioses-Salinas, D.C., Fernández Severini, M.D., Dobaradaran, S., Madadi, R., Ben-Haddad, M., 2023. Microplastics released from face masks used during the COVID-19 pandemic: A review of the characterization techniques. *TrAC - Trends Anal. Chem.* **167**, 117227. https://doi.org/10.1016/j.trac.2023.117227

10. Mak, C.W., Tsang, Y.Y., Leung, M.M.-L., Fang, J.K.-H., Chan, K.M., 2020. Microplastics from effluents of sewage treatment works and stormwater discharging into the Victoria Harbor, Hong Kong. *Mar. Pollut. Bull.* **157**. https://doi.org/10.1016/j.marpolbul.2020.111181
11. Zendehboudi, A., Mohammadi, A., Dobaradaran, S., De-la-Torre, G.E., Ramavandi, B., Hashemi, S.E., Saeedi, R., Tayebi, E.M., Vafaee, A., Darabi, A.H., 2024. Analysis of microplastics in ships ballast water and its ecological risk assessment studies from the Persian Gulf. *Mar. Pollut. Bull.* **198**, 115825. https://doi.org/10.1016/j.marpolbul.2023.115825
12. Benson, N.U., Agboola, O.D., Fred-Ahmadu, O.H., De-la-Torre, G.E., Oluwalana, A., Williams, A., 2022. Micro(nano)plastics prevalence, food web interactions and toxicity assessment in aquatic organisms: A review. *Front. Mar. Sci.* **9**, 851281. https://doi.org/10.3389/FMARS.2022.851281
13. Hüffer, T., Weniger, A.K., Hofmann, T., 2018. Sorption of organic compounds by aged polystyrene microplastic particles. *Environ. Pollut.* **236**, 218–225. https://doi.org/10.1016/j.envpol.2018.01.022
14. Gurumoorthi, K., Luis, A.J., 2023. Recent trends on microplastics abundance and risk assessment in coastal Antarctica: Regional meta-analysis. *Environ. Pollut.* **324**, 121385. https://doi.org/10.1016/j.envpol.2023.121385
15. Holm-Hansen, O., 1985. Nutrient cycles in Antarctic marine ecosystems BT - Antarctic nutrient cycles and food webs, in: Siegfried, W.R., Condy, P.R., Laws, R.M. (Eds.), *Antarctic Nutrient Cycles and Food Webs.* Springer Berlin Heidelberg, Berlin, Heidelberg, pp. 6–10.
16. Gonzalez-Pineda, M., Salvadó, H., Avila, C., 2024. Do Antarctic bivalves present microdebris? The case of Livingston Island. *Environ. Pollut.* **351**, 124086. https://doi.org/10.1016/j.envpol.2024.124086
17. Wilkie Johnston, L., Bergami, E., Rowlands, E., Manno, C., 2023. Organic or junk food? Microplastic contamination in Antarctic krill and salps. *R. Soc. Open Sci.* **10**, 221421. https://doi.org/10.1098/rsos.221421
18. Sfriso, A.A., Tomio, Y., Rosso, B., Gambaro, A., Sfriso, A., Corami, F., Rastelli, E., Corinaldesi, C., Mistri, M., Munari, C., 2020. Microplastic accumulation in benthic invertebrates in Terra Nova Bay (Ross Sea, Antarctica). *Environ. Int.* **137**, 105587. https://doi.org/10.1016/j.envint.2020.105587
19. Hodkovicova, N., Hollerova, A., Svobodova, Z., Faldyna, M., Faggio, C., 2022. Effects of plastic particles on aquatic invertebrates and fish: A review. *Environ. Toxicol. Pharmacol.* **96**, 104013. https://doi.org/10.1016/j.etap.2022.104013

20. Mancuso, M., Conti Nibali, V., Porcino, N., Branca, C., Natale, S., Smedile, F., Azzaro, M., D'Angelo, G., Bottari, T., 2023. Monitoring of anthropogenic microplastic pollution in Antarctic fish (emerald rockcod) from the Terranova Bay after a quarter of century. *Sci. Total Environ.* **904**, 167244. https://doi.org/10.1016/j.scitotenv.2023.167244
21. Zhu, W., Zhao, N., Liu, W., Guo, R., Jin, H., 2023. Occurrence of microplastics in Antarctic fishes: Abundance, size, shape, and polymer composition. *Sci. Total Environ.* **903**, 166186. https://doi.org/10.1016/j.scitotenv.2023.166186
22. Ergas, M., Figueroa, D., Paschke, K., Urbina, M.A., Navarro, J.M., Vargas-Chacoff, L., 2023. Cellulosic and microplastic fibers in the Antarctic fish *Harpagifer antarcticus* and Sub-Antarctic *Harpagifer bispinis*. *Mar. Pollut. Bull.* **194**, 115380. https://doi.org/10.1016/j.marpolbul.2023.115380
23. Geng, Q., Wang, C., Zhu, G., 2023. A novel report on the distribution of microplastics in ocellated icefish (*Chionodraco rastrospinosus*) in the northern Antarctic Peninsula. *Mar. Pollut. Bull.* **193**, 115188. https://doi.org/10.1016/j.marpolbul.2023.115188
24. Garcia-Garin, O., García-Cuevas, I., Drago, M., Rita, D., Parga, M., Gazo, M., Cardona, L., 2020. No evidence of microplastics in Antarctic fur seal scats from a hotspot of human activity in Western Antarctica. *Sci. Total Environ.* **737**. https://doi.org/10.1016/j.scitotenv.2020.140210
25. Cebuhar, J.D., Negrete, J., Rodríguez Pirani, L.S., Picone, A.L., Proietti, M., Romano, R.M., Della Védova, C.O., Casaux, R., Secchi, E.R., Botta, S., 2024. Anthropogenic debris in three sympatric seal species of the Western Antarctic Peninsula. *Sci. Total Environ.* **922**, 171273. https://doi.org/10.1016/j.scitotenv.2024.171273
26. Bessa, F., Ratcliffe, N., Otero, V., Sobral, P., Marques, J.C., Waluda, C.M., Trathan, P.N., Xavier, J.C., 2019. Microplastics in gentoo penguins from the Antarctic region. *Sci. Rep.* **9**, 1–7. https://doi.org/10.1038/s41598-019-50621-2
27. Fragão, J., Bessa, F., Otero, V., Barbosa, A., Sobral, P., Waluda, C.M., Guímaro, H.R., Xavier, J.C., 2021. Microplastics and other anthropogenic particles in Antarctica: Using penguins as biological samplers. *Sci. Total Environ.* **788**, 147698. https://doi.org/10.1016/j.scitotenv.2021.147698
28. Leistenschneider, C., Le Bohec, C., Eisen, O., Houstin, A., Neff, S., Primpke, S., Zitterbart, D.P., Burkhardt-Holm, P., Gerdts, G., 2022. No evidence of microplastic ingestion in emperor penguin chicks (*Aptenodytes forsteri*) from the Atka Bay colony (Dronning Maud

Land, Antarctica). *Sci. Total Environ.* **851**, 158314. https://doi.org/10.1016/j.scitotenv.2022.158314

29. Perold, V., Connan, M., Suaria, G., Weideman, E.A., Dilley, B.J., Ryan, P.G., 2024. Regurgitated skua pellets containing the remains of South Atlantic seabirds can be used as biomonitors of small buoyant plastics at sea. *Mar. Pollut. Bull.* **203**, 116400. https://doi.org/10.1016/j.marpolbul.2024.116400
30. Ibañez, A.E., Morales, L.M., Torres, D.S., Borghello, P., Haidr, N.S., Montalti, D., 2020. Plastic ingestion risk is related to the anthropogenic activity and breeding stage in an Antarctic top predator seabird species. *Mar. Pollut. Bull.* **157**, 111351. https://doi.org/10.1016/j.marpolbul.2020.111351
31. Lenzi, J., Bresesti, F., Lozoya, J.P., De Feo, B., Krojmal, E., Lacerot, G., Braun, C., Teixeira de Mello, F., 2022. Diet and debris ingestion of skuas on Fildes Peninsula, King George Island, Antarctica. *Mar. Pollut. Bull.* **185**, 114211. https://doi.org/10.1016/j.marpolbul.2022.114211
32. De-la-Torre, G.E., Santillán, L., Dioses-Salinas, D.C., Yenney, E., Toapanta, T., Okoffo, E.D., Kannan, G., Madadi, R., Dobaradaran, S., 2024. Assessing the current state of plastic pollution research in Antarctica: Knowledge gaps and recommendations. *Chemosphere* **355**, 141870. https://doi.org/10.1016/j.chemosphere.2024.141870
33. Suaria, G., Perold, V., Lee, J.R., Lebouard, F., Aliani, S., Ryan, P.G., 2020. Floating macro- and microplastics around the Southern Ocean: Results from the Antarctic Circumnavigation Expedition. *Environ. Int.* **136**, 105494. https://doi.org/10.1016/j.envint.2020.105494
34. Isobe, A., Uchiyama-Matsumoto, K., Uchida, K., Tokai, T., 2017. Microplastics in the Southern Ocean. *Mar. Pollut. Bull.* **114**, 623–626. https://doi.org/10.1016/j.marpolbul.2016.09.037
35. Cincinelli, A., Scopetani, C., Chelazzi, D., Lombardini, E., Martellini, T., Katsoyiannis, A., Fossi, M.C., Corsolini, S., 2017. Microplastic in the surface waters of the Ross Sea (Antarctica): Occurrence, distribution and characterization by FTIR. *Chemosphere* **175**, 391–400. https://doi.org/10.1016/j.chemosphere.2017.02.024
36. Kim, B.K., Hwang, J.H., Kim, S.K., 2023. Modeling of microplastics discharged from a station in Marian Cove, West Antarctica. *Mar. Pollut. Bull.* **186**, 114441. https://doi.org/10.1016/j.marpolbul.2022.114441
37. Zhang, S., Zhang, W., Ju, M., Qu, L., Chu, X., Huo, C., Wang, J., 2022. Distribution characteristics of microplastics in surface and subsurface Antarctic seawater. *Sci. Total Environ.* **838**, 156051. https://doi.org/10.1016/j.scitotenv.2022.156051

38. Leistenschneider, C., Wu, F., Primpke, S., Gerdts, G., Burkhardt-Holm, P., 2024. Unveiling high concentrations of small microplastics (11–500 μm) in surface water samples from the southern Weddell Sea off Antarctica. *Sci. Total Environ.* **927**, 172124. https://doi.org/10.1016/j.scitotenv.2024.172124
39. Pakhomova, S., Berezina, A., Lusher, A.L., Zhdanov, I., Silvestrova, K., Zavialov, P., van Bavel, B., Yakushev, E., 2022. Microplastic variability in subsurface water from the Arctic to Antarctica. *Environ. Pollut.* **298**, 118808. https://doi.org/10.1016/j.envpol.2022.118808
40. Munari, C., Infantini, V., Scoponi, M., Rastelli, E., Corinaldesi, C., Mistri, M., 2017. Microplastics in the sediments of Terra Nova Bay (Ross Sea, Antarctica). *Mar. Pollut. Bull.* **122**, 161–165. https://doi.org/10.1016/j.marpolbul.2017.06.039
41. Reed, S., Clark, M., Thompson, R., Hughes, K.A., 2018. Microplastics in marine sediments near Rothera Research Station, Antarctica. *Mar. Pollut. Bull.* **133**, 460–463. https://doi.org/10.1016/j.marpolbul.2018.05.068
42. Cunningham, E.M., Ehlers, S.M., Dick, J.T.A., Sigwart, J.D., Linse, K., Dick, J.J., Kiriakoulakis, K., 2020. High abundances of microplastic pollution in deep-sea sediments: Evidence from Antarctica and the Southern Ocean. *Environ. Sci. Technol.* **54**, 13661–13671. https://doi.org/10.1021/acs.est.0c03441
43. Alurralde, G., Isla, E., Fuentes, V., Olariaga, A., Maggioni, T., Rimondino, G., Tatián, M., 2022. Anthropogenic microfibres flux in an Antarctic coastal ecosystem: The tip of an iceberg? *Mar. Pollut. Bull.* **175**, 113388. https://doi.org/10.1016/j.marpolbul.2022.113388
44. Perfetti-Bolaño, A., Araneda, A., Muñoz, K., Barra, R.O., 2022. Occurrence and distribution of microplastics in soils and intertidal sediments at Fildes Bay, Maritime Antarctica. *Front. Mar. Sci.* **8**, 774055. https://doi.org/10.3389/fmars.2021.774055
45. Cunningham, E.M., Rico Seijo, N., Altieri, K.E., Audh, R.R., Burger, J.M., Bornman, T.G., Fawcett, S., Gwinnett, C.M.B., Osborne, A.O., Woodall, L.C., 2022. The transport and fate of microplastic fibres in the Antarctic: The role of multiple global processes. *Front. Mar. Sci.* **9**, 1056081. https://doi.org/10.3389/fmars.2022.1056081
46. González-Pleiter, M., Edo, C., Velázquez, D., Casero-Chamorro, M.C., Leganés, F., Quesada, A., Fernández-Piñas, F., Rosal, R., 2020. First detection of microplastics in the freshwater of an Antarctic Specially Protected Area. *Mar. Pollut. Bull.* **161**. https://doi.org/10.1016/j.marpolbul.2020.111811

47. González-Pleiter, M., Lacerot, G., Edo, C., Pablo Lozoya, J., Leganeacutes, F., Fernández-Pinãs, F., Rosal, R., Teixeira-De-Mello, F., 2021. A pilot study about microplastics and mesoplastics in an Antarctic glacier. *Cryosphere* **15**, 2531–2539. https://doi.org/10.5194/tc-15-2531-2021

48. Materić, D., Kjær, H.A., Vallelonga, P., Tison, J.L., Röckmann, T., Holzinger, R., 2022. Nanoplastics measurements in Northern and Southern polar ice. *Environ. Res.* **208**, 112741. https://doi.org/10.1016/j.envres.2022.112741

49. Chen, Q., Shi, G., Revell, L.E., Zhang, J., Zuo, C., Wang, D., Le Ru, E.C., Wu, G., Mitrano, D.M., 2023. Long-range atmospheric transport of microplastics across the southern hemisphere. *Nat. Commun.* **14**, 7898. https://doi.org/10.1038/s41467-023-43695-0

50. Coyle, R., Hardiman, G., Driscoll, K.O., 2020. Microplastics in the marine environment: A review of their sources, distribution processes, uptake and exchange in ecosystems. *Case Stud. Chem. Environ. Eng.* **2**, 100010. https://doi.org/10.1016/j.cscee.2020.100010

51. Ribeiro, V.V., Avelino Soares, T.M., De-la-torre, G.E., Casado-Coy, N., Sanz-Lazaro, C., Castro, Í.B., 2024. Microplastics in rocky shore mollusks of different feeding habits: An assessment of sentinel performance. *Environ. Pollut.* **346**, 123571. https://doi.org/10.1016/j.envpol.2024.123571

52. De-la-Torre, G.E., 2020. Microplastics: An emerging threat to food security and human health. *J. Food Sci. Technol.* **57**, 1601–1608. https://doi.org/10.1007/s13197-019-04138-1

53. Li, Y., Liu, C., Yang, H., He, W., Li, B., Zhu, X., Liu, S., Jia, S., Li, R., Tang, K.H.D., 2024. Leaching of chemicals from microplastics: A review of chemical types, leaching mechanisms and influencing factors. *Sci. Total Environ.* **906**, 167666. https://doi.org/10.1016/j.scitotenv.2023.167666

54. Torres, F.G., Dioses-Salinas, D.C., Pizarro-Ortega, C.I., De-la-Torre, G.E., 2021. Sorption of chemical contaminants on degradable and non-degradable microplastics: Recent progress and research trends. *Sci. Total Environ.* **757**, 143875. https://doi.org/10.1016/j.scitotenv.2020.143875

55. Miller, M.E., Hamann, M., Kroon, F.J., 2020. Bioaccumulation and biomagnification of microplastics in marine organisms: A review and meta-analysis of current data. *PLoS One* **15**, e0240792. https://doi.org/10.1371/journal.pone.0240792

56. Nik Mut, N.N., Na, J., Jung, J., 2024. A review on fate and ecotoxicity of biodegradable microplastics in aquatic system: Are biodegradable plastics truly safe for the environment? *Environ. Pollut.* **344**, 123399. https://doi.org/10.1016/j.envpol.2024.123399

57. Dawson, A., Huston, W., Kawaguchi, S., King, C., Cropp, R., Wild, S., Eisenmann, P., Townsend, K., Bengtson Nash, S., 2018. Uptake and depuration kinetics influence microplastic bioaccumulation and toxicity in Antarctic krill (*Euphausia superba*). *Environ. Sci. Technol.* **52**, 3195–3201. https://doi.org/10.1021/acs.est.7b05759

58. Rowlands, E., Galloway, T., Cole, M., Lewis, C., Hacker, C., Peck, V.L., Thorpe, S., Blackbird, S., Wolff, G.A., Manno, C., 2023. Scoping intergenerational effects of nanoplastic on the lipid reserves of Antarctic krill embryos. *Aquat. Toxicol.* **261**, 106591. https://doi.org/10.1016/j.aquatox.2023.106591

59. Manno, C., Peck, L. V., Corsi, I., Bergami, E., 2022. Under pressure: Nanoplastics as a further stressor for sub-Antarctic pteropods already tackling ocean acidification. *Mar. Pollut. Bull.* **174**, 113176. https://doi.org/10.1016/j.marpolbul.2021.113176

60. Oliveira de Miranda, C., Lelis Leal de Souza, J.J., Gonçalves Reynaud Schaefer, C.E., Huerta Lwanga, E., Nadal Junqueira Villela, F., 2024. Short-term impacts of polyethylene and polyacrylonitrile microplastics on soil physicochemical properties and microbial activity of a marine terrace environment in maritime Antarctica. *Environ. Pollut.* **347**, 123791. https://doi.org/10.1016/j.envpol.2024.123791

Chapter 8

Microplastics as a Vector of Hazardous Contaminants: Plastic Chemicals, Digestive Physiology and the Need for Chemical Simplification

Farhan R. Khan,[a] Ana I. Catarino,[b] Kristian Syberg,[c] Frank G.A.J. Van Belleghem,[d,e] and Nathaniel J. Clark[f,g]

[a]*Norwegian Research Center (NORCE), Nygårdsporten 112, NO-5008 Bergen, Norway (ORCID 0000-0002-9251-2972)*

[b]*Vlaams Instituut voor de Zee, Flanders Marine Institute InnovOcean site, Wandelaarkaai 7, 8400 Oostende, Belgium (ORCID 0000-0002-8796-0869)*

[c]*Department of Science and Environment, Roskilde University, Roskilde, Denmark*

[d]*Department of Environmental Sciences, Faculty of Science, Open Universiteit, 6419 AT Heerlen, the Netherlands*

[e]*Zoology: Biodiversity and Toxicology, Centre for Environmental Sciences, Hasselt University, BE 3590 Diepenbeek, Belgium*

[f]*School of Biomedical Sciences, University of Plymouth, Plymouth, PL4 8AA, UK (0000-0002-0968-503X)*

[g]*School of Health Professionals, University of Plymouth, PL4 8AA*

fakh@norceresearch.no; farhan.khan@gmx.com

Microplastic Pollution: Environmental Impact and Solutions
Edited by Julia Martín, Juan Luis Santos, Irene Aparicio, and Esteban Alonso

ISBN 978-981-5129-77-9 (Hardcover), 978-1-003-66006-4 (eBook)
www.jennystanford.com

DOI: 10.1201/9781003660064-8

Microplastics are one of today's major environmental problems, as they are a ubiquitous persistent pollutant group that has reached all parts of the environment. The chemical burden of microplastics results from the plastic-associated chemicals (chemical additives, monomers, and non-intentionally added substances) incorporated into the polymer matrix during manufacturing, and the environmental pollutants (e.g., hydrophobic organic compounds, trace metals or pharmaceuticals) that sorb to the plastic surfaces following release. Microplastics have also been shown to be readily ingested by a wide range of organisms across taxonomic groups, and it is this combination of biotic ingestion and chemical association that gives credence to the notion that microplastics may be vectors of hazardous contaminants. Therefore, the transfer of chemicals from microplastics has been the subject of much laboratory research, but owing to the complexity of exposure scenarios, results are often disparate. Moreover, while the relevance of microplastics as real-world vehicles for environmental pollutants has been questioned, recent analysis has shown that plastics can have as many as 16,000 chemicals added to them during manufacturing, many of which are of known toxicity. Furthermore, plastic-associated chemicals once released into the environment may be transformed into distinct substances that exhibit different toxicities compared to their parent compound. This chapter provides an overview of the current literature on microplastics as carriers of environmental pollutants and plastic-associated chemicals, and the emergence of transformation products as a topic of concern. The importance of digestive physiology is recognized as one of the main sites in which chemical release and desorption occurs resulting in chemicals being labile within an organism. Lastly, we explore how the wealth of literature linking microplastics and chemicals has heightened calls for chemical simplification to ensure better safety for environmental (and human) health.

8.1 Introduction

Since the rapid growth in plastic production in the 1950s and the ever-increasing levels of plastic use, as a consequence of the linear value chain (often coined as 'take-make-dispose'), plastic pollution and specifically microplastics have been distributed to every part of our planet. Microplastics (nominally defined as being <5 mm in size) have been sampled at the lowest depths of the ocean [1] to the highest mountain top [2], and everywhere in between. Microplastics have been found in all compartments of the aquatic environment in both marine waters and freshwaters—water, sediment, and biota [3]. The ingestion of microplastics by biota is well documented as over 700 marine species across taxonomic groups (seabirds, marine mammals, fish, and invertebrates) have been found with plastic particles in their digestive systems [4]. Focussing on the occurrence of microplastics in the piscine gastrointestinal tracts across all water types, a recent global meta-analysis showed that 728 species, spread over 35 orders and 178 families, have been collected and evaluated for microplastic ingestion [5]. There is also increasing evidence of the presence of microplastics within human tissue, with the discovery of microplastics in the placenta [6] and testis [7] gaining media attention worldwide. In other words, plastics, the novel entity once lauded as a technological marvel, are now literally found everywhere.

However, "microplastic" encompasses not only the typically petroleum-based polymers (among the most common include polyethylene (PE), polypropylene (PP), polystyrene (PS), and polyvinyl chloride (PVC)) and their physical characteristics of morphology, shape, and size, but also the specific chemical signature of the microplastic. The chemical burden of microplastics results from both the thousands of plastic-associated chemicals (including chemical additives, monomers, and non-intentionally added substances) used within the production process and the plethora of existing environmental pollutants, such as hydrophobic organic compounds (HOCs), trace metals, or pharmaceuticals that microplastics will inevitably encounter following environmental release [8, 9]. It is the association of plastics and chemicals that has led to much research into microplastics being carriers of toxic chemicals. This "microplastic vector-effect" [10] thus describes the influence the microplastic particles may have in transporting and depositing its chemical load into novel scenarios, for instance, in (a) new geographic locations and between environmental compartments as an "environmental-vector," (b) following the inadvertent ingestion of microplastics, chemicals are transported into organisms, i.e., "organismal-vector," and (c) chemicals hitchhiking into cells associated to nanoplastics subject to possible endocytotic or phagocytotic processes.

From a toxicological perspective, the microplastic particle represents a "triple exposure nexus" [11] consisting of the particle itself, which has been shown to cause physical injury and distress to digestive tracts following ingestion [12, 13], and the plastic-associated chemicals and environmental pollutants. Numerous studies have established that different pollutant groups sorb to microplastics [14–17]. This sorption is influenced by several factors related to the physicochemical properties of the plastic particle (including weathering status [18]), the chemical in question and the ambient environment. To date, much of the research concerning potential vector effects has been related to environmental pollutants at the organismal level, and by comparison, the leaching of plastic-associated chemicals, which can be defined as the migration of additives from the resin (polymer matrix) to the external medium, has been relatively understudied. However, there is a current rethinking of the relative importance of environmental pollutants and plastic-associated chemicals.

During the plastic manufacturing process, plastic-associated chemicals (chemical additives, monomers, and non-intentionally added substances) are incorporated into the matrix of the polymeric resin to enhance the functionality of the final application [19]. The "Chemicals in Plastics: A Technical Report" published in 2023 by the United Nations Environment Programme [20] identified more than 13,000 plastic-associated chemicals resulting from production processes. Within a year of publication, the United Nations Environment Programme (UNEP) report was superseded by the results of the PlastChem project identifying over 16,000 chemicals of which 4200 are of known concern, as they exhibit high human health and/or environmental hazard, and 400 chemicals that are present in each major plastic type [21]. To the previously defined "triple exposure nexus," we may consider adding a fourth dimension, which is the emergence of transformation products as a toxicologically relevant concern. Such products are formed when a plastic-associated chemical is released from the polymeric structure and transformed in the environment, for example by oxidation, UV or microbial degradation, into a chemical product that exhibits greater toxicity than the parent chemical. A notable example is the tyre-derived chemical 6PPD and derivative 6PPD-quinone [22].

The subsequent transfer of environmental pollutants to biota has been the subject of much research in laboratory settings [23], but consensus on the transfer and effects of chemical exposure via microplastics remains elusive. The key biotic process is the ingestion of microplastics, which change the route of uptake of the associated chemical from a waterborne to a dietary exposure [24], and following desorption within an organism's digestive system, there is potentially a greater level of internal chemical exposure [25]. However, one of the lesser-understood variables is the impact of ingesting on an organism's digestive physiology. For instance, feeding modes, gut retention times, and gut environment all play a role in whether an environmental pollutant or a plastic-associated chemical is bioaccessible and then bioavailable to the organism [8, 26]. Furthermore, the lumen of the gastrointestinal tract (GIT) is a dynamic environment that varies within and between species, adding another level of complexity to the comparability between vector studies.

This chapter provides an overview of the literature that has examined the role of microplastics as chemical vectors to biota, with a focus on aquatic organisms. The influence and interactions of microplastics with environmental pollutants and plastic-associated chemicals are evaluated particularly as the relevance of the former is being questioned while thousands of chemicals that are part of the manufacturing process are being determined by advanced analytical techniques. Furthermore, we focus on the emergent concern of transformation products that are, in many cases, an unanticipated consequence of plastic-associated chemicals entering the environment. We highlight the importance of digestive physiology as a facilitator of chemical transfer from microplastics to an organism. Lastly, given the ongoing discussion at the global level to mitigate the problems of plastic pollution, we emphasize the need to reduce the extensive number of chemicals used in plastic production, termed "chemical simplification" (Fig. 8.1).

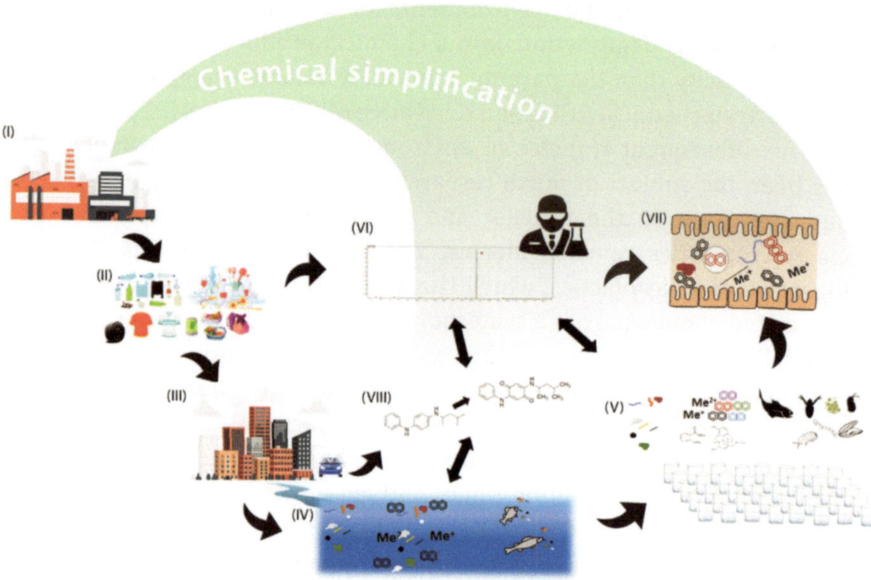

Figure 8.1 Schematic of key processes in understanding microplastics as a vector of hazardous contaminants. (I) The production process includes thousands of plastic-associated chemicals from which a vast number of products are made (II). Through use and (often improper) disposal (III), plastics degrade into a variety of microplastic forms that have been found in the environment, where they both sorb environmental pollutants and are ingested by a host of aquatic organisms (IV). Observations of both these phenomena have led to a multitude of laboratory-based microplastic vector studies combining different microplastic polymers and types, different environmental pollutants (hydrophobic organic pollutants, metals, and pharmaceuticals) and numerous test organisms (V). Plastic products have also been analysed through non-target screenings to determine the number of plastic-associated chemicals they contain (VI), and the bioaccessibility and bioavailability of these chemicals have also been tested under laboratory conditions. For both the plastic-associated chemicals and environmental pollutants, understanding the role of digestive physiology and the gut environment on the release and/or desorption of ingested microplastics is vital (VII). Recent studies have also shown that plastic-associated chemicals released into the environment can transform into structurally distinct compounds that exhibit a different toxicological profile than their parent chemical, as exemplified by 6PPD and its transformation product 6PPD-quinone (VIII). The wealth of information gathered from these research activities has heightened calls for chemical simplification within manufacturing processes.

8.2 Impacts and Relevance of Microplastics as Carriers for Environmental Pollutants

Microplastics enter the environment with an existing chemical cocktail, including HOCs (polycyclic aromatic hydrocarbons (PAHs) and polychlorinated biphenyls (PCBs)) [27, 28], trace metals [29, 30], and pharmaceuticals [31, 32], all of which have been measured on microplastics sampled from the environment. Understanding the transfer of environmental pollutants from a microplastic to the tissue of an organism is first dependent on understanding how the pollutant attaches to the microplastic and second how desorption occurs post-ingestion. This attachment has been mainly characterized by HOC–microplastic interactions, and "sorption" is generically used to cover both "adsorption," defined as the surface interaction between HOCs and microplastics, and "partition," which denotes the diffusion of HOCs into the internal space of the microplastics [16, 33]. Sorption kinetics are governed by the polymeric structure, essentially the ratio of crystalline and amorphous regions, the hydrophobicity of the HOC, and ambient environmental conditions. Polymers are composed of both ordered crystalline regions and less structured amorphous regions. Crystalline regions contain tightly packed polymer chains that are largely impermeable to solutions. Crystallinity controls properties such as hardness and density, and the degree of crystallinity can vary greatly from approximately 50% (e.g., polypropylene and nylon) to almost 0% (e.g., atactic polystyrene), but no polymer is 100% crystalline. The loosely packed amorphous regions interact more strongly with HOCs than the crystalline areas and largely control the rate of migration of HOCs into microplastics and their concentrations at equilibrium. Modelling this interaction has become a mandatory aspect of vector studies, and a hybrid Freundlich–Langmuir model that recognizes the dominance of adsorption at lower HOC concentrations and partitioning at higher concentrations could be considered most appropriate to capture both sorption processes [33]. Desorption processes are additionally dependent on the environment of the gut lumen post-ingestion, which is covered in Section 8.5. However, early investigations into desorption processes have shown that HOC desorption was 30 times greater in gut fluids than in seawater alone and that both higher temperatures and acidic pHs facilitate the desorption of HOCs [34]. The effect of pH has also been demonstrated for trace metals [35].

The scientific literature on vector studies is too vast to comprehensively cover in this chapter (also see reviews by Khan et al. [8], Hartmann et al. [36], Huang et al. [37], and Koelmans et al. [38]), yet it remains difficult to generalize the validity of the phenomenon across so many different laboratory experiments. Since the earliest investigations into the influence of microplastics on the transfer of environmental pollutants outcomes have fit broadly into three categories: (a) microplastics enhancing the bioavailability and toxicity of environmental pollutants [39, 40], (b) where the addition of microplastics to the exposure scenario resulted in negligible impacts [41], and (c) cases where microplastics reduced pollutant bioavailability [24, 42]. Despite the wealth of studies investigating vector effects in the ensuing decade, this categorization remains. Two approaches are primarily used in vector studies, first, to directly measure chemicals in tissues to determine the influence of microplastics on the exposure scenario and, second, to measure a toxicological marker of exposure as an indicator of chemical bioavailability and biological reactivity. Numerous laboratory studies have reported that environmental pollutants sorbed to or co-exposed with microplastics are bioavailable and lead to a measurable chemical transfer into tissue [43–45] and toxicological impacts based upon the assessment of biomarkers [44, 46, 47], even if tissue burdens do not correspondingly increase [48–50]. Conversely, several studies have demonstrated that at least for some measured endpoints, the role of microplastics on chemical-induced negative effects is not significant [51–54]. While most vector studies attempt to determine whether microplastics can facilitate the transfer of environmental pollutants, in the case of *Talitrus saltator* (amphipod crustacean) microplastics were shown to carry HOCs into the tissue following ingestion. However, when uncontaminated microplastics were fed, the microplastic scavenged the chemicals and reduced tissue burden through a "cleaning effect" [55].

The variation in outcomes is understandable considering that comparisons are being made across studies that have employed different polymers, environmental pollutants, organisms, experimental designs, and endpoints [26]. Multiplying across all these variables highlights the challenges in finding consensus among vector studies. This variation will most likely increase as approximately 2000 new chemicals enter the market each year [56] and novel polymers are introduced. For instance, biobased plastics may be considered a viable solution to plastic pollution, although doubts about their green credentials [57] and toxicity [58] persist. The "bioplastic" polyhydroxybutyrate behaved similarly to polyethylene as a vector of fluoranthene with both minimally affecting the oxidative stress response in *Mytilus edulis* [59]. Whether such new materials constitute toxicologically safer alternatives needs to be verified. Moreover, how we view vector studies may need revising as the combination of microplastics and environmental pollutants may resemble the interactions within a chemical mixture, often described as independent or dependent action, or additivity, synergism, or antagonism [60]. However, even when assessed through a recognized framework designed to disentangle the effects of single components within a mixture, the microplastic vector effect surprises us. The co-exposure of the marine copepod *Acartia tonsa* to polyethylene microbeads and triclosan resulted in a relatively obscure mixture effect known as potentiation, in which the microplastic, without being toxic itself, enhanced the toxicity of triclosan [60].

Despite the large number of studies published on the topic, laboratory-performed microplastic vector studies are often criticized for a perceived lack of environmental realism and relevance. Notably, the concentrations of microplastics and environmental pollutants often exceed measured levels in the environment. Additionally, the overuse of certain microplastic polymers, types such as beads or spheres, and the use of pristine particles lessen environmental realism. Similarly, short equilibrium times and methods to artificially accelerate sorption kinetics between plastics and environmental pollutants do not accurately reflect how microplastics and environmental pollutants may interact in the real world [33, 61, 62]. The use of aged or weathered particles and realistic concentrations are, therefore, desirable experimental ambitions. As for whether, microplastics are a relevant pathway for chemical transfer, and Koelmans et al. [62] have comprehensively addressed the topic. Their modelled analysis showed that of nine compartments of the ocean, including plastics, waterborne HOCs would be partitioned to the largest compartment of water (98.3%) and only 0.0002%

would be attached to plastics. Therefore, when assessing the relative importance of microplastics as chemical carriers, consideration must be given to whether experimental set-ups adequately reflect natural conditions and whether the presence of microplastics in the environment is as important as other compartments (e.g., food and water) as a vector of environmental pollutants [62].

8.3 Bioaccessibility, Bioaccumulation, and Bioavailability of Plastic-Associated Chemicals

In a recent data compilation, 16,000 chemicals were identified as potentially used in plastic materials and products, of which only 6% are currently subject to international regulations [21]. Of these plastic-associated chemicals, over 4200 are considered of concern due to their persistence, potential for bioaccumulation, mobility (the potential to spread in fresh and drinking water), and toxicity (PBMT) [21]. To assess the environmental risk of these chemicals with microplastics, it is key to determine the exposure route (via ingestion or contact) and (eco)toxicological effects. With the likely risk that exposure to a pollutant or contaminant will have adverse effects on organisms, exposure is defined as the contact of organisms with plastic-associated chemicals, in terms of concentration, duration, and frequency [8, 63]. The complex nature of plastic particles (diversity of polymers and associated chemicals, sizes, weathering state, etc.) and the current uncertainty associated with the lack of knowledge of the potential adverse effects of microplastics and associated chemicals can lead to challenges when estimating these particles' risk [64]. In general terms, microplastic exposure concentration and duration determine the magnitude of potential negative effects on organisms via changes in the state of an organism at the molecular, cellular, tissue, or individual level [63]. Additionally, understanding the extent to which organisms are exposed to microplastics-associated chemicals independent of the microplastic particle will enable a comprehensive risk characterization and assessment [64–66]. This is because released leachates (chemicals released from the polymer matrix to the environment) from microplastics can induce chemical-led toxicological effects, rather than particle-led ones if substances are made accessible to biota [8]. Efforts to quantify and assess the toxicological risks of microplastics are not yet established, and this complexity is evident by the growing understanding that microplastics incorporate a wide variety of contaminants [9]. Because of this complex nature of plastic particles (micro- and nanoplastics), authors such as Cunningham et al. [64] have suggested new alternative methodologies for risk assessment (Bayesian network relative risk model, BN-RRM) which consider the current knowledge uncertainties and account for associated chemicals and other levels of complexity of microplastic particles [65].

Some of the plastic-associated chemicals are well-known toxicants, according to the recent data compilation mentioned above [21] and may potentially increase the bioaccessibility, bioaccumulation, and bioavailability of these plastic-associated chemicals upon exposure to organisms [8]. Once microplastics are ingested, the particles and associated chemicals will be exposed to a medium that is more acidic compared to water. As discussed later in this chapter, the gastrointestinal physiology (especially the low pH) of the organism determines the release or leaching of the associated substances, and a portion of the substances will then be bioaccessible, i.e., available for uptake [8, 67]. These microplastic additives can be transferred from polymers to biological tissues in higher concentrations, in part due to their strong affinity for fatty tissues, similar to hydrophobic compounds [62]. For instance, highly lipophilic flame retardants were found to leach from plastics into the digestive fluids of birds [68]. Also, in more acidic media, elements used as pigments in plastic production (Al, Zn, Ba, Cu, Pb, Cd, Mn, Cr) can leach and sorb to the surface of microplastics, becoming bioaccessible [69].

As a portion of these substances can be accessible for absorption by organisms, some of these substances can also bioaccumulate, as they are stored in tissues, often at higher concentrations than in the environment [62]. In studies assessing the accumulation of plastic additives in wildlife, the levels of organophosphorus plastic additives accumulated in fish were mostly associated with reproductive tissues, which may have potential population and ecological consequences in the future [70]. In a review, Hermabessiere et al. [71] identified polybrominated diphenyl ethers (PBDEs), phthalates, nonylphenols (NP), bisphenol A (BPA), and antioxidants as the most common plastic-associated chemicals (additives, in this specific case) observed in marine environments and that the bioaccumulation of some of these substances has been demonstrated in both laboratory and field studies [71]. In another study, Tanaka et al. [72] demonstrated that feeding seabird chicks with microplastics whose polymeric resin was spiked with known additives resulted in accumulation at levels 101–105 times higher than the baseline [72]. However, some studies also demonstrate that bioaccumulation of plastic-associated chemicals from microplastics may be less relevant when particles are ingested rather than when substances are leached into the water, as in the case of mussels exposed to the additive hexabromocyclododecane (HBCD) [73]. A modelling study estimated that exposure to leaching of nonylphenol (NP) and bisphenol A (BPA) in the intestinal tract of lugworms and North Sea cod via microplastics ingestion would result in concentrations in these organisms that are below field concentration ranges and, therefore, is unlikely to be a relevant exposure pathway [74]. The diversity of substances and plastic particles (sizes, weathering status, polymers, morphology, etc.) increases the challenges in assessing and reporting all bioaccumulation mechanisms and requires careful assessment of each contributing factor to accurately determine the overall impact on biological and environmental systems.

Once released in the gastrointestinal environment, chemicals associated with microplastics can be bioavailable by crossing epithelial membranes and reaching systemic circulation and/or entering cells to become biologically active [8, 67]. It is at this moment that substances are processed and can alter physiological functions at the cellular level [75]. Bioaccumulation can also lead to toxic effects, especially if the substances are persistent and do not easily break down or are not readily excreted by organisms. As an example of plastic additives that are bioavailable to organisms, in the case of zebrafish exposed to PVC microplastics, the bioavailability of Pb additives was indicated by the increased expression of metal-binding proteins [76]. Exposure of water fleas to flexible PVC particles also demonstrated the bioavailability of diisononylphthalate as their life history and morphological traits were affected [77]. Bioaccessibility and bioavailability also need to be placed in a broader context, as other sources of these substances are also present in natural systems. For example, in contaminated gut environments with other food items, clean microplastics could quickly extract (i.e., clean) PCBs from other items simultaneously present in the gut, suggesting that these particles can also have an inverse role of the "vector effect" depending on the specific context [78]. In a different context, the adhesion of microorganism communities to the surfaces of microplastics in the environment can cause micromechanical stresses on the particle surfaces, enhance leaching rates, and act as an additional phase that increases the bioavailable dose of chemicals to organisms that ingest these particles [79].

Plastic-associated chemicals are present as a complex mixture, instead of isolated, and the combined effects of these mixtures are poorly known. Leachate compounds can include both metals and organic substances [76, 80], leaching rates from particular polymers are highly variable, and leachates from different products made of the same polymer can differ significantly in their additive composition and bioavailability [8]. For example, mussels exposed to leachates from car tyre rubber, polypropylene (PP), polyethylene terephthalate (PET), polystyrene (PS), and PVC affected stress biomarker parameters, indicating their bioavailability [81]. However, other studies report different effects of PVC-based products under the same assessment, with leachates from some products inducing toxicity while others not [82–84]. The weathering status of the plastic particles also influences the leaching rates and toxicity of their additives but without a consistent pattern. For example, Bejgarn et al. [82] showed that leachate toxicity increased with UV exposure for some products such as PVC packaging and biodegradable bags, but decreased for others such as garden hoses and computer housings. This highlights the complexity of bioaccessibility, bioavailability, and toxicity of plastic-associated chemicals and combined mixtures in leachates [8], emphasizing the need for a more nuanced approach to understanding chemical additives. Some plastic-associated chemicals, in particular additives, may interact with one another in unpredictable ways; others may have unknown effects, and some could be highly toxic [21] or transform into more toxic compounds (discussed in the next section). Therefore, it is crucial to investigate these interactions and effects in a simplified yet comprehensive manner to accurately evaluate the risks associated with different additives and their combined impact on the environment and organisms. From here it would be possible to propose methods for chemical simplification (as discussed at the end of this chapter).

8.4 Emergence of Transformation Products as an Environmental Concern

When plastics undergo physical, chemical, or biological degradation, plastic-associated chemicals (described in the previous section) can be released into the environment. The released chemicals can also undergo processes of change through, for example, photooxidation or biotransformation. The resulting transformation products, which are structurally diverse intermediates, may exhibit a different toxicological profile than the initial parent chemical [85]. Thus, if a decrease in toxicity caused by the formation of a transformation product is parallel to the decrease in parent chemical concentration, the transformation products are considered irrelevant. However, when toxicity increases as a result of the formation of transformation products or the decrease in toxicity is not proportional to the parent chemical concentration, then the transformation product must be identified [86, 87]. Transformation products, their generation and toxicity, have been well studied in other fields such as pharmaceuticals and personal care products, biocides, and pesticides that may be transformed as they move through wastewater treatment plants [88–90]. To date, little attention has been paid to the risks posed by the transformation products of plastic-associated chemicals. The notable exception to this concerns chemical compounds found in tyre rubber, particularly 6PPD (N-(1,3-Dimethylbutyl)-N′-phenyl-p-phenylenediamine) and hexamethoxymethylmelamine (HMMM).

The substance 6PPD in tyre rubber and its transformation product 6PPD-quinone exemplify how compounds present in low concentrations can have significant environmental impacts. 6PPD is an antioxidant and antiozonant used in vehicle tyres, accounting for 0.4% to 2% of the total tyre rubber mass [91]. This 6PPD undergoes oxidative transformation with ground-level ozone to form 6PPD-quinone and several other transformation products, including 4-nitrosodiphenylamine and $C_{18}H_{22}N_2O$ [92]. But it is 6PPD-quinone in particular that has attracted considerable attention because it has been linked to a phenomenon known as urban run-off mortality syndrome (URMS), the mass mortality of pre-spawn coho salmon in the Pacific Northwest [91, 93–96]. However, sensitivity to 6PPD-quinone varies by salmon species: Coho salmon, brown trout, and rainbow trout are very sensitive; Chinook and steelhead are moderately sensitive; while Arctic char, chum salmon, Atlantic salmon, sockeye salmon, and brown trout have lower sensitivity [97]. In a recent study, 11 species were tested for zooplankton URMS. The results showed that copepods had high tolerance in most cases, while rotifers and cladocerans (such as *Daphnia pullex*) were more sensitive, except for the small cladoceran *Corylus cornuta* and the rotifer *Keratella valga* [98]. Overall, the growing literature indicates that 6PPD-quinone is toxic to a wide range of taxa, including nematodes, mammals, and zebrafish [99, 100].

HMMM, a polymer crosslinker used in the manufacture of coatings, plastics, and tyres, is another compound of concern. HMMM is a common pollutant in urban surface waters that arises from tyre wear particles and is transported via road run-off during rainfall events, similar to 6PPD. Recent studies have found that HMMM readily transforms into several persistent compounds, many of which have been detected in aquatic environments [101–103]. While the toxicity of HMMM and its transformation products has not been investigated, it has been shown that HMMM, similar to 6PPD, is readily absorbed by lettuce and that these compounds are metabolized in the plant to transformation products, most of which were found to be more stable in lettuce leaves than the parent compounds [104].

In addition to the compounds currently under investigation, numerous transformation products are generated through various plastic degradation processes. Improper disposal methods, such as incomplete or faulty incineration or uncontrolled combustion, lead to the thermal degradation of (micro)plastics and the release of volatile organic compounds (VOCs), PCBs, dioxins, furans, PAHs, and heavy metals [105]. VOCs include a wide range of compounds such as alcohols, aldehydes, ketones, esters, and aromatic hydrocarbons, contributing to air pollution, ground-level ozone, and smog, with associated health risks, such as respiratory issues, eye irritation, and cancer [19]. PCBs, dioxins, and furans are persistent organic pollutants (POPs) that are not typically found in plastics but can be inadvertently formed during combustion or thermal degradation processes when chlorinated compounds such as PVC are present [106]. These compounds are known for their environmental persistence, bioaccumulation, and toxic effects. While PCBs, dioxins, and furans share many toxicological characteristics, such as endocrine disruption, carcinogenicity, immune system effects, developmental toxicity, and liver toxicity, their mechanisms of action and specific toxic potencies differ. Dioxins, particularly 2,3,7,8-tetrachlorodibenzo-p-dioxin (TCDD), are more toxic and act primarily through the AhR pathway [107], whereas PCBs have diverse mechanisms of action depending on the congener structure [108].

Polycyclic aromatic hydrocarbons (PAHs) such as benzo[a] pyrene, naphthalene, and anthracene are persistent, bioaccumulative, carcinogenic, and mutagenic compounds that pose significant health risks [109]. Photooxidative degradation is another process that chemically transforms plastics, releasing a range of polymers, dissolved and VOCs, that are potentially harmful [110]. Photooxidation of polyolefins such as polyethylene and polypropylene, as well as poly(styrene), results in the emission of harmful compounds such as acrolein, benzene, propanal, methyl vinyl ketone, and methyl propenyl ketone, highlighting plastic waste as an overlooked but potentially significant source of hazardous VOCs and environmental pollution [111]. Photooxidation is likely not limited to the surface of the polymer itself, but the absorbed pollutants may also undergo photooxidation and form toxic derivatives, such as the oxy-PAHs, which result from photomodification and prove to be highly toxic to *Daphnia magna* [112]. Biodegradation leads to the mineralization of plastic particles, a complex process that produces various transformation products influenced by environmental conditions, polymer types, and microorganisms [113]. The biodegradation of PET leads to the formation of terephthalic acid (TA), benzoic acid (BA), methyl acrylate (MA), mono-(2-hydroxyethyl) terephthalate (MHET), and bis-(2-hydroxyethyl) terephthalate (BHET) [114]. Although these compounds have low toxicity and are unlikely to pose a significant environmental threat, the transformation products of other plastics may present different environmental risks.

Assessing the potential risks of transformation products, particularly in the context of plastic pollution, is crucial but significantly more complex than assessing the risks based on reading across from the parent chemicals [86]. This field of research is still in its infancy, and there is much to learn about transformation pathways, ecosystem fate, and long-term implications for ecosystems and human health.

8.5 Role of Digestive Physiology

The toxicological consequences of animals ingesting microplastics and any plastic-associated chemicals or environmental pollutants are influenced by natural gastrointestinal physiology. There are two possibilities for microplastics with plastic-associated chemicals and environmental pollutants to enter the gastrointestinal tract of aquatic animals from the environment: independently or sorbed to the particle. In the latter case, microplastics and associated chemicals have been found in birds [115], where the ingested microplastics can release the chemicals within the gastrointestinal tract via desorption. The microplastics and the now desorbed chemical enter the intestine separately as a co-exposure. Here, the microplastics themselves have no impact on the trans-epithelial transport (i.e., across the gut) of the contaminant, but the net result of the microplastic vector effect is that potentially labile chemicals enter the intestine [35]. The factors that affect this adsorption in the environment are related to the properties of the plastic (polymer type, crystallinity, polarity, functional groups, and size), the medium (pH, ionic strength, salinity), and the chemical pollutant properties [116–118], with the reversal of this (desorption) likely to also be dependent on these factors. From the perspective of ingestion, it is the medium of the gut lumen that may differ from other environmental compartments, and its influence on microplastics as vectors has received less attention.

The lumen of the gastrointestinal tract is an environment with a dynamic chemistry that varies between and within species. The most notable change is that of pH. In aquatic invertebrates, the pH of the luminal gastrointestinal tract ranges from 5.4 to 7.2 between species, with an interspecies variation of about 0.5 pH or less (polychaetes worms, freshwater snails, Daphnia [119]). In comparison, the vertebrate luminal pH of the gastrointestinal tract can contain greater variation, with a pH of 2 being found in the stomach and a pH of around 8 in the intestines of carnivorous fish [119]. Notably, the pH within gastrointestinal compartments changes through the ingestion of food, with the food's composition depending on dietary habits, as well as through ontogeny. For instance, the presence of food raises the pH (less acidic) to 5–6 for up to 6 h following food intake, taking more than 24 h to decrease to around pH 2 [120]. In omnivorous fish (*Diplodus sargus*), the opposite can be observed when gastric pH starts at pH 6–7 and decreases with the ingestion of food reaching the lowest pH (~3) after 8 h, before gradually climbing back after 24 h [121]. Moreover, the pH of the luminal gastrointestinal tract changes throughout fish development (correlated in fish weight and size); the gastric pH of omnivorous *Diolodus sargus* at 0.05 g is 7, and this decreases to around pH 3 at 1 g, with no consistent changes in growing to over 100 g [121]. A similar pattern of reduced gastric pH with increasing size has also been found in *Solea senegalensis* [122]. Measurements of such pH changes in response to food consumption and ontogeny are absent from the invertebrate literature, but it is reasonable to assume that these occur to some extent due to or during the breakdown of food and the secretion from the digestive system into the gut throughout development. In comparison to pH, the ionic strength and salinity are less well described in aquatic invertebrates, with some data existing for snails that contain around 35 mmol/L of NaCl [119]. Most data are available for freshwater fish, with the stomach of rainbow trout (*Oncorhynchus mykiss*) containing higher concentrations of NaCl (140–225 mmol/L) compared to the intestinal fluid (4–170 mmol/L).

Studying the interactions of gastrointestinal chemistry *in vivo* in the presence of microplastics and their associated chemicals is challenging. A common method is to co-expose the microplastic with a chemical, allowing any sorption within the system (e.g., water/diet). However, this does not disentangle the cause/effect of the vector. A second approach is to sorb the chemicals to the microplastics and then feed them to animals, observing any changes in physiology [47]. However, this largely captures the effects on the animal and not the dynamics of the microplastics and chemical interactions within the gut lumen. To overcome this, sorption studies have been performed to simulate gastrointestinal conditions *in vitro*. Numerous studies have demonstrated that both plastic-associated chemicals and environmental pollutants separate from microplastics in intestinal fluids [123–125]. The pH is a main driver for determining the partitioning of chemicals onto the surface of microplastics for ionisable organic chemicals [126, 127] and dissolved metals [128], typically with lower pH values causing less chemical to bind to the surface of the microplastics or enhancing the desorption of them. Thus, from a fish's perspective, older individuals with a developed gastrointestinal tract (low pH) may be at a higher risk of plastic-associated chemicals and environmental pollutants dissociating from microplastics in the gut lumen, whereas juvenile fish (high pH) are more susceptible to the physical damage from the microplastic exposure. Moreover, the variation in pH along the lumen of the gastrointestinal tract, at least in fish, creates the potential for a desorption–adsorption cycle of chemicals on and off the microplastics [78], which could protect the intestinal regions from chemical exposure due to a higher pH value, which promotes the sorption to the microplastics. As ionic strength increases, the sorption of dissolved metals decreases [129], which means that at the mmol/L concentrations found in the lumen of the gastrointestinal tract, the sorption of metals may be low compared to other environments (e.g., external freshwater). Therefore, it is difficult to determine the relative contribution of each parameter to the potential for vector effects of microplastics and the associated chemicals to transfer into tissues, and a systematic approach with a focus on gastrointestinal fluids would be beneficial to assess the toxicological effects of microplastics and to better understand chemical exposures.

Other physiological parameters will also shape the behaviour of microplastics in the gastrointestinal tract with plastic-associated chemicals and environmental pollutants. The gastrointestinal tract is subject to secretions from accessory organs, such as bile salts (derived from cholesterol) to facilitate the breakdown of fat for absorption. This is a component of the gastrointestinal tract that is not frequently captured in sorption studies (pH, salinity, and ionic strength) yet remains an important aspect of organic chemical desorption. For instance, the desorption of phenanthrene from PVC and PE in the presence of bile salt (sodium taurocholate) was 1.9

and 2.2 times faster, respectively, in the relatively neutral pH of the intestine compared to seawater (without sodium taurocholate; [34]). Something similar has been found in the invertebrate digestive fluid [124], suggesting that natural organic matter (i.e., bile secretions) could also be a prominent determinant of organic chemical sorption, which is abundant in ingested food.

The digestive physiology of humans is somewhat similar to that of carnivorous fish in that it involves compartmentalized acidic and basic conditions, with the main difference being that the body temperature of mammals is constant and not poikilothermic. As in animals, microplastics have been found in the gastrointestinal tract [130, 131] and in the gastrointestinal tissue of humans [132], but the nature of any plastic-associated chemicals or environmental pollutants here remains unclear. Regardless, the passage of plastic through the (simulated) human gastrointestinal tract (mouth, stomach, and intestine) enhances the accumulation of nanoplastics (compared to the virgin nanoplastic) that are subsequently exposed to macrophage cells [133], suggesting the surface chemistry is modified and promotes uptake, but the consequence of the sorption/desorption of plastic-associated chemicals and environmental pollutants throughout the gastrointestinal tract remains largely unexplored.

The aforementioned discussion has been based on the normal gastrointestinal physiology of animals (and humans), yet it is conceivable that through the development of pathology, this environment changes, hence altering the vector effect on microplastics and associated chemicals. Microplastics have also been present in human pathologies outside of the gastrointestinal tract, including atheromas [134] and cirrhosis [135], which involve an element of tissue inflammation. Pathological inflammation in the gastrointestinal tract also affects the chemistry of the gut lumen. In humans, inflammation of the gut (as inflammatory bowel disease) reduces the intestinal lumen pH to 3–5 (from pH 7.1–7.5; [136]), making the pH similar to that of the stomach. Not only does the reduced pH increase desorption (as discussed above), but also enhances the toxicity of microplastic exposure [137], creating a scenario for increased toxicological effects. Certainly, in fish, the ingestion of microplastics alone has led to the inflammation of the intestinal wall [12], but the consequences this has on the gut lumen pH remain unexplored in animals. Therefore, there is a need to understand how such changes in digestive physiology can alter the profiles of plastic-associated chemicals, as this could contribute to and exacerbate pathological processes in animals and humans.

8.6 Need for Chemical Simplification

A recent focal area in the debate about plastic pollution has been the simplification of polymer types and product compositions [138]. The complexity of current plastic products reduces recycling efficiency, which was just around 9% in 2019 according to OECD [139]. Other factors, such as lack of waste management infrastructure and poor quality of single-use plastic products, also play important roles in the low recycling rates, but product complexity is a fundamental challenge that must be addressed to improve the environmental performance of plastics. However, reducing polymer complexity is not sufficient, since the vast amount of plastic chemicals in production today further reduces the environmental performance of end-of-life (EoL) plastics [138]. Apart from the impact of EoL performance, many plastic chemicals pose a risk to the environment and human health, as elaborated on earlier in this chapter and recently reviewed by Costa et al. [140]. To solve the problems associated with chemicals in plastics, four fundamental aspects have to be addressed: (a) well-known chemicals of concern in plastics, (b) major knowledge gaps about the use and composition of chemicals in plastics, (c) market-based barriers to chemical simplification, and (d) introduction of non-intentionally added substances (NIAS) in the plastic life cycle.

An array of plastic chemicals have been known to pose risks to the environment and human health for decades, including phthalates, per- and polyfluoroalkyl substances (PFAS), and BPAs [141]. A recent study by Wagner et al. [21] provided an overview of plastic chemicals. Of the 16,000 chemicals identified in current plastic production, more than 4200 were deemed of concern due to their hazardous properties. More than 3600 of these chemicals of concern are currently not subject to international regulation, illustrating the magnitude of the challenge with hazardous chemicals in plastics. Adequate risk assessment and regulation of hazardous plastic chemicals typically take decades, as the history of the two plastic chemicals BPA and DEHP perfectly illustrates [142]. These chemicals have been in production in high tonnage volumes for decades, and scientific evidence of their hazardous properties was first published more than 50 years ago [142]. Despite the significant body of evidence of these chemicals' hazardous properties, they are still used today in, for example, the healthcare sector [141]. The history of these problematic chemicals illustrates that regulating these thousands of chemicals one by one will take additional decades and might even be practically impossible, with severe consequences for the environment and human health.

To further add to the challenge of providing sufficient protection against harm from plastic chemicals, the PlastChem report found that the hazardous properties of approximately 9000 plastic chemicals are currently unknown [21]. This is largely due to the lack of disclosure about chemicals used in plastics, which is why the UNEP has argued that transparency on plastic chemicals should be an essential part of the forthcoming Plastic Treaty [143]. While the integration of mandatory disclosure of plastic chemicals is a vital component for the new Plastics Treaty and other relevant plastic legislations, history has shown that compliance and sufficient control are extremely challenging, mainly due to the magnitude of work required. The European chemicals legislation REACH (Registration, Evaluation, Authorisation and Restriction of Chemicals) provides an example of how strict requirements for disclosure do not necessarily lead to sufficient compliance. More than a decade after the adoption of REACH, the European Chemicals Agency (ECHA) concluded that data on chemicals needs to improve, as the majority of dossiers submitted by the industry did not have the required data. Due to the vast amount of chemicals on the European market, ECHA has reached 20% checks of the submitted dossiers [144].

A final aspect that further complicates the task of ensuring that plastics do not contain hazardous chemicals relates to the fact that NIAS often contaminate plastics [145]. These chemicals typically migrate into the plastics either during production or during the recycling process. As an example, NIAS has been shown to migrate from, for instance, microwaveable plastic food containers [146]. Given the increasing demand to keep materials in the value chain for longer, as part of the transition to a more circular economy [147], the magnitude of NIAS impacts will only become more important in the future. The plethora of known chemicals of concern, chemicals of unknown origin, and hazards and NIAS make a scientifically based risk assessment of plastics virtually impossible. This implies that this chemical complexity has to be reduced to allow for sufficient scientific assessment, which can only be achieved via chemical simplification. Such chemical simplification can be achieved through a three-step approach that includes (a) elimination of non-essential products and chemicals, (b) reduction by avoiding redundant additives, and (c) assessment and regulation based on groupings of chemicals.

First and foremost, the elimination of non-essential products and chemicals should be prioritized. The essential use concept has been implemented in international law in Decision IV/25 of the Montreal Protocol and later in the EU chemicals policy [148]. In essence, the essential use criteria prescribe that chemicals of concern should only be used if they are necessary for health and safety or are critical for the functioning of society (encompassing cultural and intellectual aspects). Furthermore, there should be no available technically and economically feasible alternatives or substitutes that are acceptable from an environmental and health perspective [149]. Concerning plastics, the implementation of the essential use criteria will be more complicated, due to the high amount of both chemicals in plastics but also plastic applications on the global market. If implemented sufficiently, the criteria could be used to drastically reduce the number of single-use products on the market, many of which are not essential for health or critical functions in society. For some categories, some uses should be phased out, whereas others remain essential. The European single-use plastic directive provides an example of such a case. Single-use plastic straws are banned with this legislation, except for medical use, where some patients might have physical conditions that deem the use of straws essential [150]. Another important reason for the myriads of plastic chemicals is the market-based mechanism that governs economies across the globe. When an additive is developed, the company typically files a patent on its use. This implies that many of the additives currently on the market have redundant properties, simply because different companies need chemicals with the same properties but must make their product unique [151]. Eliminating redundant additives can be a way to achieve chemical simplification without losing any chemical properties on the market. For example, 500 products containing 66 different phenolic antioxidants could potentially be reduced to just 8 functional groups [151]. The grouping approach is also something that Wagner et al. [21] recommend as an approach to chemical simplification. PFAS serves as a good example of a large group of chemicals with similar properties and hazardous profiles that can and should be regulated as a group [152]. The authors discuss two different approaches to grouping the approximately 4300 types of PFAS on the market. The more precautionary approach is based on grouping chemicals with similar intrinsic properties, whereas the other approach is based on a more classical risk-based approach in which chemicals with expected cumulative effects, such as those with similar modes of action, are grouped [152]. However, both approaches require a much higher degree of disclosure on where and how plastic chemicals are used. As long as the uses and effects of the majority of plastic chemicals are unknown, it will be impossible to ensure that plastic products on the market are safe. A systematic approach to simplify the chemical content in plastics will be an important step towards providing the scientific overview needed to ensure safety for the environment and human health.

8.7 Conclusion

The role of microplastics as a vector affecting the bioavailability and toxicological consequences of both plastic-associated chemicals and environmental pollutants has been a much-researched and debated aspect of plastic pollution research. There is a substantial body of work investigating the combinations of polymers, sorbed chemicals, aquatic organisms, test set-ups, and endpoints, and given the complexity of the variables involved, it is perhaps not surprising that studies often display inconsistencies either validating or invalidating vector effects. Moreover, questions are raised about the realism and relevance of such laboratory studies in explaining real-world scenarios. Thus, when placed within a wider body of work, it remains difficult to assess whether microplastics are realistic carriers of environmental pollutants. However, it is becoming increasingly clear that plastics are produced with an alarming number of plastic-associated chemicals. The bioaccessibility and bioavailability of these chemicals, once microplastics enter the environment, need to be considered, but already there is evidence demonstrating that these chemicals, many of known toxicity, can be transferred to biota and cause adverse effects. Transformation products add a further dimension to the toxicity of chemicals that are introduced into the environment via plastics. Similarly, the influence of digestive physiology needs to be fully understood, with the gastrointestinal tract being the primary location of chemical desorption and release from the polymer matrix. Again, species, microplastic, chemical, and environmental differences increase the complexity of the topic. Perhaps, the best solution for at least part of the problem is to reduce the number of chemicals in plastics to begin with, and the research conducted to date strengthens the case for such calls (Fig. 8.1). However, chemical simplification is not easy as full disclosure of chemical content from manufacturers remains absent and the essential function of chemicals can be argued for. Nonetheless, chemical simplification is a worthy goal alongside reductions in plastic production and use to limit further environmental damage.

References

1. Peng, X., Chen, M., Chen, S., Dasgupta, S., Xu, H., Ta, K., Du, M., Li, J., Guo, Z., and Bai, S. (2018). Microplastics contaminate the deepest part of the world's ocean. *Geochemical Perspectives Letters*, **9**, 1–5. https://doi.org/10.7185/geochemlet.1829
2. Napper, I. E., Davies, B. F. R., Clifford, H., Elvin, S., Koldewey, H. J., Mayewski, P. A., Miner, K. R., Potocki, M., Elmore, A. C., Gajurel, A. P., and Thompson, R. C. (2020). Reaching new heights in plastic pollution: Preliminary findings of microplastics on Mount Everest. *One Earth*, **3**(5), 621–630. https://doi.org/10.1016/j.oneear.2020.10.020
3. Mihai, F. C., Gündoğdu, S., Khan, F. R., Olivelli, A., Markley, L. A., and Van Emmerik, T. (2022). Plastic pollution in marine and freshwater environments: Abundance, sources, and mitigation. In *Emerging Contaminants in the Environment: Challenges and Sustainable Practices* (pp. 241–274). Elsevier. https://doi.org/10.1016/B978-0-323-85160-2.00016-0
4. Kühn, S., and van Franeker, J. A. (2020). Quantitative overview of marine debris ingested by marine megafauna. *Marine Pollution Bulletin*, **151**. https://doi.org/10.1016/j.marpolbul.2019.110858

5. Hossain, M. A. R., and Olden, J. D. (2022). Global meta-analysis reveals diverse effects of microplastics on freshwater and marine fishes. *Fish and Fisheries*, **23**(6), 1439–1454. https://doi.org/10.1111/faf.12701
6. Ragusa, A., Svelato, A., Santacroce, C., Catalano, P., Notarstefano, V., Carnevali, O., Papa, F., Rongioletti, M. C. A., Baiocco, F., Draghi, S., D'Amore, E., Rinaldo, D., Matta, M., and Giorgini, E. (2021). Plasticenta: First evidence of microplastics in human placenta. *Environment International*, **146**. https://doi.org/10.1016/j.envint.2020.106274
7. Zhao, Q., Zhu, L., Weng, J., Jin, Z., Cao, Y., Jiang, H., and Zhang, Z. (2023). Detection and characterization of microplastics in the human testis and semen. *Science of the Total Environment*, **877**. https://doi.org/10.1016/j.scitotenv.2023.162713
8. Khan, F. R., Patsiou, D., and Catarino, A. I. (2021). Pollutants bioavailability and toxicological risk from microplastics. In *Handbook of Microplastics in the Environment* (pp. 1–40). Springer International Publishing. https://doi.org/10.1007/978-3-030-10618-8_19-1
9. Rochman, C. M., Brookson, C., Bikker, J., Djuric, N., Earn, A., Bucci, K., Athey, S., Huntington, A., McIlwraith, H., Munno, K., Frond, H. De, Kolomijeca, A., Erdle, L., Grbic, J., Bayoumi, M., Borrelle, S. B., Wu, T., Santoro, S., Werbowski, L. M., Zhu, X., Giles, R. K., Hamilton, B. M., Thaysen, C., Kaura, A., Klasios, N., Ead, L., Kim, J., Sherlock, C., Ho, A., and Hung, C. (2019). Rethinking microplastics as a diverse contaminant suite. *Environmental Toxicology and Chemistry*, **38**(4), 703–711. https://doi.org/10.1002/etc.4371
10. Syberg, K., Khan, F. R., Selck, H., Palmqvist, A., Banta, G. T., Daley, J., Sano, L., and Duhaime, M. B. (2015). Microplastics: Addressing ecological risk through lessons learned. *Environmental Toxicology and Chemistry*, **34**(5), 945–953. https://doi.org/10.1002/etc.2914
11. Alijagic, A., Suljević, D., Fočak, M., Sulejmanović, J., Šehović, E., Särndahl, E., and Engwall, M. (2024). The triple exposure nexus of microplastic particles, plastic-associated chemicals, and environmental pollutants from a human health perspective. In *Environment International* (Vol. 188). Elsevier Ltd. https://doi.org/10.1016/j.envint.2024.108736
12. Mbugani, J. J., Machiwa, J. F., Shilla, D. A., Joseph, D., Kimaro, W. H., and Khan, F. R. (2022). Impaired growth performance of Wami tilapia juveniles (*Oreochromis urolepis*) (Norman, 1922) due to microplastic induced degeneration of the small intestine. *Microplastics*, **1**(3), 334–345. https://doi.org/10.3390/microplastics1030025
13. Mbugani, J. J., Machiwa, J. F., Shilla, D. A., Kimaro, W., Joseph, D., and Khan, F. R. (2022). Histomorphological damage in the small intestine

of Wami tilapia (*Oreochromis urolepis*) (Norman, 1922) exposed to microplastics remain long after depuration. *Microplastics*, **1**(2), 240–253. https://doi.org/10.3390/microplastics1020017

14. Frias, J. P. G. L., Sobral, P., and Ferreira, A. M. (2010). Organic pollutants in microplastics from two beaches of the Portuguese coast. *Marine Pollution Bulletin*, **60**(11), 1988–1992. https://doi.org/10.1016/j.marpolbul.2010.07.030
15. Holmes, L. A., Turner, A., and Thompson, R. C. (2014). Interactions between trace metals and plastic production pellets under estuarine conditions. *Marine Chemistry*, **167**, 25–32. https://doi.org/10.1016/j.marchem.2014.06.001
16. Hüffer, T., and Hofmann, T. (2016). Sorption of non-polar organic compounds by micro-sized plastic particles in aqueous solution. *Environmental Pollution*, **214**, 194–201. https://doi.org/10.1016/j.envpol.2016.04.018
17. Santos, L. H. M. L. M., Rodríguez-Mozaz, S., and Barceló, D. (2021). Microplastics as vectors of pharmaceuticals in aquatic organisms: An overview of their environmental implications. *Case Studies in Chemical and Environmental Engineering*, **3**. https://doi.org/10.1016/j.cscee.2021.100079
18. Hüffer, T., Weniger, A. K., and Hofmann, T. (2018). Data on sorption of organic compounds by aged polystyrene microplastic particles. *Data in Brief*, **18**, 474–479. https://doi.org/10.1016/j.dib.2018.03.053
19. Hahladakis, J. N., Velis, C. A., Weber, R., Iacovidou, E., and Purnell, P. (2018). An overview of chemical additives present in plastics: Migration, release, fate and environmental impact during their use, disposal and recycling. *Journal of Hazardous Materials*, **344**, 179–199. https://doi.org/10.1016/j.jhazmat.2017.10.014
20. UNEP. (2023). *Chemicals in Plastics a Technical Report*. United Nations Environment Programme and Secretariat of the Basel, Rotterdam and Stockholm Conventions.
21. Wagner, M., Monclús, L., Arp, H. P. H., Groh, K. J., Løseth, M. E., Mucke, J., Wang, Z., Wolf, R., and Zimmerman, L. (2024). *State of the science on plastic chemicals: Identifying and addressing chemicals and polymers of concern*. https://doi.org/http://dx.doi.org/10.5281/zenodo.10701706.
22. Chen, X., He, T., Yang, X., Gan, Y., Qing, X., Wang, J., and Huang, Y. (2023). Analysis, environmental occurrence, fate and potential toxicity of tire wear compounds 6PPD and 6PPD-quinone. *Journal of Hazardous Materials*, **452**. Elsevier B.V. https://doi.org/10.1016/j.jhazmat.2023.131245

23. Mouneyrac, C., Lagarde, F., Châtel, A., Khan, F. R., Syberg, K., and Palmqvist, A. (2017). The role of laboratory experiments in the validation of field data. *Comprehensive Analytical Chemistry*, **75**, 241–273. https://doi.org/10.1016/bs.coac.2016.10.005

24. Khan, F. R., Syberg, K., Shashoua, Y., and Bury, N. R. (2015). Influence of polyethylene microplastic beads on the uptake and localization of silver in zebrafish (*Danio rerio*). *Environmental Pollution*, **206**, 73–79. https://doi.org/10.1016/j.envpol.2015.06.009

25. Sørensen, L., Rogers, E., Altin, D., Salaberria, I., and Booth, A. M. (2020). Sorption of PAHs to microplastic and their bioavailability and toxicity to marine copepods under co-exposure conditions. *Environmental Pollution*, **258**. https://doi.org/10.1016/j.envpol.2019.113844

26. Khan, F. R., Catarino, A. I., and Clark, N. J. (2022). The ecotoxicological consequences of microplastics and co-contaminants in aquatic organisms: A mini-review. *Emerging Topics in Life Sciences*, **6**(4), 339–348. https://doi.org/10.1042/ETLS20220014

27. Fred-Ahmadu, O. H., Tenebe, I. T., Ayejuyo, O. O., and Benson, N. U. (2022). Microplastics and associated organic pollutants in beach sediments from the Gulf of Guinea (SE Atlantic) coastal ecosystems. *Chemosphere*, **298**. https://doi.org/10.1016/j.chemosphere.2022.134193

28. Syberg, K., Knudsen, C. M. H., Tairova, Z., Khan, F. R., Shashoua, Y., Geertz, T., Pedersen, H. B., Sick, C., Mortensen, J., Strand, J., and Palmqvist, A. (2020). Sorption of PCBs to environmental plastic pollution in the North Atlantic Ocean: Importance of size and polymer type. *Case Studies in Chemical and Environmental Engineering*, **2**. https://doi.org/10.1016/j.cscee.2020.100062

29. Carbery, M., MacFarlane, G. R., O'Connor, W., Afrose, S., Taylor, H., and Palanisami, T. (2020). Baseline analysis of metal(loid)s on microplastics collected from the Australian shoreline using citizen science. *Marine Pollution Bulletin*, **152**. https://doi.org/10.1016/j.marpolbul.2020.110914

30. Vedolin, M. C., Teophilo, C. Y. S., Turra, A., and Figueira, R. C. L. (2018). Spatial variability in the concentrations of metals in beached microplastics. *Marine Pollution Bulletin*, **129**(2), 487–493. https://doi.org/10.1016/j.marpolbul.2017.10.019

31. McDougall, L., Thomson, L., Brand, S., Wagstaff, A., Lawton, L. A., and Petrie, B. (2022). Adsorption of a diverse range of pharmaceuticals to polyethylene microplastics in wastewater and their desorption in environmental matrices. *Science of the Total Environment*, **808**. https://doi.org/10.1016/j.scitotenv.2021.152071

32. Wagstaff, A., Lawton, L. A., and Petrie, B. (2022). Polyamide microplastics in wastewater as vectors of cationic pharmaceutical drugs. *Chemosphere*, **288**. https://doi.org/10.1016/j.chemosphere.2021.132578

33. Velez, J. F. M., Shashoua, Y., Syberg, K., and Khan, F. R. (2018). Considerations on the use of equilibrium models for the characterisation of HOC-microplastic interactions in vector studies. *Chemosphere*, **210**, 359–365. https://doi.org/10.1016/j.chemosphere.2018.07.020

34. Bakir, A., Rowland, S. J., and Thompson, R. C. (2014). Enhanced desorption of persistent organic pollutants from microplastics under simulated physiological conditions. *Environmental Pollution*, **185**, 16–23. https://doi.org/10.1016/j.envpol.2013.10.007

35. Khan, F. R., Boyle, D., Chang, E., and Bury, N. R. (2017). Do polyethylene microplastic beads alter the intestinal uptake of Ag in rainbow trout (*Oncorhynchus mykiss*)? Analysis of the MP vector effect using in vitro gut sacs. *Environmental Pollution*, **231**, 200–206. https://doi.org/10.1016/j.envpol.2017.08.019

36. Hartmann, N. B., Rist, S., Bodin, J., Jensen, L. H. S., Schmidt, S. N., Mayer, P., Meibom, A., and Baun, A. (2017). Microplastics as vectors for environmental contaminants: Exploring sorption, desorption, and transfer to biota. *Integrated Environmental Assessment and Management*, **13**(3), 488–493. Wiley-Blackwell. https://doi.org/10.1002/ieam.1904

37. Huang, W., Song, B., Liang, J., Niu, Q., Zeng, G., Shen, M., Deng, J., Luo, Y., Wen, X., and Zhang, Y. (2021). Microplastics and associated contaminants in the aquatic environment: A review on their ecotoxicological effects, trophic transfer, and potential impacts to human health. *Journal of Hazardous Materials*, **405**. https://doi.org/10.1016/j.jhazmat.2020.124187

38. Koelmans, A. A., Diepens, N. J., and Mohamed Nor, N. H. (2022). *Weight of Evidence for the Microplastic Vector Effect in the Context of Chemical Risk Assessment* (M. S. Bank, Ed.; pp. 155–197). https://doi.org/10.1007/978-3-030-78627-4_6

39. Oliveira, M., Ribeiro, A., Hylland, K., and Guilhermino, L. (2013). Single and combined effects of microplastics and pyrene on juveniles (0+ group) of the common goby *Pomatoschistus microps* (Teleostei, Gobiidae). *Ecological Indicators*, **34**, 641–647. https://doi.org/10.1016/j.ecolind.2013.06.019

40. Rochman, C. M., Kurobe, T., Flores, I., and Teh, S. J. (2014). Early warning signs of endocrine disruption in adult fish from the ingestion of polyethylene with and without sorbed chemical pollutants from the

marine environment. *Science of the Total Environment*, **493**, 656–661. https://doi.org/10.1016/j.scitotenv.2014.06.051

41. Besseling, E., Wegner, A., Foekema, E. M., Van Den Heuvel-Greve, M. J., and Koelmans, A. A. (2013). Effects of microplastic on fitness and PCB bioaccumulation by the lugworm Arenicola marina (L.). *Environmental Science and Technology*, **47**(1), 593–600. https://doi.org/10.1021/es302763x
42. Chua, E. M., Shimeta, J., Nugegoda, D., Morrison, P. D., and Clarke, B. O. (2014). Assimilation of polybrominated diphenyl ethers from microplastics by the marine amphipod, *Allorchestes compressa*. *Environmental Science and Technology*, **48**(14), 8127–8134. https://doi.org/10.1021/es405717z
43. González-Soto, N., Hatfield, J., Katsumiti, A., Duroudier, N., Lacave, J. M., Bilbao, E., Orbea, A., Navarro, E., and Cajaraville, M. P. (2019). Impacts of dietary exposure to different sized polystyrene microplastics alone and with sorbed benzo[a]pyrene on biomarkers and whole organism responses in mussels *Mytilus galloprovincialis*. *Science of the Total Environment*, **684**, 548–566. https://doi.org/10.1016/j.scitotenv.2019.05.161
44. Lu, K., Qiao, R., An, H., and Zhang, Y. (2018). Influence of microplastics on the accumulation and chronic toxic effects of cadmium in zebrafish (*Danio rerio*). *Chemosphere*, **202**, 514–520. https://doi.org/10.1016/j.chemosphere.2018.03.145
45. O'Donovan, S., Mestre, N. C., Abel, S., Fonseca, T. G., Carteny, C. C., Cormier, B., Keiter, S. H., and Bebianno, M. J. (2018). Ecotoxicological effects of chemical contaminants adsorbed to microplastics in the clam *Scrobicularia plana*. *Frontiers in Marine Science*, **5**(APR). https://doi.org/10.3389/fmars.2018.00143
46. Batel, A., Borchert, F., Reinwald, H., Erdinger, L., and Braunbeck, T. (2018). Microplastic accumulation patterns and transfer of benzo[a]pyrene to adult zebrafish (*Danio rerio*) gills and zebrafish embryos. *Environmental Pollution*, **235**, 918–930. https://doi.org/10.1016/j.envpol.2018.01.028
47. Rainieri, S., Conlledo, N., Larsen, B. K., Granby, K., and Barranco, A. (2018). Combined effects of microplastics and chemical contaminants on the organ toxicity of zebrafish (*Danio rerio*). *Environmental Research*, **162**, 135–143. https://doi.org/10.1016/j.envres.2017.12.019
48. Ašmonaitė, G., Tivefälth, M., Westberg, E., Magnér, J., Backhaus, T., and Carney Almroth, B. (2020). Microplastics as a vector for exposure to hydrophobic organic chemicals in fish: A comparison of two polymers

and silica particles spiked with three model compounds. *Frontiers in Environmental Science*, **8**. https://doi.org/10.3389/fenvs.2020.00087

49. Granby, K., Rainieri, S., Rasmussen, R. R., Kotterman, M. J. J., Sloth, J. J., Cederberg, T. L., Barranco, A., Marques, A., and Larsen, B. K. (2018). The influence of microplastics and halogenated contaminants in feed on toxicokinetics and gene expression in European seabass (*Dicentrarchus labrax*). *Environmental Research*, **164**, 430–443. https://doi.org/10.1016/j.envres.2018.02.035

50. Magara, G., Elia, A. C., Syberg, K., and Khan, F. R. (2018). Single contaminant and combined exposures of polyethylene microplastics and fluoranthene: Accumulation and oxidative stress response in the blue mussel, *Mytilus edulis*. *Journal of Toxicology and Environmental Health - Part A: Current Issues*, **81**(16), 761–773. https://doi.org/10.1080/15287394.2018.1488639

51. Beiras, R., Muniategui-Lorenzo, S., Rodil, R., Tato, T., Montes, R., López-Ibáñez, S., Concha-Graña, E., Campoy-López, P., Salgueiro-González, N., and Quintana, J. B. (2019). Polyethylene microplastics do not increase bioaccumulation or toxicity of nonylphenol and 4-MBC to marine zooplankton. *Science of the Total Environment*, **692**, 1–9. https://doi.org/10.1016/j.scitotenv.2019.07.106

52. Beiras, R., and Tato, T. (2019). Microplastics do not increase toxicity of a hydrophobic organic chemical to marine plankton. *Marine Pollution Bulletin*, **138**, 58–62. https://doi.org/10.1016/j.marpolbul.2018.11.029

53. Norland, S., Vorkamp, K., Bogevik, A. S., Koelmans, A. A., Diepens, N. J., Burgerhout, E., Hansen, Ø. J., Puvanendran, V., and Rønnestad, I. (2021). Assessing microplastic as a vector for chemical entry into fish larvae using a novel tube-feeding approach. *Chemosphere*, **265**. https://doi.org/10.1016/j.chemosphere.2020.129144

54. Wen, B., Jin, S. R., Chen, Z. Z., Gao, J. Z., Liu, Y. N., Liu, J. H., and Feng, X. S. (2018). Single and combined effects of microplastics and cadmium on the cadmium accumulation, antioxidant defence and innate immunity of the discus fish (*Symphysodon aequifasciatus*). *Environmental Pollution*, **243**, 462–471. https://doi.org/10.1016/j.envpol.2018.09.029

55. Scopetani, C., Cincinelli, A., Martellini, T., Lombardini, E., Ciofini, A., Fortunati, A., Pasquali, V., Ciattini, S., and Ugolini, A. (2018). Ingested microplastic as a two-way transporter for PBDEs in *Talitrus saltator*. *Environmental Research*, **167**, 411–417. https://doi.org/10.1016/j.envres.2018.07.030

56. Brander, S. M. (2022). Rethinking our chemical legacy and reclaiming our planet. *One Earth*, **5**(4), 316–319. https://doi.org/10.1016/j.oneear.2022.03.020

57. Tong, H., Zhong, X., Duan, Z., Yi, X., Cheng, F., Xu, W., and Yang, X. (2022). Micro- and nanoplastics released from biodegradable and conventional plastics during degradation: Formation, aging factors, and toxicity. *Science of the Total Environment*, **833**. https://doi.org/10.1016/j.scitotenv.2022.155275

58. Uribe-Echeverría, T., and Beiras, R. (2022). Acute toxicity of bioplastic leachates to *Paracentrotus lividus* sea urchin larvae. *Marine Environmental Research*, **176**. https://doi.org/10.1016/j.marenvres.2022.105605

59. Magara, G., Khan, F. R., Pinti, M., Syberg, K., Inzirillo, A., and Elia, A. C. (2019). Effects of combined exposures of fluoranthene and polyethylene or polyhydroxybutyrate microplastics on oxidative stress biomarkers in the blue mussel (*Mytilus edulis*). *Journal of Toxicology and Environmental Health - Part A: Current Issues*, **82**(10), 616–625. https://doi.org/10.1080/15287394.2019.1633451

60. Syberg, K., Nielsen, A., Khan, F. R., Banta, G. T., Palmqvist, A., and Jepsen, P. M. (2017). Microplastic potentiates triclosan toxicity to the marine copepod *Acartia tonsa* (Dana). *Journal of Toxicology and Environmental Health - Part A: Current Issues*, **80**(23–24), 1369–1371. https://doi.org/10.1080/15287394.2017.1385046

61. Bakir, A., O'Connor, I. A., Rowland, S. J., Hendriks, A. J., and Thompson, R. C. (2016). Relative importance of microplastics as a pathway for the transfer of hydrophobic organic chemicals to marine life. *Environmental Pollution*, **219**, 56–65. https://doi.org/10.1016/j.envpol.2016.09.046

62. Koelmans, A. A., Bakir, A., Burton, G. A., and Janssen, C. R. (2016). Microplastic as a vector for chemicals in the aquatic environment: Critical review and model-supported reinterpretation of empirical studies. *Environmental Science and Technology*, **50**(7), 3315–3326. https://doi.org/10.1021/acs.est.5b06069

63. Amiard, J.-C., and Amiard-Triquet, C. (2015). Conventional risk assessment of environmental contaminants. In C. Amiard-Triquet, J.-C. Amiard, and C. Mouneyrac (Eds.), *Aquatic Ecotoxicology* (pp. 25–49). Academic Press. https://doi.org/https://doi.org/10.1016/B978-0-12-800949-9.00002-4

64. Cunningham, B. E., Sharpe, E. E., Brander, S. M., Landis, W. G., and Harper, S. L. (2023). Critical gaps in nanoplastics research and their connection to risk assessment. *Frontiers in Toxicology*, **5**. https://doi.org/10.3389/ftox.2023.1154538

65. Catarino, A. I., Patsiou, D., Summers, S., Everaert, G., Henry, T. B., and Gutierrez, T. (2023). Challenges and recommendations in experimentation and risk assessment of nanoplastics in aquatic organisms. *TrAC - Trends in Analytical Chemistry*, **167**. https://doi.org/10.1016/j.trac.2023.117262

66. Thornton Hampton, L. M., Brander, S. M., Coffin, S., Cole, M., Hermabessiere, L., Koelmans, A. A., and Rochman, C. M. (2022). Characterizing microplastic hazards: Which concentration metrics and particle characteristics are most informative for understanding toxicity in aquatic organisms? *Microplastics and Nanoplastics*, **2**(1). https://doi.org/10.1186/s43591-022-00040-4

67. Umeh, A. C., Naidu, R., Owojori, O. J., and Semple, K. T. (2020). Bioavailability and bioaccessibility of hydrophobic organic contaminants in soil and associated desorption-based measurements. In J. J. Ortega-Calvo and J. R. Parsons (Eds.), *Bioavailability of Organic Chemicals in Soil and Sediment* (pp. 293–350). Springer International Publishing. https://doi.org/10.1007/698_2020_521

68. Guo, H., Zheng, X., Ru, S., Luo, X., and Mai, B. (2019). The leaching of additive-derived flame retardants (FRs) from plastics in avian digestive fluids: The significant risk of highly lipophilic FRs. *Journal of Environmental Sciences (China)*, **85**, 200–207. https://doi.org/10.1016/j.jes.2019.06.013

69. Catrouillet, C., Davranche, M., Khatib, I., Fauny, C., Wahl, A., and Gigault, J. (2021). Metals in microplastics: Determining which are additive, adsorbed, and bioavailable. *Processes & Impacts*, **4**. https://doi.org/10.1039/D1EM00017Aï

70. Peng, X., Chen, G., Fan, Y., Zhu, Z., Guo, S., Zhou, J., and Tan, J. (2021). Lifetime bioaccumulation, gender difference, tissue distribution, and parental transfer of organophosphorus plastic additives in freshwater fish. *Environmental Pollution*, **280**. https://doi.org/10.1016/j.envpol.2021.116948

71. Hermabessiere, L., Dehaut, A., Paul-Pont, I., Lacroix, C., Jezequel, R., Soudant, P., and Duflos, G. (2017). Occurrence and effects of plastic additives on marine environments and organisms: A review. *Chemosphere*, **182**, 781–793. https://doi.org/10.1016/j.chemosphere.2017.05.096

72. Tanaka, K., Watanuki, Y., Takada, H., Ishizuka, M., Yamashita, R., Kazama, M., Hiki, N., Kashiwada, F., Mizukawa, K., Mizukawa, H., Hyrenbach, D., Hester, M., Ikenaka, Y., and Nakayama, S. M. M. (2020). In vivo accumulation of plastic-derived chemicals into seabird tissues. *Current Biology*, **30**(4), 723–728. https://doi.org/10.1016/j.cub.2019.12.037

73. Jang, M., Shim, W. J., Han, G. M., Cho, Y., Moon, Y., and Hong, S. H. (2021). Relative importance of aqueous leachate versus particle ingestion as uptake routes for microplastic additives (hexabromocyclododecane) to mussels. *Environmental Pollution*, **270**. https://doi.org/10.1016/j.envpol.2020.116272

74. Koelmans, A. A., Besseling, E., and Foekema, E. M. (2014). Leaching of plastic additives to marine organisms. *Environmental Pollution*, **187**, 49–54. https://doi.org/10.1016/j.envpol.2013.12.013

75. Semple, K. T., Doick, K. J., Jones, K. C., Burauel, P., Crave, A., and Harms, H. (2004). Peer reviewed: Defining bioavailability and bioaccessibility of contaminated soil and sediment is complicated. *Environmental Science & Technology*, **38**(12), 228A–231A. https://doi.org/10.1021/es040548w

76. Boyle, D., Catarino, A. I., Clark, N. J., and Henry, T. B. (2020). Polyvinyl chloride (PVC) plastic fragments release Pb additives that are bioavailable in zebrafish. *Environmental Pollution*, **263**. https://doi.org/10.1016/j.envpol.2020.114422

77. Schrank, I., Trotter, B., Dummert, J., Scholz-Böttcher, B. M., Löder, M. G. J., and Laforsch, C. (2019). Effects of microplastic particles and leaching additive on the life history and morphology of *Daphnia magna*. *Environmental Pollution*, **255**. https://doi.org/10.1016/j.envpol.2019.113233

78. Mohamed Nor, N. H., and Koelmans, A. A. (2019). Transfer of PCBs from Microplastics under simulated gut fluid conditions is biphasic and reversible. *Environmental Science and Technology*. https://doi.org/10.1021/acs.est.8b05143

79. Rummel, C. D., Jahnke, A., Gorokhova, E., Kühnel, D., and Schmitt-Jansen, M. (2017). Impacts of biofilm formation on the fate and potential effects of microplastic in the aquatic environment. *Environmental Science & Technology Letters*, **4**(7), 258–267. https://doi.org/10.1021/acs.estlett.7b00164

80. Capolupo, M., Gunaalan, K., Booth, A. M., Sørensen, L., Valbonesi, P., and Fabbri, E. (2021). The sub-lethal impact of plastic and tire rubber leachates on the Mediterranean mussel *Mytilus galloprovincialis*. *Environmental Pollution*, **283**. https://doi.org/10.1016/j.envpol.2021.117081

81. Capolupo, M., Sørensen, L., Jayasena, K. D. R., Booth, A. M., and Fabbri, E. (2020). Chemical composition and ecotoxicity of plastic and car tire rubber leachates to aquatic organisms. *Water Research*, **169**. https://doi.org/10.1016/j.watres.2019.115270

82. Bejgarn, S., MacLeod, M., Bogdal, C., and Breitholtz, M. (2015). Toxicity of leachate from weathering plastics: An exploratory screening study with *Nitocra spinipes*. *Chemosphere*, **132**, 114–119. https://doi.org/10.1016/j.chemosphere.2015.03.010

83. Oliviero, M., Tato, T., Schiavo, S., Fernández, V., Manzo, S., and Beiras, R. (2019). Leachates of micronized plastic toys provoke embryotoxic effects upon sea urchin *Paracentrotus lividus*. *Environmental Pollution*, **247**, 706–715. https://doi.org/10.1016/j.envpol.2019.01.098

84. Zimmermann, L., Dierkes, G., Ternes, T. A., Völker, C., and Wagner, M. (2019). Benchmarking the in vitro toxicity and chemical composition of plastic consumer products. *Environmental Science & Technology*, **53**(19), 11467–11477. https://doi.org/10.1021/acs.est.9b02293

85. Kotthoff, L., Keller, J., Lörchner, D., Mekonnen, T. F., and Koch, M. (2019). Transformation products of organic contaminants and residues: Overview of current simulation methods. *Molecules*, **24**(4). https://doi.org/10.3390/molecules24040753

86. Escher, B. I., and Fenner, K. (2011). Recent advances in environmental risk assessment of transformation products. *Environmental Science & Technology*, **45**(9), 3835–3847. https://doi.org/10.1021/es1030799

87. Wilkinson, J., Hooda, P. S., Barker, J., Barton, S., and Swinden, J. (2017). Occurrence, fate and transformation of emerging contaminants in water: An overarching review of the field. *Environmental Pollution*, **231**, 954–970. Elsevier Ltd. https://doi.org/10.1016/j.envpol.2017.08.032

88. Belfroid, A. C., Van Drunen, M., Beek, M. A., Schrap, S. M., Van Gestel, C. A. M., and Van Hattum, B. (1998). Relative risks of transformation products of pesticides for aquatic ecosystems. In *The Science of the Total Environment* (Vol. 222).

89. Deeb, A. A., Stephan, S., Schmitz, O. J., and Schmidt, T. C. (2017). Suspect screening of micropollutants and their transformation products in advanced wastewater treatment. *Science of the Total Environment*, **601–602**, 1247–1253. https://doi.org/10.1016/j.scitotenv.2017.05.271

90. Parezanović, G. Š., Lalic-Popovic, M., Golocorbin-Kon, S., Vasovic, V., Milijašević, B., Al-Salami, H., and Mikov, M. (2019). Environmental transformation of pharmaceutical formulations: A scientific review. *Archives of Environmental Contamination and Toxicology*, **77**(2), 155–161. https://doi.org/10.1007/s00244-019-00630-z

91. Tian, Z., Zhao, H., Peter, K. T., Gonzalez, M., Wetzel, J., Wu, C., Hu, X., Prat, J., Mudrock, E., and Hettinger, R. (2021). A ubiquitous tire rubber–derived chemical induces acute mortality in coho salmon. *Science*, **371**(6525), 185–189.

92. Hu, X., Zhao, H. N., Tian, Z., Peter, K. T., Dodd, M. C., and Kolodziej, E. P. (2022). Transformation product formation upon heterogeneous ozonation of the tire rubber antioxidant 6PPD (N-(1,3-dimethylbutyl)-N′-phenyl-p-phenylenediamine). *Environmental Science and Technology Letters*, **9**(5), 413–419. https://doi.org/10.1021/acs.estlett.2c00187

93. Chow, M. I., Lundin, J. I., Mitchell, C. J., Davis, J. W., Young, G., Scholz, N. L., and McIntyre, J. K. (2019). An urban stormwater runoff mortality syndrome in juvenile coho salmon. *Aquatic Toxicology*, **214**. https://doi.org/10.1016/j.aquatox.2019.105231

94. French, B. F., Baldwin, D. H., Cameron, J., Prat, J., King, K., Davis, J. W., McIntyre, J. K., and Scholz, N. L. (2022). Urban roadway runoff is lethal to juvenile coho, steelhead, and chinook salmonids, but not congeneric sockeye. *Environmental Science and Technology Letters*. https://doi.org/10.1021/acs.estlett.2c00467

95. McIntyre, J. K., Prat, J., Cameron, J., Wetzel, J., Mudrock, E., Peter, K. T., Tian, Z., Mackenzie, C., Lundin, J., Stark, J. D., King, K., Davis, J. W., Kolodziej, E. P., and Scholz, N. L. (2021). Treading water: Tire wear particle leachate recreates an urban runoff mortality syndrome in coho but not chum salmon. *Environmental Science and Technology*, **55**(17), 11767–11774. https://doi.org/10.1021/acs.est.1c03569

96. Tian, Z., Gonzalez, M., Rideout, C. A., Zhao, H. N., Hu, X., Wetzel, J., Mudrock, E., James, C. A., McIntyre, J. K., and Kolodziej, E. P. (2022). 6PPD-quinone: Revised toxicity assessment and quantification with a commercial standard. *Environmental Science & Technology Letters*, **9**(2), 140–146.

97. Khan, F. R., Rødland, E. S., Kole, P. J., Van Belleghem, F. G. A. J., Jaén-Gil, A., Hansen, S. F., and Gomiero, A. (2024). An overview of the key topics related to the study of tire particles and their chemical leachates: From problems to solutions. *TrAC - Trends in Analytical Chemistry*, **172**. https://doi.org/10.1016/j.trac.2024.117563

98. Li, J., Xu, J., and Jiang, X. (2023). Urban runoff mortality syndrome in zooplankton caused by tire wear particles. *Environmental Pollution*, **329**. https://doi.org/10.1016/j.envpol.2023.121721

99. Jiang, Y., Wang, C., Ma, L., Gao, T., and Wāng, Y. (2024). Environmental profiles, hazard identification, and toxicological hallmarks of emerging tire rubber-related contaminants 6PPD and 6PPD-quinone. *Environment International*, **187**. https://doi.org/10.1016/j.envint.2024.108677

100. Zoroufchi Benis, K., Behnami, A., Minaei, S., Brinkmann, M., McPhedran, K. N., and Soltan, J. (2023). Environmental occurrence and toxicity of 6PPD quinone: An emerging tire rubber-derived chemical: A review. *Environmental Science and Technology Letters*, **10**(10), 815–823. https://doi.org/10.1021/acs.estlett.3c00521

101. Alhelou, R., Seiwert, B., and Reemtsma, T. (2019). Hexamethoxymethylmelamine: A precursor of persistent and mobile contaminants in municipal wastewater and the water cycle. *Water Research*, **165**. https://doi.org/10.1016/j.watres.2019.114973

102. Foscari, A., Seiwert, B., Zahn, D., Schmidt, M., and Reemtsma, T. (2024). Leaching of tire particles and simultaneous biodegradation of leachables. *Water Research*, **253**. https://doi.org/10.1016/j.watres.2024.121322

103. Johannessen, C., and Metcalfe, C. D. (2022). The occurrence of tire wear compounds and their transformation products in municipal wastewater and drinking water treatment plants. *Environmental Monitoring and Assessment*, **194**(10). https://doi.org/10.1007/s10661-022-10450-9

104. Castan, S., Sherman, A., Peng, R., Zumstein, M. T., Wanek, W., Hüffer, T., and Hofmann, T. (2023). Uptake, metabolism, and accumulation of tire wear particle-derived compounds in lettuce. *Environmental Science and Technology*, **57**(1), 168–178. https://doi.org/10.1021/acs.est.2c05660

105. Sarkar, S., Diab, H., and Thompson, J. (2023). Microplastic pollution: Chemical characterization and impact on wildlife. *International Journal of Environmental Research and Public Health*, **20**(3). MDPI. https://doi.org/10.3390/ijerph20031745

106. Katami, T., Yasuhara, A., Okuda, T., and Shibamoto, T. (2002). Formation of PCDDs, PCDFs, and coplanar PCBs from polyvinyl chloride during combustion in an incinerator. *Environmental Science & Technology*, **36**(6), 1320–1324. https://doi.org/10.1021/es0109904

107. Swanson, H. (2011). Dioxin response elements and regulation of gene transcription. In *The AH Receptor in Biology and Toxicology* (pp. 81–91). https://doi.org/https://doi.org/10.1002/9781118140574.ch5

108. Stenberg, M., Hamers, T., Machala, M., Fonnum, F., Stenius, U., Lauy, A. A., van Duursen, M. B. M., Westerink, R. H. S., Fernandes, E. C. A., and Andersson, P. L. (2011). Multivariate toxicity profiles and QSAR modeling of non-dioxin-like PCBs: An investigation of in vitro screening data from ultra-pure congeners. *Chemosphere*, **85**(9), 1423–1429. https://doi.org/10.1016/j.chemosphere.2011.08.019

109. Kim, K. H., Jahan, S. A., Kabir, E., and Brown, R. J. C. (2013). A review of airborne polycyclic aromatic hydrocarbons (PAHs) and their human health effects. *Environment International*, **60**, 71–80. https://doi.org/10.1016/j.envint.2013.07.019

110. Walsh, A. N., Reddy, C. M., Niles, S. F., McKenna, A. M., Hansel, C. M., and Ward, C. P. (2021). Plastic formulation is an emerging control of its photochemical fate in the ocean. *Environmental Science and Technology*, **55**(18), 12383–12392. https://doi.org/10.1021/acs.est.1c02272

111. Lomonaco, T., Manco, E., Corti, A., La Nasa, J., Ghimenti, S., Biagini, D., Di Francesco, F., Modugno, F., Ceccarini, A., Fuoco, R., and Castelvetro, V. (2020). Release of harmful volatile organic compounds (VOCs) from photo-degraded plastic debris: A neglected source of environmental pollution. *Journal of Hazardous Materials*, **394**. https://doi.org/10.1016/j.jhazmat.2020.122596

112. Lampi, M. A., Gurska, J., McDonald, K. I. C., Xie, F., Huang, X. D., Dixon, D. G., and Greenberg, B. M. (2006). Photoinduced toxicity of polycyclic aromatic hydrocarbons to *Daphnia magna*: Ultraviolet-mediated effects and the toxicity of polycyclic aromatic hydrocarbon photoproducts. *Environmental Toxicology and Chemistry*, **25**(4), 1079–1087. https://doi.org/10.1897/05-276R.1

113. Zhai, X., Zhang, X. H., and Yu, M. (2023). Microbial colonization and degradation of marine microplastics in the plastisphere: A review. *Frontiers in Microbiology*, **14**. https://doi.org/10.3389/fmicb.2023.1127308

114. Gong, J., Kong, T., Li, Y., Li, Q., Li, Z., and Zhang, J. (2018). Biodegradation of microplastic derived from poly(ethylene terephthalate) with bacterial whole-cell biocatalysts. *Polymers*, **10**(12). https://doi.org/10.3390/polym10121326

115. Sühring, R., Baak, J. E., Letcher, R. J., Braune, B. M., de Silva, A., Dey, C., Fernie, K., Lu, Z., Mallory, M. L., Avery-Gomm, S., and Provencher, J. F. (2022). Co-contaminants of microplastics in two seabird species from the Canadian Arctic. *Environmental Science and Ecotechnology*, **12**. https://doi.org/10.1016/j.ese.2022.100189

116. Fred-Ahmadu, O. H., Bhagwat, G., Oluyoye, I., Benson, N. U., Ayejuyo, O. O., and Palanisami, T. (2020). Interaction of chemical contaminants with microplastics: Principles and perspectives. *Science of the Total Environment*, **706**. https://doi.org/10.1016/j.scitotenv.2019.135978

117. Martín, J., Santos, J. L., Aparicio, I., and Alonso, E. (2022). Microplastics and associated emerging contaminants in the environment:

Analysis, sorption mechanisms and effects of co-exposure. *Trends in Environmental Analytical Chemistry*, **35**. https://doi.org/10.1016/j.teac.2022.e00170

118. Torres, F. G., Dioses-Salinas, D. C., Pizarro-Ortega, C. I., and De-la-Torre, G. E. (2021). Sorption of chemical contaminants on degradable and non-degradable microplastics: Recent progress and research trends. *Science of the Total Environment*, **757**. https://doi.org/10.1016/j.scitotenv.2020.143875

119. Van Der Zande, M., Jemec Kokalj, A., Spurgeon, D. J., Loureiro, S., Silva, P. V., Khodaparast, Z., Drobne, D., Clark, N. J., Van Den Brink, N. W., Baccaro, M., Van Gestel, C. A. M., Bouwmeester, H., and Handy, R. D. (2020). The gut barrier and the fate of engineered nanomaterials: A view from comparative physiology. *Environmental Science: Nano*, **7**(7), 1874–1898. https://doi.org/10.1039/d0en00174k

120. Bucking, C., and Wood, C. M. (2009). The effect of postprandial changes in pH along the gastrointestinal tract on the distribution of ions between the solid and fluid phases of chyme in rainbow trout. *Aquaculture Nutrition*, **15**(3), 282–296. https://doi.org/10.1111/j.1365-2095.2008.00593.x

121. Yúfera, M., Moyano, F. J., Astola, A., Pousão-Ferreira, P., and Martínez-Rodríguez, G. (2012). Acidic digestion in a teleost: Postprandial and circadian pattern of gastric pH, pepsin activity, and pepsinogen and proton pump mRNAs expression. *PLoS ONE*, **7**(3). https://doi.org/10.1371/journal.pone.0033687

122. Yúfera, M., and Darías, M. J. (2007). Changes in the gastrointestinal pH from larvae to adult in Senegal sole (*Solea senegalensis*). *Aquaculture*, **267**(1–4), 94–99. https://doi.org/10.1016/j.aquaculture.2007.02.009

123. Bao, Z. Z., Chen, Z. F., Lu, S. Q., Wang, G., Qi, Z., and Cai, Z. (2021). Effects of hydroxyl group content on adsorption and desorption of anthracene and anthrol by polyvinyl chloride microplastics. *Science of the Total Environment*, **790**. https://doi.org/10.1016/j.scitotenv.2021.148077

124. Coffin, S., Huang, G. Y., Lee, I., and Schlenk, D. (2019). Fish and seabird gut conditions enhance desorption of estrogenic chemicals from commonly-ingested plastic items. *Environmental Science and Technology*, **53**(8), 4588–4599. https://doi.org/10.1021/acs.est.8b07140

125. Wu, P., Tang, Y., Jin, H., Song, Y., Liu, Y., and Cai, Z. (2020). Consequential fate of bisphenol-attached PVC microplastics in water and simulated intestinal fluids. *Environmental Science and Ecotechnology*, **2**. https://doi.org/10.1016/j.ese.2020.100027

126. Wang, Z., Ding, J., Razanajatovo, R. M., Huang, J., Zheng, L., Zou, H., Wang, Z., and Liu, J. (2022). Sorption of selected pharmaceutical compounds on polyethylene microplastics: Roles of pH, aging, and competitive sorption. *Chemosphere*, **307**. https://doi.org/10.1016/j.chemosphere.2022.135561

127. Zhang, H., Wang, J., Zhou, B., Zhou, Y., Dai, Z., Zhou, Q., Chriestie, P., and Luo, Y. (2018). Enhanced adsorption of oxytetracycline to weathered microplastic polystyrene: Kinetics, isotherms and influencing factors. *Environmental Pollution*, **243**, 1550–1557. https://doi.org/10.1016/j.envpol.2018.09.122

128. Ahechti, M., Benomar, M., El Alami, M., and Mendiguchía, C. (2022). Metal adsorption by microplastics in aquatic environments under controlled conditions: Exposure time, pH and salinity. *International Journal of Environmental Analytical Chemistry*, **102**(5), 1118–1125. https://doi.org/10.1080/03067319.2020.1733546

129. Zou, J., Liu, X., Zhang, D., and Yuan, X. (2020). Adsorption of three bivalent metals by four chemical distinct microplastics. *Chemosphere*, **248**. https://doi.org/10.1016/j.chemosphere.2020.126064

130. Schwabl, P., Köppel, S., Königshofer, P., Bucsics, T., Trauner, M., Reiberger, T., and Liebmann, B. (2019). Detection of various microplastics in human stool. *Annals of Internal Medicine*, **171**(7), 453–457. https://doi.org/10.7326/M19-0618

131. Yan, Z., Liu, Y., Zhang, T., Zhang, F., Ren, H., and Zhang, Y. (2022). Analysis of microplastics in human feces reveals a correlation between fecal microplastics and inflammatory bowel disease status. *Environmental Science & Technology*, **56**(1), 414–421. https://doi.org/10.1021/acs.est.1c03924

132. Ibrahim, Y. S., Tuan Anuar, S., Azmi, A. A., Wan Mohd Khalik, W. M. A., Lehata, S., Hamzah, S. R., Ismail, D., Ma, Z. F., Dzulkarnaen, A., Zakaria, Z., Mustaffa, N., Tuan Sharif, S. E., and Lee, Y. Y. (2021). Detection of microplastics in human colectomy specimens. *JGH Open*, **5**(1), 116–121. https://doi.org/10.1002/jgh3.12457

133. Brouwer, H., Porbahaie, M., Boeren, S., Busch, M., and Bouwmeester, H. (2024). The in vitro gastrointestinal digestion-associated protein corona of polystyrene nano- and microplastics increases their uptake by human THP-1-derived macrophages. *Particle and Fibre Toxicology*, **21**(1). https://doi.org/10.1186/s12989-024-00563-z

134. Marfella, R., Prattichizzo, F., Sardu, C., Fulgenzi, G., Graciotti, L., Spadoni, T., D'Onofrio, N., Scisciola, L., La Grotta, R., Frigé, C., Pellegrini, V., Municinò, M., Siniscalchi, M., Spinetti, F., Vigliotti, G., Vecchione,

C., Carrizzo, A., Accarino, G., Squillante, A., Spaziano, G., Mirra, D., Esposito, R., Altieri, S., Falco, G., Fenti, A., Galoppo, S., Canzano, S., Sasso, F. C., Matacchione, G., Olivieri, F., Ferraraccio, F., Panarese, I., Paolisso, P., Barbato, E., Lubritto, C., Balestrieri, M. L., Mauro, C., Caballero, A. E., Rajagopalan, S., Ceriello, A., D'Agostino, B., Iovino, P., and Paolisso, G. (2024). Microplastics and nanoplastics in atheromas and cardiovascular events. *New England Journal of Medicine*, **390**(10), 900–910. https://doi.org/10.1056/nejmoa2309822

135. Horvatits, T., Tamminga, M., Liu, B., Sebode, M., Carambia, A., Fischer, L., P€, K., Huber, S., and Fischer, E. K. (2022). Microplastics detected in cirrhotic liver tissue. *EbioMedicine*, **82**, 104147. doi: 10.1016/j.ebiom.2022.104147

136. Nugent, S. G., Kumar, D., Rampton, D. S., and Evans, D. F. (2001). Intestinal luminal pH in inflammatory bowel disease: Possible determinants and implications for therapy with aminosalicylates and other drugs. *Gut*, **48**(4), 571. https://doi.org/10.1136/gut.48.4.571

137. Piccardo, M., Provenza, F., Grazioli, E., Cavallo, A., Terlizzi, A., and Renzi, M. (2020). PET microplastics toxicity on marine key species is influenced by pH, particle size and food variations. *Science of the Total Environment*, **715**. https://doi.org/10.1016/j.scitotenv.2020.136947

138. OECD. (2022). *Plastic pollution is growing relentlessly as waste management and recycling fall short* Press release, OECD.

139. Vogt, B. D., Stokes, K. K., and Kumar, S. K. (2021). Why is recycling of postconsumer plastics so challenging? *ACS Applied Polymer Materials*, **3**(9), 4325–4346. https://doi.org/10.1021/acsapm.1c00648

140. Costa, J. P. da, Avellan, A., Mouneyrac, C., Duarte, A., and Rocha-Santos, T. (2023). Plastic additives and microplastics as emerging contaminants: Mechanisms and analytical assessment. *TrAC - Trends in Analytical Chemistry*, **158**. https://doi.org/10.1016/j.trac.2022.116898

141. Ramos, T. M., Christensen, T. B., Bour, A., Almroth, B. C., Kristensen, D. M., Selck, H., and Syberg, K. (2023). A not so circular healthcare economy: A review of challenges with plastic associated chemicals. *TrAC - Trends in Analytical Chemistry*, **166**. https://doi.org/10.1016/j.trac.2023.117191

142. Warner, G. R., and Flaws, J. A. (2018). Bisphenol A and phthalates: How environmental chemicals are reshaping toxicology. *Toxicological Sciences*, **166**(2), 246–249. https://doi.org/10.1093/toxsci/kfy232

143. UNEP. (2022). *Global Plastics Treaty: Transparency Requirement for Chemical Constituents in Plastic is a Must*. https://apps1.unep.org/resolutions/uploads/submission_chemical_transparency.pdf

144. ECHA. (2019). *ECHA checked over 20 % of REACH registration dossiers for compliance.* https://Echa.Europa.Eu/Sv/-/Echa-Checked-over-20-of-Reach-Registration-Dossiers-for-Compliance-1. https://echa.europa.eu/sv/-/echa-checked-over-20-of-reach-registration-dossiers-for-compliance-1

145. Horodytska, O., Cabanes, A., and Fullana, A. (2020). Non-intentionally added substances (NIAS) in recycled plastics. *Chemosphere*, **251**. https://doi.org/10.1016/j.chemosphere.2020.126373

146. He, Y. J., Qin, Y., Zhang, T. L., Zhu, Y. Y., Wang, Z. J., Zhou, Z. S., Xie, T. Z., and Luo, X. D. (2021). Migration of (non-) intentionally added substances and microplastics from microwavable plastic food containers. *Journal of Hazardous Materials*, **417**. https://doi.org/10.1016/j.jhazmat.2021.126074

147. Neves, S. A., and Marques, A. C. (2022). Drivers and barriers in the transition from a linear economy to a circular economy. *Journal of Cleaner Production*, **341**. https://doi.org/10.1016/j.jclepro.2022.130865

148. European Commission. (2024). Guiding criteria and principles for the essential use concept in EU legislation dealing with chemicals, *Communication from the Commission*. https://doi.org/10.2779/529713

149. Aanesen, M., Ahi, J. C., Abate, T. G., Khan, F. R., de Vries, F. P., Kite-Powell, H., and Beaumont, N. J. (2024). Insights from international environmental legislation and protocols for the global plastic treaty. *Scientific Reports*, **14**(1). https://doi.org/10.1038/s41598-024-53099-9

150. European Union. (2019). *Directive (EU) 2019/904 of the European Parliament and of the Council of 5 June 2019 on the reduction of the impact of certain plastic products on the environment.* https://eur-lex.europa.eu/legal-content/EN/TXT/PDF/?uri=CELEX:32019L0904

151. Orndoff, D., Beymer-Farris, B., Wood, M., Sadler, J., Ternes, M. E., Law, W., Hester, T., Miller, K. M., and Seay, J. (2023). A review of additive usage in polymer manufacturing: Case study phenolic antioxidants. *Pre-Print.* https://doi.org/10.21203/rs.3.rs-3639945/v1

152. Cousins, I. T., Dewitt, J. C., Glüge, J., Goldenman, G., Herzke, D., Lohmann, R., Miller, M., Ng, C. A., Scheringer, M., Vierke, L., and Wang, Z. (2020). Strategies for grouping per- and polyfluoroalkyl substances (PFAS) to protect human and environmental health. *Environmental Science: Processes and Impacts*, **22**(7), 1444–1460. https://doi.org/10.1039/d0em00147c

Chapter 9

Microplastics as Potential Vector of Antibiotics in Aquatic Media: Environmental Implications

Carmen Mejías, Marina Arenas, Julia Martín, Juan Luis Santos, Irene Aparicio, and Esteban Alonso
Departamento de Química Analítica, Escuela Politécnica Superior, Universidad de Sevilla. C/ Virgen de África, 7, E-41011 Seville, Spain
jbueno@us.es

9.1 Introduction

Microplastics (MPs) are plastic particles that have a particle size smaller than 5 mm [1] and, due to their low biodegradability, have gained the attention of the scientific community and regulatory authorities, thanks to their potential risks to organisms and ecosystems. Some studies [1–3] have reported that MPs, because of their small particle size, large specific surface area, and strong hydrophobicity, could adsorb and concentrate organic pollutants from the aquatic and soil media. Moreover, during the co-exposure of MPs and organic pollutants, MPs could transfer organic pollutants into organisms and promote their bioaccumulation [4].

Microplastic Pollution: Environmental Impact and Solutions
Edited by Julia Martín, Juan Luis Santos, Irene Aparicio, and Esteban Alonso

ISBN 978-981-5129-77-9 (Hardcover), 978-1-003-66006-4 (eBook)
www.jennystanford.com

DOI: 10.1201/9781003660064-9

In the last few decades, the accumulation of pharmaceuticals in aquatic environments has received growing attention. Among them, antibiotics are considered a big threat to human health due to their widespread presence in the environment, which facilitates the development of antibiotic-resistant genes [5] and promotes the widespread existence of antibiotic-resistant bacteria [6]. Moreover, antibiotics are one of the most problematic organic pollutants in association with MPs [5, 7–11]. Some studies have shown that MPs could serve as carriers or vectors of gene transfer by facilitating gene exchange between different bacteria and introducing and supporting pathogenicity and antibiotic resistance in the environment.

Therefore, this chapter reviews the distribution and occurrence of both MPs and antibiotics in the aquatic environment, the factors influencing the sorption of antibiotics onto MPs, as well as their adsorption mechanisms. Finally, interactive toxicity to aquatic organisms has been also considered.

9.2 Distribution and Occurrence of MPs and Antibiotics in the Aquatic Environment

The distribution and occurrence of MPs and antibiotics in the aquatic environment have become a major concern in recent years due to their potential impact on ecosystems and human health [12]. MPs have been found in different aquatic environments worldwide, including marine environments, rivers, lakes, and even groundwater (Fig. 9.1).

Figure 9.1 Aquatic environments where MPs have been found.

Table 9.1 summarizes the abundance and type of MPs reported in different aquatic environments over the most recent years. An increasing trend can be observed in their abundance detected in different aquatic environments, with generally higher concentrations in rivers and groundwater, followed by lakes and, finally, marine environments.

Table 9.1 Occurrence of MPs in different aquatic environments

Aquatic Environment	MP Type	Abundance (particles/L)	Year	Ref.
Seawater	PE, PP	0.00125 ± 0.00088	2020	[13]
Seawater	PE	0.00130 ± 0.00098	2021	[14]
Seawater	PET, PP, PS, PVC	0.000048–0.0033	2022	[15]
Seawater	PE, PET, PS	0.103 ± 0.0983	2023	[16]
River	PP	0.912–3.395	2020	[17]
River	PE, PVC	0.133–5.467	2021	[18]
River	PA, PE, PET PP, PS, PVC	2.3–21.05	2022	[19]
Lake	PET	0.0009 ± 0.0006	2020	[20]
Lake	PE, PP	0.0119–0.0612	2021	[21]
Lake	PET	0.117–0.423	2022	[22]
Lake	PE, PET, PP	8.770 ± 0.270	2023	[23]
Groundwater	PE, PP	3.0–23.0	2021	[24]
Groundwater	PE, PET, PS	17.0–44.0	2022	[25]
Groundwater	PS	34.0–64.0	2023	[26]

PA: polyamide; PE: polyethylene; PET: polyethylene terephthalate; PP: polypropylene; PS: polystyrene; PVC: polyvinylchloride

Plastic represents approximately 60–80% of the total global waste, and almost 10% of the production ends up in the oceans [27]. The main sources of plastic entering the ocean include beaches [5] and terrestrial sources such as rivers [28], stormwater run-off [29], sewage discharges [30], wind-driven transport of terrestrial waste [31], and also maritime activities such as fishing and waste discharged due to maritime traffic [32, 33]. The highest concentrations were found in groundwater (64.0 particles/L). These

MPs can originate from a variety of sources, such as the breakdown of larger plastic debris [17, 30], the abrasion of synthetic textiles [34], and the fragmentation of plastic products [19, 35]. Once in the aquatic environment, MPs can persist for long periods and can be ingested by a wide range of aquatic organisms, including fish [36], shellfish [37], and plankton [38]. This ingestion can lead to blockages in digestive systems, oxidative and pathological stress, growth inhibition, and reproductive disorders [13]. Furthermore, due to distinctive characteristics such as hydrophobicity, MPs can adsorb a greater amount of contaminants [39], including persistent organic pollutants such as pharmaceutical products. This heightened adsorption increases the toxic effects of MPs on organisms [40].

Similarly, antibiotics have been detected in the same aquatic environments due to their widespread use in human and veterinary medicine [41]. Table 9.2 presents the concentration of several antibiotics documented in different aquatic environments in recent years. The highest concentrations were found in rivers (76,800 ng/L).

Table 9.2 Occurrence of antibiotics in different aquatic environments

Aquatic Environment	Antibiotic	Concentration (ng/L)	Year	Ref.
Seawater	AMOX, AZI, CAP, CIP, ENR, ERI, FF, LCM, NOR, OFL, RTM, SDZ, SMM, TMP	1.60–49.2	2020	[42]
Seawater	CIP, DOTC, ENR, NOR, OFL, OTC, SAR, SDZ, SMM, SMX, SMZ SQX, TC, TMP	0.35–301	2021	[43]
Seawater	CIP, DAN, DIF, ENO, ENR, FLE, FLU, LOM, NOR, OFL, ORB, OXO, PEF, SAR, SPA	18.0–68.0	2022	[44]
Seawater	AMOX, AMP, CIP, DAN, ERI, OFL, SDZ, SMX, SMZ, SPD, STZ	0.14–70.9	2023	[45]
River	AZI, CEP, CIP, CLI, NOR, SDZ, SMX, TMP	20.0–344	2020	[46]

Aquatic Environment	Antibiotic	Concentration (ng/L)	Year	Ref.
River	CIP, ENR, SMX, TMP	40.2–2188	2021	[47]
River	CIP, ERI, LOM, NOR, OTC, SDZ, SMX, SMZ, SPD, TC, TMP	1.38–54.9	2022	[48]
River	AMOX, AMP, AZM, CIP, DOTC, ERI, GEN, MET, NOR, OFL, SMX, SPD, SMZO, TC, TMP	8–76,800	2023	[49]
Lake	CAP, ENR, ERI, FF, RTM, SDZ, SMM, SMX, SMZ, TMP	1.06–66.3	2020	[50]
Lake	CIP, CTC, ENR, ERI, NOR, OFL, OTC, RTM, SDZ, SMZ, SMX, TC	0.04–95.4	2021	[51]
Lake	AMP, CIP, DIF, ERI, MTC, NOR, OFL, PNC, SDZ, SMX, SMZ, TC, TMP	1.0–892	2022	[52]
Lake	CTC, DOTC, ERI, ENR, NOR, OFL, RTM, SDZ, SME, SMX, TC	0.13–259	2023	[53]
Groundwater	CIP, DIF, ENR, LOM, NOR, OFL	1.8–84.9	2020	[54]
Groundwater	CTC, DOTC, OTC, SMX, SMZ, TC	21.0–365	2021	[55]
Groundwater	AZM, CIP, ENR, ERI, NOR, OFL, SMX, SMZ, TMP	0.47–20.5	2022	[56]

AMOX: amoxicillin; AMP: ampicillin; AZM: azithromycin; CAP: chloramphenicol; CEP: cephalexin; CIP: ciprofloxacin; CLI: clindamycin; CTC: chlortetracycline; DAN: danofloxacin; DIF: difloxacin; DOTC: doxycycline; ERI: erythromycin; ENO: enoxacin; ENR: enrofloxacin; FF: florfenicol; FLE: fleroxacin; FLU: flumequine; GEN: gentamicin; LOM: lomefloxacin; LCM: leucomycin; MET: metronidazole; MTC: metacycline; NOR: norfloxacin; OFL: ofloxacin; OTC: oxytetracycline; ORB: orbifloxacin; OXO: oxolinic acid; PNC: penicillin; PEF: pefloxacin; RTM: roxithromycin; SAR: sarafloxacin; SDZ: sulfadiazine; SME: sulfameter; SMM: sulfamonomethoxine; SMX: sulfamethoxazole; SMZ: sulfamethazine; SMZO: sulfamethizole; SPA: sparfloxacin; SPD: sulfapyridine; SQX: sulfaquinoxaline; STZ: sulfathiazole; TC: tetracycline; TMP: trimethoprim

Antibiotics originate from various sources, including manufacturing facilities, discharge from hospitals and households, livestock farming, as well as run-off from aquaculture and agricultural sites [57]. Once in the aquatic environment, antibiotics can persist for extended periods and can exert selective pressure on bacteria, leading to the development and spread of antibiotic-resistant bacteria [46, 51]. This resistance poses a significant threat to public health as it can render antibiotics ineffective in treating bacterial infections in humans and animals [50].

9.3 Sorption of Antibiotics

To study the ability of MPs to act as vectors of antibiotics in the environment, laboratory-scale tests are carried out to study the effect of different parameters. The main thing to start is to characterize the MP through different characterization techniques to know in depth its physicochemical characteristics. Common material characterization techniques are used to obtain information on the morphological and structural characteristics of the MP. They include nitrogen adsorption–desorption isotherms, scanning electron microscopy, transmission electron microscopy, Fourier transform infrared spectroscopy, X-ray diffraction, zeta potential and point of zero charge analyses. The main factors affecting the sorption of antibiotics onto MPs can be divided into three categories: (a) MPs weathering factors, (b) physicochemical properties of antibiotics, and (c) environmental factors. To understand their effects on the adsorption, the most common approach is to perform batch experiments in which all variables are kept constant except for one, which is the one that is modified. Figure 9.2 shows a general outline of the tests that are performed and the variables that are usually studied.

Table 9.3 summarizes 39 laboratory studies that have investigated the sorption of antibiotics onto virgin and aged MPs. As shown in Table 9.3, polyethylene (PE) ($n = 19$) was the most common polymer studied in laboratory experiments followed by polystyrene (PS) ($n = 18$), polyvinylchloride (PVC) ($n = 13$), polypropylene (PP) ($n = 9$), polyamide (PA) ($n = 8$), polyethylene terephtalate (PET) ($n = 8$), polylactic acid (PLA) ($n = 5$), polyoxymethyelene (POM) ($n = 1$), and polybutylene succinate (PBS) ($n = 1$). As can be observed, fluoroquinolones, sulfonamides, and tetracyclines were the most studied antibiotic groups.

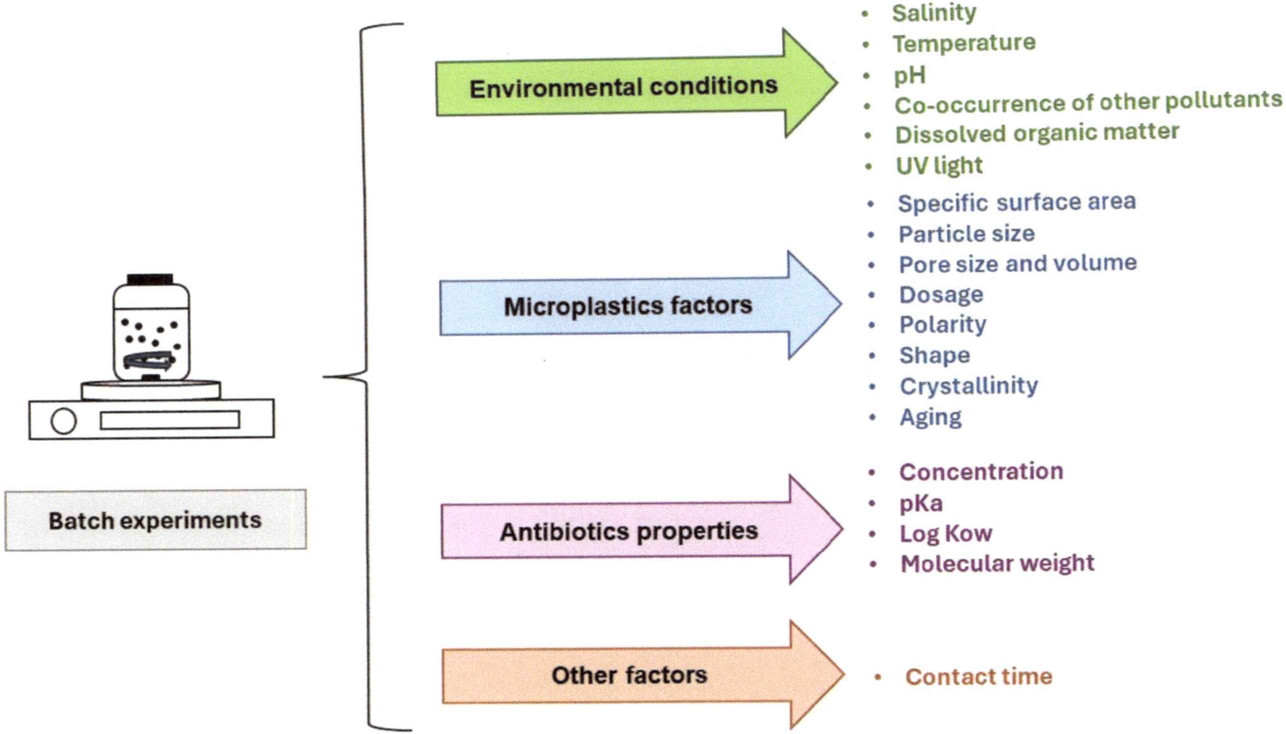

Figure 9.2 Scheme of typical batch adsorption experiments.

Table 9.3 Sorption studies of antibiotics onto MPs

MP Type	Antibiotics	Variables Tested	Type of Interaction	Ref.
PE, PS, PA, PP, PVC	AMOX, CIP, SDZ, TC, TMP	Freshwater and seawater systems	Porous structure and hydrogen bonding	[11]
PE	CIP	pH, salinity, and DOM	Hydrophobic and electrostatic interactions	[1]
PS	OTC	pH, salinity, and DOM	Electrostatic interaction, H-bonding, and multivalent cationic bridging	[58]
PLA, PVC	CIP, TC	Ageing and desorption	Hydrogen, π-π bonding and electrostatic interactions	[59]
PE	CTC, OTC, TC	pH and salinity	Non-bond interactions	[60]
PET, PLA, POM, PS	AZM, CLA	Desorption and effect on cyanobacterium	Hydrophobic interactions	[61]
PET, PLA, PP	SMX	Ageing, pH, salinity, and DOM	Hydrophobic interaction or electrostatic interaction	[62]
PP, PE, PVC	CIP, ENR, NOR	pH and salinity	Hydrophobic interaction and electrostatic interaction	[63]
PP	CAP, CIP, ENR, NOR, OFL, SMZ, SMX, SRZ, STZ, TC	Ageing	Hydrophobic and hydrogen bonding interaction for fresh PP and electrostatic interaction and hydrogen bonding for aged	[64]
PE, PET, PP, PS, PVC	AMOX, VAN	pH and temperature	Electrostatic interactions	[65]
PE	SMX	-	Hydrophobicity and electrostatic interactions	[66]

MP Type	Antibiotics	Variables Tested	Type of Interaction	Ref.
PE, PS, PBS	NOR	Temperature, pH, DOM, heavy metal ions and salinity	π-π conjugation, intermolecular hydrogen bonds, ion exchange, and electrostatic interactions	[8]
PE	LEV, (*R*)-OFL	Particle size, pH, salinity, DOM, and ageing	Hydrogen-bond interactions	[67]
PS	ERI, SMZ, TYL	Ageing	Hydrogen-bond interactions	[68]
PE, PP, PS, PVC	TYL	pH, salinity	Electrostatic interaction, surface complexation, and hydrophobic interaction	[69]
PS, PE	CPC, SMZ, SMX	Freshwater and simulated seawater	Hydrophobic, van der Waals, and electrostatic interactions	[70]
PET, PE, PA, PS, PVC, PP	SMZ	pH, salinity	Electrostatic and van der Waals interaction	[71]
PA, PE, PET, PS, PVC, PP	SMX	pH, salinity	Electrostatics interactions	[72]
PA	TC	Metal ions, salinity, pH, temperature, particle size	van der Waals forces	[73]
PS, PVC	CIP	Salinity, pH, ageing, desorption	Partitioning, electrostatic interactions, and intermolecular hydrogen bonding	[74]
PE	TC	DOM, phthalates, ageing, Cu	Complexation, H-bond, electrostatic interaction, and hydrophobic interaction	[9]

(*Continued*)

Table 9.3 (*Continued*)

MP Type	Antibiotics	Variables Tested	Type of Interaction	Ref.
PS, PVC	CIP	Ageing, salinity, pH, DOM, cations, particle size	Electrostatic interaction and hydrogen bonding	[75]
PE	SMX	pH, DOM, and salinity	van der Waals interaction	[76]
PS	CIP	Particle size, pH	Electrostatic interactions, hydrogen bonding, and hydrophobic interactions	[77]
PE, PS, PVC	TC	Salinity, metal ions, particle size	Hydrogen bonding, surface properties	[78]
PE, PP, PS	TC	pH, DOM, salinity	Hydrophobic interactions, electrostatic interactions, and π-π interactions	[79]
PE	TC	pH, Cu, desorption	Hydrophobic interaction, electrostatic interactions, and π-π interactions	[80]
PS	NOR levofloxacin	pH, DOM, salinity	Polar interaction, electrostatic interaction, and hydrogen bonding	[81]
PA	SMX	Ageing	Electrostatic interactions	[82]
PVC	TC	pH, salinity, Cu	Hydrophobic interactions	[83]
PLA, PE	CIP, NOR	Ageing, temperature, pH, DOM, salinity, desorption	Hydrogen bonding, π-π conjugation, ion exchange, and electrostatic interactions	[84]
PET	AMOX	Ageing	Hydrogen-bond interactions	[85]

MP Type	Antibiotics	Variables Tested	Type of Interaction	Ref.
PLA	CIP	Desorption, ageing time, particle size, and dosage	Hydrogen-bond interactions	[86]
PVC	CLA, CIP	Temperature, pH	Hydrophobic and electrostatic interactions	[87]
PA	SMX	Desorption, pH, salinity, gastrointestinal adsorption	Hydrogen-bond and electrostatic interactions	[88]
PA, PVC	CIP, TC	Heavy metals and pH	Hydrophobic interactions, electrostatic interactions, and negative charge-assisted hydrogen bonds	[89]
PS, PET	TC	Ageing, Cu, pH, desorption, temperature	Physical interactions	[90]
PE, PET	AZM, CLA, ERI, RTM, desmethyl CLA	pH, salinity, DOM, particle size, dosage, real water samples	Hydrophobic interactions, pore filling, and hydrogen bonding	[91]
PA, PS	SMX, SMZ, SDZ, acetyl SMX, acetyl SMZ, acetyl SDZ	pH, salinity, DOM, particle size, dosage, real water samples	Hydrophobic interactions, pore filling, and hydrogen bonding	[92]

AMOX: amoxicillin; AZM: azithromycin; CAP: chloramphenicol; CEP: cephalexin; CIP: ciprofloxacin; CLA: clarithromycin; CPC: cephalosporin C; CTC: chlortetracycline; DOM: dissolved organic matter; ERI: erythromycin; ENR: enrofloxacin; LEV: levofloxacin; NOR: norfloxacin, OFL: ofloxacin; OTC: oxytetracycline; PA: polyamide; PBS: polybutylene succinate; PE: polyethylene; PET: polyethylene terephthalate; PLA: polylactic acid; POM: polyoxymethylene; PP: polypropylene; PS: polystyrene; PVC: polyvinylchloride; RTM: roxithromycin; SDZ: sulfadiazine; SMX: sulfamethoxazole; SMZ: sulfamethazine; SRZ: sulfamerazine; STZ: sulfathiazole; TC: tetracycline; TMP: trimethoprim; TYL: tylosin; VAN: vancomycin

9.3.1 Polymer Factors

The main factors affecting the sorption of antibiotics onto MPs can be MP surface such as surface functional groups as well as textural characteristics. Several characteristics of MPs, such as size, specific surface area, pore size distribution, pore volume, dosage, polarity, shape, and degree of crystallinity as well as ageing, affect antibiotic adsorption. Therefore, the elucidation of those characteristics is essential to further evaluate adsorption. In addition, MP ageing changes the MP surface, thus also affecting the adsorption process. Zheng et al. [87] indicated that the reduction in MP particle size, which is highly related to specific surface area, increased the adsorption of ciprofloxacin (CIP) onto PLA MPs. Pore size distribution and pore volume are the main characteristics affecting the pore-filling mechanism. In addition, polarity and degree of crystallinity are crucial to elucidate the adsorption mechanism. It is essential to study the ageing of MPs because their structure can be modified due to UV radiation and water currents. Xiong et al. [77] studied the influence of ageing on the adsorption of CIP onto PS MPs and revealed that adsorption capacity at equilibrium increased from 0.15 to 4.07 mg/g after weathering.

9.3.2 Antibiotic Factors

The main factors affecting the sorption of antibiotics onto MPs can be the physicochemical properties of antibiotics such as hydrophobicity, molecular weight, size, and pK_a. These properties are essential as they are the main factors affecting the capacity of antibiotics to form hydrogen bonding with polar MPs (such as PA) or hydrophobic interactions with non-polar MPs (such as PE). Tong et al. [90] concluded that due to pK_a, zwitterionic forms of tetracycline (TC) and CIP prevailed in aqueous solutions within the pH ranges of 4–7 and 6–8, respectively. At pH values outside these ranges, TC and CIP predominantly existed in their cationic or anionic forms. Consequently, the highest adsorption of CIP onto PA and PVC was observed at pH 6, likely due to the minimal electrostatic repulsion between the MP surface and the CIP zwitterion at a pH near CIP pK_a (pK_{a1} = 6.05). This indicates that electrostatic interactions are crucial for CIP adsorption by MPs. In contrast, the lowest adsorption capacity for TC on MPs occurred at pH 6, with increased adsorption at both higher and lower pH levels. Alkaline conditions favoured TC adsorption more than acidic conditions, suggesting that the pH dependence of TC adsorption onto MPs cannot be solely attributed to electrostatic interactions.

9.3.3 Environmental Factors

The main factors affecting the sorption of antibiotics onto MPs can be environmental conditions such as salinity, temperature, pH, co-occurrence of other pollutants, and dissolved organic matter (DOM). Therefore, these variables are usually tested in adsorption assays, as can be seen from Table 9.3. Salinity is an environmental factor that could influence the adsorption of different pollutants. The salinity effect was studied to simulate the salinity in aquatic environments, such as rivers, lakes, and oceans. In most adsorption cases, an increase in ionic strength will reduce the adsorption capacity of antibiotics on MPs due to the competition of ionic strength with antibiotics for adsorption sites, to the less formation of hydrogen bonds for the substitution of H^+ for inorganic exchangeable cations, and an increase in chlorine atom increases the cohesion energy, thus making the diffusion between materials more difficult. Environmental water samples possess a different range of pH. Solution pH affects both the surface charge of the adsorbent material and the speciation of the antibiotic molecules. When MPs and antibiotics have the same charge, they repeal, but when they have different charges, they attract. The co-occurrence of other pollutants and DOM can compete with antibiotics for the adsorption-active sites on the MP surface and for interacting with antibiotics, preventing their adsorption on MPs.

9.4 Interactions between Microplastics and Antibiotics: Sorption Mechanisms

The adsorption mechanism is used to describe the types of molecular interactions taking place during the adsorption process. Figure 9.3 represents the main adsorption mechanisms. As can be seen, the main adsorption mechanisms are hydrophobic interactions, electrostatic interactions, partitioning, π-π bonding, pore filling, hydrogen bonding, and van der Waals forces. The adsorption mechanism depends on antibiotics and MPs' physicochemical characteristics. Several adsorption mechanisms can occur simultaneously; for example, Wang et al. [80] reported that the antibiotic TC adsorbed on PE MPs by hydrophobic interaction, electrostatic interactions, and π-π interactions. Common material characterization techniques are used to obtain information on the morphological and structural characteristics of the MP and to elucidate the type of interaction between the MPs and antibiotics. They include nitrogen adsorption–desorption isotherms, scanning electron microscopy, transmission electron microscopy, Fourier transform infrared spectroscopy, X-ray diffraction, zeta potential, and point of zero charge analyses. Therefore, to further evaluate adsorption mechanisms between MPs and antibiotics, these techniques should be employed. Unfortunately, this evaluation should be performed for every antibiotic with every MP as no suitable prediction tool is available.

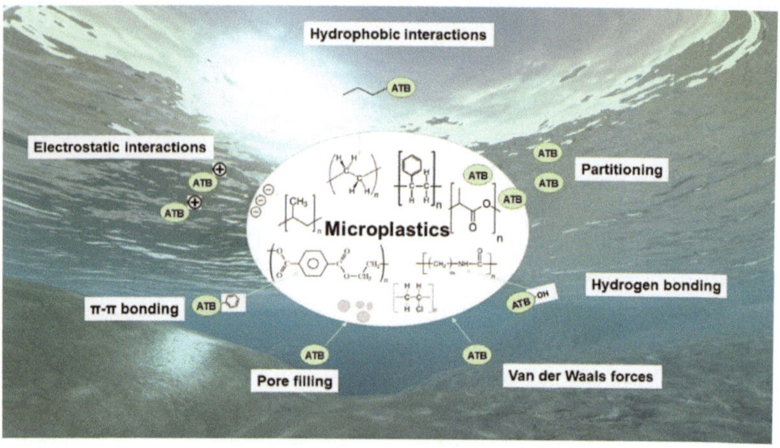

Figure 9.3 Adsorption mechanism of antibiotics onto MPs.

9.5 Interactive Toxicity

MPs, as shown by the sorption studies mentioned earlier, can act as vectors for antibiotics, which lead to higher concentrations of these pharmaceuticals on their surfaces compared to the surrounding environment [4]. Once MPs with sorbed antibiotics at these higher concentrations are ingested by aquatic organisms, they increase antibiotic bioaccumulation along the food chain. Consequently, antibiotic toxicity and adverse effects can be significantly enhanced [93].

More recently, there has also been increasing concern about the interactive toxicity of MPs and antibiotics. Despite their potential interaction, little research has been conducted on the combined toxicity of antibiotics and MPs in aquatic systems. The effects produced when antibiotics are attached to the surface of MPs can be additive, antagonistic, or synergistic. Understanding the overall toxicity of this mixture is still limited, as the specific effects of these interactions remain unclear [94]. Figure 9.4 shows a scheme of interactive toxicity tests of MPs and antibiotics on different aquatic organisms.

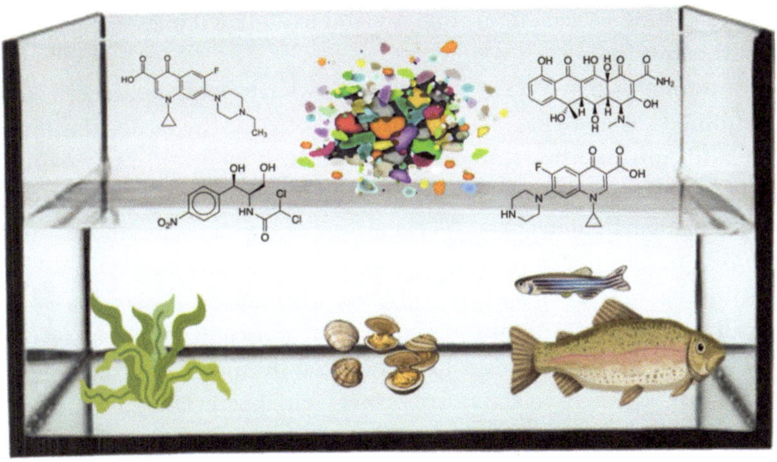

Figure 9.4 Scheme of interactive toxicity tests of MPs and antibiotics conducted on algae, bivalves, and fish.

According to some research, the presence of both MPs and antibiotic pollutants in aquatic ecosystems could lead to further damage. Feng et al. [95] reported increased acute toxicity of PS in the presence of TC on studies performed in marine microalgae. Li et al. [96] reported that the co-exposure of PS and chloramphenicol induced higher toxicity than the single exposure in freshwater algae, increasing photosynthetic inhibition, oxidative stress, and cell membrane damage. Zhou et al. [97] found that the immunotoxicity of oxytetracycline and florfenicol to bivalve was potentially aggravated by the co-presence of PS. Banaee et al. [98] concluded that PE could have synergistic effects on enrofloxacin toxicity in freshwater fish, as co-exposure resulted in more pronounced toxicological responses in parameters such as oxidative biomarkers than exposure to enrofloxacin itself.

Other authors, in contrast, have reported antagonistic effects on co-exposure to MPs and antibiotics for certain aquatic organisms, leading to a decreased interactive toxicity. Zhu et al. [99] assessed, through oxidative stress assays in marine microalgae, that the joint toxicity of PVC and triclosan was reduced due to the high adsorption capacity of these MPs, which lowers the concentration of triclosan in water. Guo et al. [100] studied the toxicity of CIP in the presence of PS, finding a toxicity reduction in the digestive glands of freshwater bivalves. Zhang et al. [101] reported that the presence of PS could mitigate the neurotoxicity and oxidative damage caused by roxithromycin in freshwater fish.

Although these recent investigations have been conducted, the combined toxic effects of MPs and antibiotics at the molecular level are still not fully understood and seem to depend on interactions between MPs and antibiotics. In addition to the combined toxic effects, there are particular concerns regarding the potential contribution of MPs to the development of antibiotic-resistant bacteria. When antibiotics are adsorbed by MPs, the adsorption process creates localized regions with elevated antibiotic concentrations, exposing bacteria to sub-lethal doses for prolonged periods. This continuous exposure causes selective pressure on bacterial populations, facilitating the survival and proliferation of antibiotic-resistant strains [94].

9.6 Final Considerations

The interactions between MPs and antibiotics have become a cause of concern. It seems appropriate to conduct laboratory experiments and parallel field experiments using real environmental samples such as seawater, surface water, or wastewater to evaluate real environmental adsorption of antibiotics on MPs as well as multisolute systems. As fibres are the main MPs found in the environment, more adsorption experiments should be performed using this shape of MPs. The shape of MPs has not been evaluated previously, so differences in the adsorption of pellets and fibres should be performed. Antibiotic metabolites could be more toxic than their parent compounds; therefore, adsorption studies including antibiotic metabolites should be considered in the future in addition to antibiotic chirality.

Acknowledgements

This work was financially supported by the Ministerio de Ciencia e Innovación-Agencia Estatal de Investigación (grant number: PID2020-117641RB-I00, funded by MCIN/AEI/10.13039/501100011033) and by Junta de Andalucía, Consejería de Economía, Conocimiento, Empresas y Universidad (Project No. P20_00556). C Mejías and M Arenas gratefully acknowledge the University of Seville and the Spanish Ministerio de Universidades for their predoctoral contracts (grant numbers VIPPIT-US2021II.2A and FPU20/00540, respectively).

References

1. Atugoda, T., Vithanage, M., Wijesekara, H., Bolan, N., Sarmah, A.K., Bank, M.S., You, S., and Ok, Y.S. (2021). Interactions between microplastics, pharmaceuticals and personal care products: Implications for vector transport. *Environ. Int.*, **149**, 106367.
2. Ahmad, M., Li, J.L., Wang, P.D., Hozzein, W.N., and Li, W.J. (2020). Environmental perspectives of microplastic pollution in the aquatic environment: A review. *Mar. Life Sci. Technol.*, **2**, 414–430.
3. Arienzo, M., Ferrara, L., and Trifuoggi, M. (2021). The dual role of microplastics in marine environment: Sink and vectors of pollutants. *J. Mar. Sci. Eng.*, **9**, 642.
4. Xiang, Y., Jiang, L., Zhou, Y., Luo, Z., Zhi, D., Yang, J., and Lam, S.S. (2022). Microplastics and environmental pollutants: Key interaction and toxicology in aquatic and soil environments. *J. Hazard. Mater.*, **422**, 126843.
5. Stenger, K.S., Wikmark, O.G., Bezuidenhout, C.C., and Molale-Tom, L.G. (2021). Microplastics pollution in the ocean: Potential carrier of resistant bacteria and resistance genes. *Environ. Pollut.*, **291**, 118130.

6. Gui, B., Xu, X., Zhang, S., Wang, Y., Li, C., Zhang, D., Su, L., and Zhao, Y. (2021). Prediction of organic compounds adsorbed by polyethylene and chlorinated polyethylene microplastics in freshwater using QSAR. *Environ. Res.*, **197**, 111001.
7. Elgarahy, A.M., Akhdhar, A., and Elwakeel, K.Z. (2021). Microplastics prevalence, interactions, and remediation in the aquatic environment: A critical review. *J. Environ. Chem. Eng.*, **9**, 106224.
8. Sun, M., Yang, Y., Huang, M., Fu, S., Hao, Y., Hu, S., Lai, D., and Zhao, L. (2022). Adsorption behaviors and mechanisms of antibiotic Norfloxacin on degradable and nondegradable microplastics. *Sci. Total Environ.*, **807**, 151042.
9. Wang, Y., Yang, Y., Liu, X., Zhao, J., Liu, R., and Xing, B. (2021). Interaction of microplastics with antibiotics in aquatic environment: Distribution, adsorption, and toxicity. *Environ. Sci. Technol.*, **55**, 15579–15595.
10. Syranidou, E., and Kalogerakis, N. (2022). Interactions of microplastics, antibiotics and antibiotic resistant genes within WWTPs. *Sci. Total Environ.*, **804**, 150141.
11. Li, J., Zhang, K., and Zhang, H. (2018). Adsorption of antibiotics on microplastics. *Environ. Pollut.*, **237**, 460–467.
12. Khare, T., Mathur, V., and Kumar, V. (2024). Agro-ecological microplastics enriching the antibiotic resistance in aquatic environment. *Curr. Opin. Environ. Sci. Heal.*, **37**, 100534.
13. Robin, R.S., Karthik, R., Purvaja, R., Ganguly, D., Anandavelu, I., Mugilarasan, M., and Ramesh, R. (2020). Holistic assessment of microplastics in various coastal environmental matrices, southwest coast of India. *Sci. Total Environ.*, **703**, 134947.
14. Expósito, N., Rovira, J., Sierra, J., Folch, J., and Schuhmacher, M. (2021). Microplastics levels, size, morphology and composition in marine water, sediments and sand beaches. Case study of Tarragona Coast (Western Mediterranean). *Sci. Total Environ.*, **786**, 147453.
15. Abelouah, M.R., Ben-Haddad, M., Hajji, S., De-la-Torre, G.E., Aziz, T., Oualid, J.A., Banni, M., and Ait Alla, A. (2022). Floating microplastics pollution in the central Atlantic Ocean of Morocco: Insights into the occurrence, characterization, and fate. *Mar. Pollut. Bull.*, **182**, 113969.
16. Liu, B., Lu, Y., Deng, H., Huang, H., Wei, N., Jiang, Y., Jiang, Y., Liu, L., Sun, K., and Zheng, H. (2023). Occurrence of microplastics in the seawater and atmosphere of the South China Sea: Pollution patterns and interrelationship. *Sci. Total Environ.*, **889**, 164173.

17. Zhou, G., Wang, Q., Zhang, J., Li, Q., Wang, Y., Wang, M., and Huang, X. (2020). Distribution and characteristics of microplastics in urban waters of seven cities in the Tuojiang River Basin, China. *Environ. Res.*, **189**, 109893.

18. Buwono, N.R., Risjani, Y., and Soegianto, A. (2021). Distribution of microplastic in relation to water quality parameters in the Brantas River, East Java, Indonesia. *Environ. Technol. Innov.*, **24**, 101915.

19. Bian, P., Liu, Y., Zhao, K., Hu, Y., Zhang, J., Kang, L., and Shen, W. (2022). Spatial variability of microplastic pollution on surface of rivers in a mountain-plain transitional area: A case study in the Chin Ling-Wei River Plain, China. *Ecotoxicol. Environ. Saf.*, **232**, 113298.

20. Alfonso, M.B., Scordo, F., Seitz, C., Mavo Manstretta, G.M., Ronda, A.C., Arias, A.H., Tomba, J.P., Silva, L.I., Perillo, G.M.E., and Piccolo, M.C. (2020). First evidence of microplastics in nine lakes across Patagonia (South America). *Sci. Total Environ.*, **733**, 139385.

21. Bertoldi, C., Lara, L.Z., Mizushima, F.A. d. L., Martins, F.C.G., Battisti, M.A., Hinrichs, R., and Fernandes, A.N. (2021). First evidence of microplastic contamination in the freshwater of Lake Guaíba, Porto Alegre, Brazil. *Sci. Total Environ.*, **759**, 143503.

22. Warrier, A.K., Kulkarni, B., Amrutha, K., Jayaram, D., Valsan, G., and Agarwal, P. (2022). Seasonal variations in the abundance and distribution of microplastic particles in the surface waters of a Southern Indian lake. *Chemosphere*, **300**, 134556.

23. Ephsy, D., and Raja, S. (2023). Characterization of microplastics and its pollution load index in freshwater Kumaraswamy Lake of Coimbatore, India. *Environ. Toxicol. Pharmacol.*, **101**, 104207.

24. K, M.B., Natesan, U., R, V., R, P.K., R, R., and S, S. (2021). Spatial distribution of microplastic concentration around landfill sites and its potential risk on groundwater. *Chemosphere*, **277**, 130263.

25. Wu, B., Li, L.W., Zu, Y.X., Nan, J., Chen, X.Q., Sun, K., and Li, Z.L. (2022). Microplastics contamination in groundwater of a drinking-water source area, Northern China. *Environ. Res.*, **214**, 114048.

26. Gong, X., Tian, L., Wang, P., Wang, Z., Zeng, L., and Hu, J. (2023). Microplastic pollution in the groundwater under a Bedrock Island in the South China Sea. *Environ. Res.*, **239**, 117277.

27. Avio, C.G., Gorbi, S., and Regoli, F. (2017). Plastics and microplastics in the oceans: From emerging pollutants to emerged threat. *Mar. Environ. Res.*, **128**, 2–11.

28. Jiang, C., Yin, L., Li, Z., Wen, X., Luo, X., Hu, S., Yang, H., Long, Y., Deng, B., Huang, L., and Liu, Y. (2019). Microplastic pollution in the rivers of the Tibet Plateau. *Environ. Pollut.*, **249**, 91–98.

29. Gola, D., Kumar Tyagi, P., Arya, A., Chauhan, N., Agarwal, M., Singh, S.K., and Gola, S. (2021). The impact of microplastics on marine environment: A review. *Environ. Nanotechnology, Monit. Manag.*, **16**, 100552.

30. Ahmed, M.B., Rahman, M.S., Alom, J., Hasan, M.S., Johir, M.A.H., Mondal, M.I.H., Lee, D.Y., Park, J., Zhou, J.L., and Yoon, M.H. (2021). Microplastic particles in the aquatic environment: A systematic review. *Sci. Total Environ.*, **775**, 145793.

31. Auta, H.S., Emenike, C.U., and Fauziah, S.H. (2017). Distribution and importance of microplastics in the marine environment: A review of the sources, fate, effects, and potential solutions. *Environ. Int.*, **102**, 165–176.

32. Cooper, D.A., and Corcoran, P.L. (2010). Effects of mechanical and chemical processes on the degradation of plastic beach debris on the Island of Kauai, Hawaii. *Mar. Pollut. Bull.*, **60**, 650–654.

33. Ribic, C.A., Sheavly, S.B., Rugg, D.J., and Erdmann, E.S. (2010). Trends and drivers of marine debris on the Atlantic Coast of the United States 1997–2007. *Mar. Pollut. Bull.*, **60**, 1231–1242.

34. Sait, S.T.L., Sørensen, L., Kubowicz, S., Vike-Jonas, K., Gonzalez, S.V., Asimakopoulos, A.G., and Booth, A.M. (2021). Microplastic fibres from synthetic textiles: Environmental degradation and additive chemical content. *Environ. Pollut.*, **268**, 115745.

35. Enfrin, M., Lee, J., Gibert, Y., Basheer, F., Kong, L., and Dumée, L.F. (2020). Release of hazardous nanoplastic contaminants due to microplastics fragmentation under shear stress forces. *J. Hazard. Mater.*, **384**, 121393.

36. Alomar, C., and Deudero, S. (2017). Evidence of microplastic ingestion in the Shark Galeus Melastomus Rafinesque, 1810 in the continental shelf off the Western Mediterranean Sea. *Environ. Pollut.*, **223**, 223–229.

37. Li, Q., Ma, C., Zhang, Q., and Shi, H. (2021). Microplastics in shellfish and implications for food safety. *Curr. Opin. Food Sci.*, **40**, 192–197.

38. Mao, Y., Ai, H., Chen, Y., Zhang, Z., Zeng, P., Kang, L., Li, W., Gu, W., He, Q., and Li, H. (2018). Phytoplankton response to polystyrene microplastics: Perspective from an entire growth period. *Chemosphere*, **208**, 59–68.

39. Mao, Y., Li, H., Gu, W., Yang, G., Liu, Y., and He, Q. (2020). Distribution and characteristics of microplastics in the Yulin River, China: Role of environmental and spatial factors. *Environ. Pollut.*, **265**, 115033.

40. Wang, F., Wong, C.S., Chen, D., Lu, X., Wang, F., and Zeng, E.Y. (2018). Interaction of toxic chemicals with microplastics: A critical review. *Water Res.*, **139**, 208–219.

41. Sharma, K., Thakur, I.S., and Kaushik, G. (2021). Occurrence and distribution of pharmaceutical compounds and their environmental impacts: A review. *Bioresour. Technol. Reports*, **16**, 100841.

42. Lu, S., Lin, C., Lei, K., Wang, B., Xin, M., Gu, X., Cao, Y., Liu, X., Ouyang, W., and He, M. (2020). Occurrence, spatiotemporal variation, and ecological risk of antibiotics in the water of the semi-enclosed urbanized Jiaozhou Bay in Eastern China. *Water Res.*, **184**, 116187.

43. Han, Q.F., Song, C., Sun, X., Zhao, S., and Wang, S.G. (2021). Spatiotemporal distribution, source apportionment and combined pollution of antibiotics in natural waters adjacent to mariculture areas in the Laizhou Bay, Bohai Sea. *Chemosphere*, **279**, 130381.

44. Zhu, Y., He, P., Hu, H., Qi, M., Li, T., Zhang, X., Guo, Y., Wu, W., Lan, Q., Yang, C., and Jin, H. (2022). Determination of quinolone antibiotics in environmental water using automatic solid-phase extraction and isotope dilution ultra-performance liquid chromatography tandem mass spectrometry. *J. Chromatogr. B Anal. Technol. Biomed. Life Sci.*, **1208**, 123390.

45. Semerjian, L., Aissaoui, S., Shanableh, A., Okoh, A., Elhadi, R., Mousa, M., Alhameed, R.A., Hassan, J.A.J., Akhtar, I., and Semreen, M.H. (2023). Occurrence, spatial and seasonal variations of emerging contaminants in the aquatic environment of Sharjah, United Arab Emirates. *Chemosphere*, **345**, 140426.

46. Arsand, J.B., Hoff, R.B., Jank, L., Bussamara, R., Dallegrave, A., Bento, F.M., Kmetzsch, L., Falção, D.A., do Carmo Ruaro Peralba, M., de Araujo Gomes, A., and Pizzolato, T.M. (2020). Presence of antibiotic resistance genes and its association with antibiotic occurrence in Dilúvio River in Southern Brazil. *Sci. Total Environ.*, **738**, 139781.

47. Biswas, P., and Vellanki, B.P. (2021). Occurrence of emerging contaminants in highly anthropogenically influenced River Yamuna in India. *Sci. Total Environ.*, **782**, 146741.

48. Tang, J., Zhang, J., Su, L., Jia, Y., and Yang, Y. (2022). Bioavailability and trophic magnification of antibiotics in aquatic food webs of Pearl River, China: Influence of physicochemical characteristics and biotransformation. *Sci. Total Environ.*, **820**, 153285.

49. Oharisi, O.-O.L., Ncube, S., Nyoni, H., Madikizela, M.L., Olowoyo, O.J., and Maseko, B.R. (2023). Occurrence and prevalence of antibiotics in wastewater treatment plants and effluent receiving rivers in South Africa using UHPLC-MS determination. *J. Environ. Manage.*, **345**, 118621.
50. Liang, X., Guan, F., Chen, B., Luo, P., Guo, C., Wu, G., Ye, Y., Zhou, Q., and Fang, H. (2020). Spatial and seasonal variations of antibiotic resistance genes and antibiotics in the surface waters of Poyang Lake in China. *Ecotoxicol. Environ. Saf.*, **196**, 110543.
51. Wang, Y., Lu, S., Liu, X., Chen, J., Han, M., Wang, Z., and Guo, W. (2021). Profiles of antibiotic resistance genes in an inland salt-lake Ebinur Lake, Xinjiang, China: The relationship with antibiotics, environmental factors, and microbial communities. *Ecotoxicol. Environ. Saf.*, **221**, 112427.
52. Zhou, Q., Liu, G., Arif, M., Shi, X., and Wang, S. (2022). Occurrence and risk assessment of antibiotics in the surface water of Chaohu Lake and its tributaries in China. *Sci. Total Environ.*, **807**, 151040.
53. Jiang, L., Zhai, W., Wang, J., Li, G., Zhou, Z., Li, B., and Zhuo, H. (2023). Antibiotics and antibiotic resistance genes in the water sources of the Wuhan Stretch of the Yangtze River: Occurrence, distribution, and ecological risks. *Environ. Res.*, **239**, 117295.
54. Ding, G., Chen, G., Liu, Y., Li, M., and Liu, X. (2020). Occurrence and risk assessment of fluoroquinolone antibiotics in reclaimed water and receiving groundwater with different replenishment pathways. *Sci. Total Environ.*, **738**, 139802.
55. Gros, M., Catalán, N., Mas-Pla, J., Čelić, M., Petrović, M., and Farré, M.J. (2021). Groundwater antibiotic pollution and its relationship with dissolved organic matter: Identification and environmental implications. *Environ. Pollut.*, **289**, 151011.
56. Arun, S., Xin, L., Gaonkar, O., Neppolian, B., Zhang, G., and Chakraborty, P. (2022). Antibiotics in sewage treatment plants, receiving water bodies and groundwater of Chennai city and the suburb, South India: Occurrence, removal efficiencies, and risk assessment. *Sci. Total Environ.*, **851**, 158195.
57. Fayaz, T., Renuka, N., and Ratha, S.K. (2024). Antibiotic occurrence, environmental risks, and their removal from aquatic environments using microalgae: Advances and future perspectives. *Chemosphere*, **349**, 140822.
58. Zhang, H., Wang, J., Zhou, B., Zhou, Y., Dai, Z., Zhou, Q., Chriestie, P., and Luo, Y. (2018). Enhanced adsorption of oxytetracycline to weathered

microplastic polystyrene: Kinetics, isotherms and influencing factors. *Environ. Pollut.*, **243**, 1550–1557.

59. Fan, X., Zou, Y., Geng, N., Liu, J., Hou, J., Li, D., Yang, C., and Li, Y. (2021). Investigation on the adsorption and desorption behaviors of antibiotics by degradable MPs with or without UV ageing process. *J. Hazard. Mater.*, **401**, 123363.
60. Chen, Y., Li, J., Wang, F., Yang, H., and Liu, L. (2021). Adsorption of tetracyclines onto polyethylene microplastics: A combined study of experiment and molecular dynamics simulation. *Chemosphere*, **265**, 129133.
61. González-Pleiter, M., Pedrouzo-Rodríguez, A., Verdú, I., Leganés, F., Marco, E., Rosal, R., and Fernández-Piñas, F. (2021). Microplastics as vectors of the antibiotics azithromycin and clarithromycin: Effects towards freshwater microalgae. *Chemosphere*, **268**, 128824.
62. Kong, F., Xu, X., Xue, Y., Gao, Y., Zhang, L., Wang, L., Jiang, S., and Zhang, Q. (2021). Investigation of the adsorption of sulfamethoxazole by degradable microplastics artificially aged by chemical oxidation. *Arch. Environ. Contam. Toxicol.*, **81**, 155–165.
63. Puckowski, A., Cwięk, W., Mioduszewska, K., Stepnowski, P., and Białk-Bielińska, A. (2021). Sorption of pharmaceuticals on the surface of microplastics. *Chemosphere*, **263**, 127976.
64. Yao, J., Wen, J., Li, H., and Yang, Y. (2022). Surface functional groups determine adsorption of pharmaceuticals and personal care products on polypropylene microplastics. *J. Hazard. Mater.*, **423**, 127131.
65. Godoy, V., Prata, J.C., Blázquez, G., Almendros, A.I., Duarte, A.C., Rocha-Santos, T., Calero, M., and Martín-Lara, M.Á. (2020). Effects of distance to the sea and geomorphological characteristics on the quantity and distribution of microplastics in beach sediments of Granada (Spain). *Sci. Total Environ.*, **746**, 142023.
66. Razanajatovo, R.M., Ding, J., Zhang, S., Jiang, H., and Zou, H. (2018). Sorption and desorption of selected pharmaceuticals by polyethylene microplastics. *Mar. Pollut. Bull.*, **136**, 516–523.
67. Chen, Y., Qian, Y., Shi, Y., Wang, X., Tan, X., and An, D. (2022). Accumulation of chiral pharmaceuticals (Ofloxacin or Levofloxacin) onto polyethylene microplastics from aqueous solutions. *Sci. Total Environ.*, **823**, 153765.
68. Ding, L., Mao, R., Ma, S., Guo, X., and Zhu, L. (2020). High temperature depended on the ageing mechanism of microplastics under different environmental conditions and its effect on the distribution of organic pollutants. *Water Res.*, **174**, 115634.

69. Guo, X., Pang, J., Chen, S., and Jia, H. (2018). Sorption properties of tylosin on four different microplastics. *Chemosphere*, **209**, 240–245.
70. Guo, X., and Wang, J. (2019). Sorption of antibiotics onto aged microplastics in freshwater and seawater. *Mar. Pollut. Bull.*, **149**, 110511.
71. Guo, X., Liu, Y., and Wang, J. (2019). Sorption of sulfamethazine onto different types of microplastics: A combined experimental and molecular dynamics simulation study. *Mar. Pollut. Bull.*, **145**, 547–554.
72. Guo, X., Chen, C., and Wang, J. (2019). Sorption of sulfamethoxazole onto six types of microplastics. *Chemosphere*, **228**, 300–308.
73. Lin, L., Tang, S., Wang, X.S., Sun, X., Han, Z., and Chen, Y. (2020). Accumulation mechanism of tetracycline hydrochloride from aqueous solutions by nylon microplastics. *Environ. Technol. Innov.*, **18**, 100750.
74. Liu, G., Zhu, Z., Yang, Y., Sun, Y., Yu, F., and Ma, J. (2019). Sorption behavior and mechanism of hydrophilic organic chemicals to virgin and aged microplastics in freshwater and seawater. *Environ. Pollut.*, **246**, 26–33.
75. Xiong, Y., Zhao, J., Li, L., Wang, Y., Dai, X., Yu, F., and Ma, J. (2020). Interfacial interaction between micro/nanoplastics and typical PPCPs and nanoplastics removal via electrosorption from an aqueous solution. *Water Res.*, **184**, 116100.
76. Xu, B., Liu, F., Brookes, P.C., and Xu, J. (2018). The sorption kinetics and isotherms of sulfamethoxazole with polyethylene microplastics. *Mar. Pollut. Bull.*, **131**, 191–196.
77. Yilimulati, M., Wang, L., Ma, X., Yang, C., and Habibul, N. (2021). Adsorption of ciprofloxacin to functionalized nano-sized polystyrene plastic: Kinetics, thermochemistry and toxicity. *Sci. Total Environ.*, **750**, 142370.
78. Yu, F., Yang, C., Huang, G., Zhou, T., Zhao, Y., and Ma, J. (2020). Interfacial interaction between diverse microplastics and tetracycline by adsorption in an aqueous solution. *Sci. Total Environ.*, **721**, 137729.
79. Xu, B., Liu, F., Brookes, P.C., and Xu, J. (2018). Microplastics play a minor role in tetracycline sorption in the presence of dissolved organic matter. *Environ. Pollut.*, **240**, 87–94.
80. Wang, F., Zhang, M., Sha, W., Wang, Y., Hao, H., Dou, Y., and Li, Y. (2020). Sorption behavior and mechanisms of organic contaminants to nano and microplastics. *Molecules*, **25**, 1827.
81. Zhang, H., Liu, F.-F., Wang, S.-C., Huang, T.-Y., Li, M.-R., Zhu, Z.-L., and Liu, G.-Z. (2020). Sorption of fluoroquinolones to nanoplastics as affected by surface functionalization and solution chemistry. *Environ. Pollut.*, **262**, 114347.

82. Wang, K., Wang, K., Chen, Y., Liang, S., Zhang, Y., Guo, C., Wang, W., and Wang, J. (2023). Desorption of sulfamethoxazole from polyamide 6 microplastics: Environmental factors, simulated gastrointestinal fluids, and desorption mechanisms. *Ecotoxicol. Environ. Saf.*, **264**, 115400.

83. Liang, Y., Ying, C., Zhu, J., Zhou, Q., Sun, K., Tian, Y., and Li, J. (2023). Effects of salinity, PH, and Cu(II) on the adsorption behaviors of tetracycline onto polyvinyl chloride microplastics: A site energy distribution analysis. *Water (Switzerland)*, **15**, 1925.

84. Liang, J., Wu, J., Zeng, Z., Li, M., Liu, W., and Zhang, T. (2023). Behavior and mechanisms of ciprofloxacin adsorption on aged polylactic acid and polyethlene microplastics. *Environ. Sci. Pollut. Res.*, **30**, 62938–62950.

85. Lionetto, F., Esposito Corcione, C., Messa, F., Perrone, S., Salomone, A., and Maffezzoli, A. (2023). The sorption of amoxicillin on engineered polyethylene terephthalate microplastics. *J. Polym. Environ.*, **31**, 1383–1397.

86. Zheng, M., Wu, P., Li, L., Yu, F., and Ma, J. (2023). Adsorption/desorption behavior of ciprofloxacin on aged biodegradable plastic PLA under different exposure conditions. *J. Environ. Chem. Eng.*, **11**, 109256.

87. Osman, D., Uyanık, İ., Mıhçıokur, H., and Özkan, O. (2023). Evaluation of ciprofloxacin (CIP) and clarithromycin (CLA) adsorption with weathered PVC microplastics. *J. Environ. Sci. Heal. Part A Toxic/Hazardous Subst. Environ. Eng.*, **58**, 498–505.

88. Zhang, X., Liu, L., Chen, X., Li, J., Chen, J., and Wang, H. (2023). The fate and risk of microplastic and antibiotic sulfamethoxazole coexisting in the environment. *Environ. Geochem. Health*, **45**, 2905–2915.

89. Tong, F., Liu, D., Zhang, Z., Chen, W., Fan, G., Gao, Y., Gu, X., and Gu, C. (2023). Heavy metal-mediated adsorption of antibiotic tetracycline and ciprofloxacin on two microplastics: Insights into the role of complexation. *Environ. Res.*, **216**, 114716.

90. Wang, H., Qiu, C., Song, Y., Bian, S., Wang, Q., Chen, Y., and Fang, C. (2022). Adsorption of tetracycline and Cd(II) on polystyrene and polyethylene terephthalate microplastics with ultraviolet and hydrogen peroxide aging treatment. *Sci. Total Environ.*, **845**, 157109.

91. Mejías, C., Martín, J., Martín-Pozo, L., Santos, J.L., Aparicio, I., and Alonso, E. (2024). Adsorption of macrolide antibiotics and a metabolite onto polyethylene terephthalate and polyethylene microplastics in aquatic environments. *Antibiotics*, **13**, 408.

92. Mejías, C., Martín, J., Santos, J.L., Aparicio, I., and Alonso, E. (2024). Implications of polystyrene and polyamide microplastics in the adsorption of sulfonamide antibiotics and their metabolites in water matrices. *Aquat. Toxicol.*, **271**, 106934.
93. Zhang, Y., Xu, X., Xu, J., Li, Z., Cheng, L., Fu, J., and Sun, W. (2024). When antibiotics encounter microplastics in aquatic environments: Interaction, combined toxicity, and risk assessments. *Sci. Total Environ.*, **929**, 172455.
94. Zhuang, S., and Wang, J. (2023). Interaction between antibiotics and microplastics: Recent advances and perspective. *Sci. Total Environ.*, **897**, 165414.
95. Feng, L.J., Shi, Y., Li, X.Y., Sun, X.D., Xiao, F., Sun, J.W., Wang, Y., Liu, X.Y., Wang, S.G., and Yuan, X.Z. (2020). Behavior of tetracycline and polystyrene nanoparticles in estuaries and their joint toxicity on marine microalgae *Skeletonema costatum*. *Environ. Pollut.*, **263**, 114453.
96. Li, N., Zeng, Z., Zhang, Y., Zhang, H., Tang, N., Guo, Y., Lu, L., Li, X., Zhu, Z., Gao, X., and Liang, J. (2023). Higher toxicity induced by co-exposure of polystyrene microplastics and chloramphenicol to *Microcystis aeruginosa*: Experimental study and molecular dynamics simulation. *Sci. Total Environ.*, **866**, 161375.
97. Zhou, W., Tang, Y., Du, X., Han, Y., Shi, W., Sun, S., Zhang, W., Zheng, H., and Liu, G. (2021). Fine polystyrene microplastics render immune responses more vulnerable to two veterinary antibiotics in a bivalve species. *Mar. Pollut. Bull.*, **164**, 111995.
98. Banaee, M., Faraji, J., Amini, M., Multisanti, C.R., and Faggio, C. (2023). Rainbow trout (*Oncorhynchus mykiss*) physiological response to microplastics and enrofloxacin: Novel pathways to investigate microplastic synergistic effects on pharmaceuticals. *Aquat. Toxicol.*, **261**, 106627.
99. Zhu, Z.-L., Wang, S.-C., Zhao, F.-F., Wang, S.-G., Liu, F.-F., and Liu, G.-Z. (2019). Joint toxicity of microplastics with triclosan to marine microalgae *Skeletonema costatum*. *Environ. Pollut.*, **246**, 509–517.
100. Guo, X., Cai, Y., Ma, C., Han, L., and Yang, Z. (2021). Combined toxicity of micro/nano scale polystyrene plastics and ciprofloxacin to *Corbicula fluminea* in freshwater sediments. *Sci. Total Environ.*, **789**, 147887.
101. Zhang, S., Ding, J., Razanajatovo, R.M., Jiang, H., Zou, H., and Zhu, W. (2019). Interactive effects of polystyrene microplastics and roxithromycin on bioaccumulation and biochemical status in the freshwater fish red tilapia (*Oreochromis niloticus*). *Sci. Total Environ.*, **648**, 1431–1439.

Chapter 10

Effects of Petroleum and Bio-micro(nano)plastics on Aquatic Organisms

Márcio Barreto, Isabel Lopes, and Miguel Oliveira
Centre for Environmental and Marine Studies (CESAM),
Department of Biology, University of Aveiro, 3810-193 Aveiro, Portugal
marcio.barreto@ua.pt

Plastics are integral to modern life but have led to significant environmental challenges, particularly in aquatic ecosystems, where micro(nano)plastics (MNPs) pose a growing concern. This chapter reviews the effects of both fossil- and bio-micro(nano)plastics on aquatic organisms, focusing on their ecotoxicological effects. While bioplastics are often marketed as eco-friendly alternatives, due to their expected biodegradability, emerging evidence indicates that these materials may degrade into micro- and nanoplastics and release harmful by-products such as additives, oligomers, and monomers. These degradation products have the potential to cause a wide range of pernicious effects on aquatic life. Overall, the available data

Microplastic Pollution: Environmental Impact and Solutions
Edited by Julia Martín, Juan Luis Santos, Irene Aparicio, and Esteban Alonso

ISBN 978-981-5129-77-9 (Hardcover), 978-1-003-66006-4 (eBook)
www.jennystanford.com

DOI: 10.1201/9781003660064-10

show that bioplastics may exhibit similar or, in some cases, greater toxicity than fossil-based plastics. Furthermore, the incomplete biodegradation of bioplastics in natural environments may lead to persistent pollution, challenging their sustainability claims. The findings suggest a cautious approach to the widespread adoption of bioplastics, emphasizing the need for further investigation into their true environmental impact before they are considered a sustainable alternative to conventional fossil-based plastics. Key knowledge gaps are identified, particularly regarding the long-term impacts of micro(nano)bioplastics (MNBPs) on aquatic ecosystems and food webs. Future research should include the assessment of the effects of MNBP degradation products at environmentally relevant concentrations and how their effects may be modulated by climate change variables.

10.1 Introduction

Plastics are ubiquitous materials that have revolutionized human life and various industries due to their durability, versatility, and cost-effectiveness. However, their widespread use has led to significant environmental challenges, particularly in aquatic ecosystems. Among these challenges, micro(nano)nanoplastics (MNPs), here defined as particles smaller than 1 mm [1], have emerged as an issue of critical concern due to their potential toxicity and persistence in the environment. MNPs, mainly produced as unintended by-products of commercial life cycle and from poor waste management of plastic products, end up in environmental matrices, like aquatic compartments, where they are exposed to diverse abiotic and biotic factors (e.g., mechanical weathering, UV radiation, and biodegradation processes) that promote their fragmentation and degradation into smaller particles [2]. Additionally, micro- and nano-sized particles can be deliberately released into the environment through their application in manifold human activities (e.g., personal care products and antifouling activities) [2, 3].

The continuous and growing trend in plastic consumption has promoted increased plastic pollution. Despite the first reports in the 1970s of the presence of microplastics and their potential effects, the last decade has seen a boom in publications reporting the presence and effects of microplastics (MPs) on different biota [4]. Still, little is known about their modes of action. Recent published articles point nanoplastics (NPs), particles smaller than 1000 nm, as a potentially more pernicious counterpart of plastic pollution [5]. Due to their small size and high surface area, NPs exhibit high reactivity, may be able to penetrate biological membranes and cause cellular damage, oxidative stress, inflammation, and ultimately pernicious effects at the individual level [6, 7]. Their chemical composition, including additives and sorbed environmental contaminants, may further exacerbate their toxicity [8]. MNPs can also bioaccumulate and potentially biomagnify through food webs, posing greater risks at higher trophic levels, including humans [2, 9].

Plastic pollution is an utmost serious environmental issue, as plastic products littered in aquatic compartments may fragment over extended periods into micro- and nano-sized particles, potentially becoming permanently embedded in the environment without feasible remediation strategies. Based on this, the European Union (EU) has been at the forefront of efforts to employ strategies (e.g., European Green Deal, Circular Economy Action Plan, Bioeconomy and Plastic strategies) to reduce plastic pollution in an attempt to prevent plastics from reaching the environmental matrices. Some of the EU initiatives aim to transition from a linear economy, characterized by the "take-make-dispose" trend, to a circular economy that emphasizes sustainability and resource efficiency [10, 11]. Thus, the promotion of a more sustainable alternative to traditional petroleum-based plastics (e.g., polystyrene (PS), polyethylene (PE), polypropylene (PP), and polyethylene terephthalate (PET)) is considered a major approach to tackle plastic resource use and environmental impact, fostering a circular economy where products are kept in use for as long as possible and are capable of biodegrading at the end of their life cycles [10, 11]. In this sense, bioplastics appear to be a promising approach. It should be highlighted that the definition of bioplastics is not consensual, and this terminology can be misleading. In this chapter, bioplastics are defined as a product fully or partially derived from a renewable/natural resource (e.g., plants), products that are both biobased and biodegradable, or products that are biodegradable despite being from non-renewable (i.e., fossil fuels) sources [12]. One of the key pathways identified by the EU for promoting a circular economy in plastics is the renewable material pathway, which focuses on biobased plastics. The use of bioplastics derived from renewable biological sources, rather than fossil fuels, reduces the dependence on non-renewable resources and can help mitigate the environmental impact associated with the extraction and processing of fossil fuels. By using renewable feedstocks, bioplastics support the EU's objectives of reducing greenhouse gas emissions and fostering sustainable agricultural practices [13]. This pathway emphasizes the importance of selecting sustainable feedstocks and ensuring that bioplastic production does not compete with food production or lead to deforestation. By prioritizing secondary raw materials such as biowaste and by-products over primary raw materials, the EU aims to create a more sustainable and circular supply chain for plastics where bioplastics may have significant environmental benefits [13].

However, the currently available data support the idea that bioplastics may not be an ultimate solution to plastic pollution, as they present several disadvantages, including weaker physicochemical properties compared to petrochemical plastics, where materials may need to be reinforced with the use of a wider variety of additives (e.g., plasticizers, flame retardants, catalysers, among others) to improve their characteristics/performance [14, 15]. While bioplastics are labelled as fully or almost fully biodegradable, most biodegradation tests occur in laboratory-controlled environments (e.g., [16]) where variables such as temperature, humidity, and bacterial populations and strains can be easily manipulated. In contrast, when these materials reach environmental matrices, they encounter scenarios drastically different from well-defined artificial laboratory ones and may not completely biodegrade. Consequently, the formation of micro(nano)bioplastics (MNBPs) is highly predictable owing to the potential incomplete mineralization of these materials in environmental scenarios [17].

In recent years, the number of studies assessing the effects of MNBPs on aquatic biota has been increasing, but the studies have been addressing different research questions, often using non-standardized methods. Thus, there is the need to summarize what is already known, and how this knowledge was obtained, for a better understanding of the risks associated with the presence of these particles in the environment. Therefore, this chapter summarizes the available and most recent peer-reviewed information about the biological effects of MNBPs on aquatic organisms identifying knowledge gaps and research lines.

10.2 Methods

An overview of the most recent articles comparing the pernicious effects of MNBPs and MNPs on aquatic organisms was performed. Thus, a bibliographic search was performed on the Scopus Database to retrieve articles addressing the effects of MNBPs and their petroleum-based counterparts. The bibliographic search was performed using a combination of words associated with Boolean operators (e.g., AND/OR) and their appropriate truncation (i.e., *). The combinations used were "biodegradable*" OR "biopolymer*" OR "bioplastic*" AND "microplastic*" AND "effect*" AND "aquatic" and the same combination was applied for NPs to search within the article title, abstract, and keywords. This search was limited to research articles published between 2022 and 2024 (June) that address the effects of MNBPs on aquatic organisms. This time period was selected to include the latest scientific reports that consider the need for the characterization of particles and provide complementary information to other reviews that addressed this issue for papers before 2022 [12]. Scientific articles regarding *in vitro* studies, or that did not include particle characterization, considered of utmost importance for a proper interpretation and regulatory use of ecotoxicological data, especially for particles generated through top-down or bottom-up approaches, were excluded from the literature review [18]. The review search and screening guidelines were conducted using the Preferred Reporting Items of Systematic Reviews and Meta-Analysis (PRISMA). The search criteria retrieved 94 articles, and 19 were kept for further analysis after the titles and abstracts of the candidate publications were read and screened for their relevance concerning the research topic. Afterwards, the citations and references of each selected article were analysed to minimize the loss of relevant information, as depicted in Fig. 10.1, and 12 more relevant articles were added to the review after citation search, totalling 25 articles. After careful analysis, the key criteria (i.e., type of polymers tested, shape and size, concentrations, relevant methodological approaches, and main findings) are summarized in Table 10.1, to be discussed further in Section 10.4.

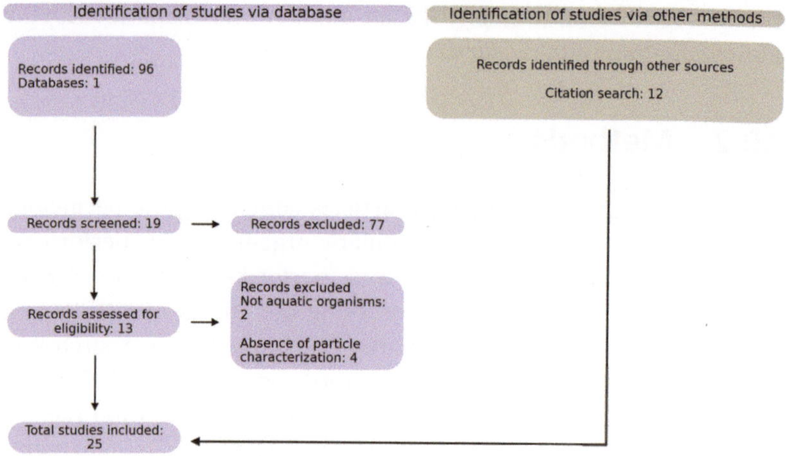

Figure 10.1 PRISMA flow diagram of the bibliographic search and study selection process, according to Page et al. [19].

10.3 Bioplastics: Chemical Nature, Properties, and Applications

The purpose of this section is to briefly synthesize the relevant features of the most commercialized bioplastics, regarding their strengths and limitations on industrial applications. According to the European Bioplastics Market reports, five main polymers can be highlighted: polylactic acid (PLA), polyhydroxyalkanoates (PHA), polybutylene succinate (PBS), polycaprolactone (PCL), and polybutylene adipate-co-terephthalate (PBAT) [20]. A schematic representation of bioplastics and their sources is depicted in Fig. 10.2.

Efforts are being made to discover new polymers for total or partial replacement of conventional plastics. Polylactic acid (PLA), an aliphatic polyester derived from lactic acid, primarily sourced through bacterial fermentation of sugars and produced from the ring-opening polymerization of a cyclic lactide dimer [21, 22], appears as a promising polymer. Recognized as safe by the US Food and Drug Administration, PLA is commonly used in food service products such as disposable coffee cups. PLA has several attractive properties, including being recyclable, industrially compostable, and having good thermal processability, allowing it to be processed by various methods such as injection moulding and film extrusion [23]. PLA is also one of the few bioplastics that can be comparable with the most produced petroleum-based plastics, such as PE, PS, PP, and PVC, in terms of physicochemical properties [24]. However, PLA has limitations, such as high brittleness, low toughness, relatively high cost, and a slow biodegradation rate. Its poor melt strength and stability also limit its use in flexible packaging [25]. PLA is compostable only under industrial conditions, requiring high temperatures and moisture, and does not degrade in home composting or aquatic environments [22, 25, 26]. Despite these drawbacks, PLA has similar barrier properties to PET, making it a potential substitute in rigid packaging, such as beverage bottles. Leading producers of PLA include NatureWorks LLC and Corbion®, with NatureWorks being the largest global producer [21].

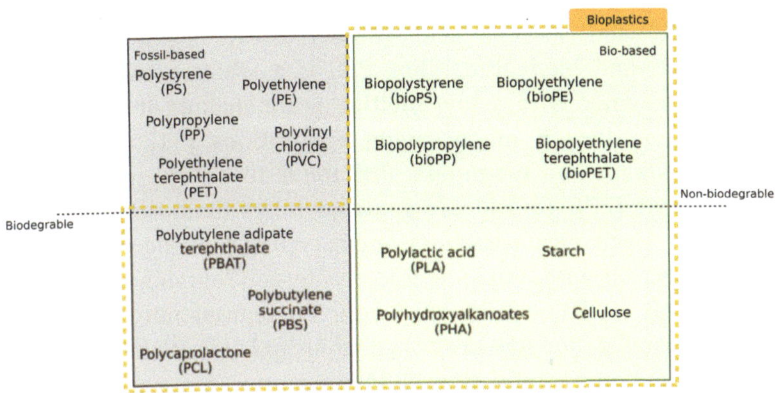

Figure 10.2 Examples of plastics categorized by their biodegradability and the sources of their monomers.

Polyhydroxyalkanoates (PHAs) are a family of biobased and biodegradable aliphatic polyesters produced through bacterial fermentation. They are classified based on their carbon atom length into short-chain, medium-chain, and long-chain PHAs, including the well-known polyhydroxybutyrate (PHB) [27]. Companies in China, Japan, and South Korea are leading in large-scale PHA production (especially PHB), with capacities ranging from 5000 to 50,000 tons per year. The amorphous forms of PHA, such as those produced by CJ BIO, offer better flexibility compared to the more brittle crystalline forms [21]. PHAs seem to be fully compostable and degrade in various environments, including rivers, oceans, and making them suitable for home composting [28]. However, PHAs also have limitations, including high production costs, which are up to 10 times higher than conventional plastics, and mechanical weaknesses such as brittleness, low impact strength, and low melt viscosity [29]. Additionally, the high thermal stability of PHAs hampers recycling, especially when mixed with other plastics such as PET [30]. Despite these challenges, PHAs are used in bioplastics and have potential use in various sectors (e.g., food and beverage materials and packaging, medical devices, and agriculture). To enhance their performance, strategies such as using plasticizers, copolymerization, blending with other polymers such as PLA or polybutylene succinate, and adding fillers are employed to improve their properties.

Polybutylene succinate (PBS) is an aliphatic thermoplastic polyester made through the condensation polymerization of butanediol and succinic acid. Known for its rigidity, ductility, toughness, and biodegradability, PBS is a white, semicrystalline, flexible polymer with a low melting point making it a potential replacement for PE in packaging applications [31]. Despite its heat resistance and toughness, PBS tends to be opaque or semi-transparent [22]. To address some of PBS's limitations, such as high cost, mechanical weakness, and recycling challenges, it is often copolymerized with adipic acid to create poly(butylene succinate-co-adipate) (PBSA), which has better biodegradability due to its lower crystallization and faster hydrolysis [32, 33]. While PBS is not suitable for home composting, PBSA is, making it more appealing for various applications. Both PBS and PBSA share similar mechanical properties with low-density PE, suitable for film extrusion. PBS and PBSA are mainly produced from petroleum-based resources, though efforts are underway to increase their biobased content. PBS has been available commercially under the name BioPBS™, produced by PTT MCC Biochem. Despite their attractive properties, the high cost of PBS and PBSA remains a barrier to wider market acceptance [34].

Polycaprolactone (PCL) is a semicrystalline aliphatic polyester produced through the ring-opening polymerization of ε-caprolactone [35]. PCL is characterized by a lower stiffness and tensile strength, but greater stretchability when compared to other bioplastics. Although it is petroleum-based, under industrial composting conditions, PCL biodegrades more effectively than PLA [36]. Its low melting point, high biodegradability, and compatibility with other bioplastics make PCL suitable for applications in food packaging [37]. However, its high cost and challenging processability limit its broader use in packaging and food service products.

Polybutylene adipate terephthalate (PBAT) is an aliphatic–aromatic biodegradable copolyester that has gained commercial popularity over the past two decades. It is produced through the copolymerization of adipic acid, 1,4-butanediol, and terephthalic acid monomers. PBAT is valued for its unique combination of flexibility, moderate stiffness, and biodegradability [38]. These properties make PBAT like low-density PE (LDPE) suitable for applications such as flexible packaging, stretch films, shopping bags, and mulch films [38, 39]. While PBAT degrades well in soil and compost, it does not break down in seawater. Despite its advantageous properties, PBAT lacks sufficient strength and rigidity for certain industrial applications. To overcome this, it is often blended with other stronger, brittle biodegradable polymers such as PLA and starch to enhance its mechanical properties. Though PBAT is fully petroleum-based, efforts are underway to replace some of its petroleum-based monomers with biobased ones [40]. For example, replacing adipic acid with succinic acid from biomass produces poly(butylene succinate-co-terephthalate), which has around 35% biobased content and better biodegradability [41].

10.4 Biological Effects of Micro(nano)-bioplastics in Aquatic Organisms

As the use of bioplastics continues to grow, concerns about their environmental impact, particularly in aquatic ecosystems, have emerged. Understanding how MNBPs affect various species, ranging from cyanobacteria and microalgae to more complex organisms such as fish and molluscs, is critical for assessing their ecological risks. This section explores the biological effects of MNBPs compared to traditional petroleum-based MNPs on key aquatic species, highlighting the complex interactions and potential consequences for marine and freshwater ecosystems. Figure 10.3 depicts some of the reported ecotoxicological effects of MNBPs on aquatic organisms.

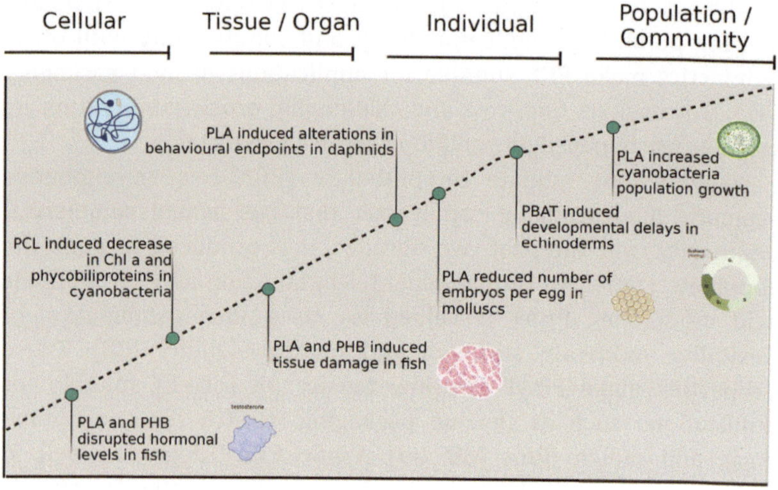

Figure 10.3 Reported ecotoxicological effects across various aquatic taxa at different biological levels of organization.

10.4.1 Cyanobacteria and Microalgae

When comparing the biological effects of micro(nano)bioplastics and their petroleum-based counterparts on cyanobacteria and microalgae, several distinct biological effects emerge. Out of five articles addressing the effects of MNBPs on cyanobacteria and microalgae, four of them cover the biological effects of PLA MPs.

PLA MPs, ranging between 2 and 100 μm, generally induced pernicious effects like other petroleum-based MPs. For example, it has been shown able to induce a decrease in cell density, chlorophyll a (Chl a), carotenoids, and phycobiliproteins content (PBPs, which help to capture light energy and transfer it to the photosystem), affecting overall health and growth of organisms (see Table 10.1) after short-term exposures (e.g., 96 h) [42] at high concentrations (i.e., 50–200 mg/L). However, over longer exposure periods (i.e., 12, 28, 63 days), at high concentrations, PLA and PCL MPs potentiate the growth and metabolism of cyanobacteria and microalgae, increasing cell density, cell metabolism, Chl a and PBPs [43–45]. Moreover, PLA MPs increased cell density and growth compared to control and PS MPs treatment [42]. This could be attributed to a panoply of factors, but potential biodegradation and increased

Table 10.1 Summary of key information of research studies reporting toxic effects of micro(nano)bioplastics on aquatic organisms

Groups	Species	Exposure Time [hours (days)]	MNPs				Main findings[d]	Ref.
			Bioplastic Type (μm)[a]	Petroleum-based Plastic Type(μm)[a]	Tested Concentration	Characterization Techniques[c]		
Cyanobacteria	*Mycrocystis aeruginosa*	1512 (63)	Spherical PLA (2.56)	none	10–200 mg/L	SEM, LPSA, and X-ray photoelectron spectroscopy	↑ cell density ↑ cell metabolism ↓ MDA content ↓ SOD enzyme activity ↑ DIC at day 39 to 51; at day 63 ↓ DIC ↓Carbonic anhydrase	[44]
		288 (12)	Spherical PLA and PCL (2 and 100)	Spherical PS (2 and 100)	0.1–10 mg/L	SEM and leachate analysis	<u>96 h</u> ↓ cell density ↓ Chl a ↓ PBPs for PS2 and PS100; ↑ PBPs for PLA2 and PLA100 ↑MC-LR synthesis <u>12 days</u> - cell density for PS2, PS100 and PCL2	[42]

(*Continued*)

Table 10.1 (*Continued*)

Groups	Species	Exposure Time [hours (days)]	MNPs				Main findings[d]	Ref.
			Bioplastic Type (μm)[a]	Petroleum-based Plastic Type(μm)[a]	Tested Concentration	Characterization Techniques[c]		
							↑ cell density for PCL100, PLA2, and PLA100 ↑ Chl a content ↓ PBPs for PS2, PS100 and PCL2; ↑ PBPs for PCL100, PLA2, and PLA100 ↑MC-LR synthesis	
	Anabaena sp. *Synechococcus* sp.	72 (3)	Spherical PCL (< 0.1)	None	91 mg/L	μ-FTIR, DLS, ELS, SEM, differential scanning calorimetry, MALDI-TOF/TOF	↓ Chl a content in both cyanobacteria; higher severity for *Synechococcus* sp. ↑ ROS production at 24 h and 72 h for both cyanobacteria - hydrogen peroxide activity for *Anabaena* sp.; ↑ for *Synechococcus* sp. ↓ Membrane potential at 24 and 72 h for *Synechococcus* sp.; ↓ at 72 h only for *Anabaena* sp.	[46]

Groups	Species	Exposure Time [hours (days)]	MNPs				Main findings[d]	Ref.
			Bioplastic Type (μm)[a]	Petroleum-based Plastic Type(μm)[a]	Tested Concentration	Characterization Techniques[c]		
Cyanobacteria/ Chlorophyta	*Mycrocystis aeruginosa* *Scenedesmus obliquus*	672 (28)	Spherical PLA (2.56)	None	10–200 mg/L	SEM, FTIR	↑ cell density for both species at 50 and 200 mg L^{-1} In co-culture ↓ cell density of *M. aeruginosa* and ↓ the inhibition effect over *S. obliquus* ↓ MC-LR synthesis both in mono and co-culture ↑ Extracellular inorganic carbon	[45]
	Chlorella vulgaris	264 (11)	Spherical PLA and PBS (57.41) (53.33)	Spherical PE and PA (77.75) (59.88)	10–1000 mg/L	DLS, SEM, LPSA, IR-FTIR, energy dispersive spectroscopy	↑ IR PE at 1000 mg/L (47.24%) and PA at 100 mg/L (40.36%); ↑ IR PLA at 100 mg/L (47.95%) and PBS (34.25%) ↓ Chl a and carotenoid at 10 mg/L for PLA and PBS, and at 100 mg/L for PA	[43]

(Continued)

Table 10.1 (*Continued*)

Groups	Species	Exposure Time [hours (days)]	MNPs				Main findings[d]	Ref.
			Bioplastic Type (μm)[a]	Petroleum-based Plastic Type(μm)[a]	Tested Concentration	Characterization Techniques[c]		
Cnidaria	*Hydra viridissima*	96 (4) 192 (8) 288 (12)	Spherical PHB (~0.2)	None	0.001–10 mg/L	DLS, SEM, ELS	96 h - Mortality, feeding rate, and malformations were not significantly different from control 8 days - Mortality and malformations were not significantly different from control ↑ Feeding rate in all treatments, except for 1 mg/L 12 days - Mortality ↑ Malformations ↑ Feeding rate	[48]

Groups	Species	Exposure Time [hours (days)]	MNPs				Main findings[d]	Ref.
			Bioplastic Type (μm)[a]	Petroleum-based Plastic Type(μm)[a]	Tested Concentration	Characterization Techniques[c]		
Cnidaria/ Crustacea	*Aurelia* sp. *Artemia franciscana*	24 (1)	Irregular PLA (50–500)	Fibers PVDF (50–500)	1–100 mg/L	SEM, UATR-FTIR	Particle uptake was observed of both MPs in both organisms only at 100 mg/L - Mortality for both organisms - *Artemia* swimming speed ↓ Pulsation frequency of *Aurelia* sp. ephyrae in both MPs exposure treatments	[50]
Cnidaria/ Chordata	*Hydra viridissima* *Danio rerio*	96 (4)	Irregular PLA (0.078)	Irregular PP, PS, and LDPE (0.142) (0.122) (0.0091)	0.001–100 mg/L	FTIR, DLS, ELS	↑ Mortality in PP and LDPE exposure to *Hydra veridissima* at 100 mg/L ↑ Malformations in PP, LDPE, and PS exposure at 10 and 100 mg/L	[49]

(*Continued*)

Table 10.1 (*Continued*)

Groups	Species	Exposure Time [hours (days)]	MNPs				Main findings[d]	Ref.
			Bioplastic Type (μm)[a]	Petroleum-based Plastic Type(μm)[a]	Tested Concentration	Characterization Techniques[c]		
							- Mortality and malformations in PLA exposure treatments ↓ Feeding rate in PLA compared to other treatments ↓ Regenerative capacity in LDPE 10 mg/L ↓ Behavioural and locomotor endpoints in all plastic treatments in both light and dark stimulation	
Mollusca	*Ilyanassa obsoleta*	744 (31)	Irregular PLA (~50)	Irregular PVC, PET, PS, and PP mixture (~50)	75–75000 particles/kg	SEM	↑ Morphological alterations in tissues in all MP treatments ↑ Nº of intercellular gaps in extreme concentrations of MPs and PLA treatment	[51]

Groups	Species	Exposure Time [hours (days)]	MNPs				Main findings[d]	Ref.
			Bioplastic Type (µm)[a]	Petroleum-based Plastic Type(µm)[a]	Tested Concentration	Characterization Techniques[c]		
							↑ Distribution of granules in extreme concentrations of MPs and PLA treatment ↑ Nº of mature hemocytes in extreme concentrations of MPs and PLA treatment ↓ Nº of embryos per egg capsule only in PLA treatment	
	Mytilus coruscus	336 (14)	Irregular PLA (5)	Spherical PS (5)	0.31–5.51 mg/L	SEM, µ-FTIR	↓ Absorption efficiency in all PLA treatments and only in medium and high concentrations of PS ↓ Clearance rate in the highest concentration of both MPs <u>Gills</u>	[52]

(*Continued*)

Table 10.1 (*Continued*)

Groups	Species	Exposure Time [hours (days)]	MNPs				Main findings[d]	Ref.
			Bioplastic Type (μm)[a]	Petroleum-based Plastic Type(μm)[a]	Tested Concentration	Characterization Techniques[c]		
							↓ LDH activity at medium concentration of PS and medium and high concentration of PLA ↓ PK activity at high concentrations of both MPs - SDH activity ↑ CAT activity in medium and high PS concentrations ↑ GSH activity in high PS concentrations Hemolymph ↓ LDH activity at low concentrations of both MPs	

Groups	Species	Exposure Time [hours (days)]	MNPs				Main findings[d]	Ref.
			Bioplastic Type (μm)[a]	Petroleum-based Plastic Type(μm)[a]	Tested Concentration	Characterization Techniques[c]		
							↑ PK activity at medium and high concentrations of PS and medium concentrations of PLA ↓ SDH activity at high concentrations of PS and low and high concentrations of PLA - CAT activity ↑ GPx activity in low concentrations of PLA and high concentrations of PS ↓ GST activity in all MP treatments	
Crustacea	*Tigriopus japonicus*	48 (2)	Irregular PBAT/PLA (10–40)	Irregular PE (10–40)	20 and 500 mg/L	SEM	MPs treatment did not influence survivability	[53]

(*Continued*)

Table 10.1 (*Continued*)

Groups	Species	Exposure Time [hours (days)]	MNPs				Main findings[d]	Ref.
			Bioplastic Type (μm)[a]	Petroleum-based Plastic Type(μm)[a]	Tested Concentration	Characterization Techniques[c]		
	Eurytemora affinis	336 (14)	Irregular PBAT (1–10)	Irregular LDPE (1–10)	300 μg/L	DLS	Simpson, Shannon, and Equitability indices (α diversity) over the five generations had no significant alterations Microbial community structure was impacted independently of MP type. PBAT had more impact on microbial community structure than LDPE Generation was a significant variable in the observed effects	[54]

Groups	Species	Exposure Time [hours (days)]	MNPs				Main findings[d]	Ref.
			Bioplastic Type (μm)[a]	Petroleum-based Plastic Type(μm)[a]	Tested Concentration	Characterization Techniques[c]		
		336 (14)	Irregular PLA (2–10)	None	400 μg/L	DLS, ELS	↓ Total population and clutch size in all generations compared to control ↓ Egg diameter in all generations, excepting the recovery (F5) ↑ Premature hatching was observed over generations ↓ Prosome length in F1 and F3 compared to control ↓ Prosome width and volume in F1, F3, and F4 compared to control ↓ Lipid droplets only in F4 compared to control	[55]

(Continued)

Table 10.1 (*Continued*)

Groups	Species	Exposure Time [hours (days)]	MNPs				Main findings[d]	Ref.
			Bioplastic Type (μm)[a]	Petroleum-based Plastic Type(μm)[a]	Tested Concentration	Characterization Techniques[c]		
	Daphnia magna	n.a	Engineered PLA, and fragmented PLA from different items (~0.27) (~0.17)	None	0.1–10 mg/L	NTA, DLS, FTIR, TEM	↑ Survivability when exposed to plastic cup PLA breakdown NPs - Survivability when exposed to dialysed and non-dialysed PLA NPs	[56]
		504 (21)	Irregular PLA (160)	none	0.124–12.5 mg/L	SEM, FTIR	↓ SOD activity dependent of exposure time, significant differences at 14-day-old individuals ↑ CAT activity 21-day-old individuals ↑ GPx activity at 14-day-old individuals and ↓ at 21-day-old individuals - Carbohydrates content	[57]

Groups	Species	Exposure Time [hours (days)]	MNPs				Main findings[d]	Ref.
			Bioplastic Type (μm)[a]	Petroleum-based Plastic Type(μm)[a]	Tested Concentration	Characterization Techniques[c]		
							↓ Protein content at 14-day-old individuals ↑ Lipid content at 14-day-old individuals and ↑ caloric content ↑ Distance moved and swimming speed at 21-day-old individuals	
		48 (2)	Irregular PLA (50)	Irregular PE (50)	0.5–20 mg/L	FE-SEM, ATR-FTIR, X-ray photoelectron spectroscopy	PLA MPs general acute toxicity was slightly higher than PE ↑ ROS content was observed in PE and PLA treatment ↑ LPO was observed in PE and PLA treatment ↑ SOD activity only in PLA treatment	[58]

(*Continued*)

Table 10.1 (*Continued*)

Groups	Species	Exposure Time [hours (days)]	MNPs				Main findings[d]	Ref.
			Bioplastic Type (µm)[a]	Petroleum-based Plastic Type(µm)[a]	Tested Concentration	Characterization Techniques[c]		
		48 (2) 504 (21)	Irregular PLA (1–80)	Irregular PET (1–80)	1–20 mg/L	FE-SEM, ATR-FTIR	48-h acute assay Computed EC50 values for PLA were slightly lower than PET not significantly different 21-day chronic assay ↓ Survivability in 5 mg/L PLA treatment, compared to control and PET treatment ↑ Deformities in 5 mg/L PLA treatment ↓ Offsprings per female in 5 mg/L in both MP treatments ↓ *vtg* relative expression in PLA treatments compared to PET treatments	[59]

Groups	Species	Exposure Time [hours (days)]	MNPs				Main findings[d]	Ref.
			Bioplastic Type (μm)[a]	Petroleum-based Plastic Type(μm)[a]	Tested Concentration	Characterization Techniques[c]		
							↑ Adenylate kinase relative expression in 1 mg/L PLA treatment only ↓ *ak* relative expression in 5 mg/L PLA treatment only	
	Dapnhia magna	24 (1)	Irregular PLA and PHB (<60)	Irregular HDPE (<60)	150 mg/L	SEM	Standard plastics ↓ In basal locomotor activity when exposed to PHB treatment compared to control ↓ In visual motor response in PHB and PLA treatments compared to control ↓ Maximal photomotor response in PLA treatment	[60]

(*Continued*)

Table 10.1 (*Continued*)

Groups	Species	Exposure Time [hours (days)]	MNPs				Main findings[d]	Ref.
			Bioplastic Type (μm)[a]	Petroleum-based Plastic Type(μm)[a]	Tested Concentration	Characterization Techniques[c]		
							↓ All standard plastics inhibited feeding Commercially available plastics ↑ In basal motor response in PHB-1, and ↓ PHB-3(3) ↓ Visual motor response in PHB-1, PHB-2(3), PLA-4(4), PLA-5, PLA-6, PLA-7, PLA8, and HDPE-9 treatments ↓ Maximal photomotor response in PHB-2(3), PHB-3(3), PLA-4(3), PLA-6 and ↑ in PHB-1 and KAO treatments ↓ PHB-2(3), PHB-3(3), PLA-4(3), PLA-5, PLA 7, and PLA 8 inhibited feeding activity	

Groups	Species	Exposure Time [hours (days)]	MNPs				Main findings[d]	Ref.
			Bioplastic Type (μm)[a]	Petroleum-based Plastic Type(μm)[a]	Tested Concentration	Characterization Techniques[c]		
	Diaphanosoma celebensis	24 (1)	Irregular PLA, β-cyclodextrin grafted PLA and PBAT (2–10)	Irregular LDPE (2–10)	0.5–10 mg/L	SEM, DLS, and ELS	Reproduction was affected at all concentrations, independent of MP type ↓ ROS production in all treatments, with the most significant decrease observed in LDPE ↑ SOD activity in PLA treatment, ↓ in CD-PLA and PBAT, - in LDPE treatment ↑ CAT activity in PLA and LDPE treatments and ↓ in CD-PLA and PBAT treatments ↑ *cat1* relative expression in CD-PLA and ↓ in LDPE treatments	[61]

(*Continued*)

Table 10.1 (*Continued*)

Groups	Species	Exposure Time [hours (days)]	MNPs				Main findings[d]	Ref.
			Bioplastic Type (μm)[a]	Petroleum-based Plastic Type(μm)[a]	Tested Concentration	Characterization Techniques[c]		
							↑ *CuZnsod1* relative expression in CD-PLA, PBAT, and LDPE treatments ↑ *Mnsod* relative expression in PLA, CD-PLA, and PBAT treatments ↑ P-gp relative fluorescence in all treatments, with the highest increase in PLA and CD-PLA ↑ MRP relative fluorescence in CD-PLA and PBAT, ↓ LDPE treatments	

Groups	Species	Exposure Time [hours (days)]	MNPs				Main findings[d]	Ref.
			Bioplastic Type (μm)[a]	Petroleum-based Plastic Type(μm)[a]	Tested Concentration	Characterization Techniques[c]		
Echinodermata	*Paracentrotus lividus*	48 (2)	Spherical PHB, Irregular PBS, PBSA, PLA, and PCL (0.64) (193) (207.70) (335) (164.90)	none	1–10 mg/L	SEM, FTIR, differential scanning calorimetry	- Egg fertilization in any treatment - PBS and PBSA did not induce effects on morphological endpoints ↑ Time for first cleavage and malformations in 10 mg/L of PHB and PLA ↑ Time to develop and malformations in 5 and 10 mg/L of PCL <u>Stress-related genes relative expression</u> ↓ Relative expression of *arf1*, *MTase*, *hif1a*, *parp-1*, *sdh*, *p53*, *ChE*, *cyp-2ui*, *gst*, *gapdh*, *pks*, and *sult1* in 5 mg/L PCL, PHB, and PLA treatments	[62]

(Continued)

Table 10.1 (*Continued*)

Groups	Species	Exposure Time [hours (days)]	MNPs				Main findings[d]	Ref.
			Bioplastic Type (μm)[a]	Petroleum-based Plastic Type(μm)[a]	Tested Concentration	Characterization Techniques[c]		
							↓ *ercc3, hsp56, hsp60, hsp70, nf-kB, and p38 MAPK* only in PLA and PHB treatment Development-related genes relative expression ↑ *cm-k* in PCL, PHB, and PLA treatments and *tak1* only in PHB and PCL ↓ *bp10, nodalm hh, m-vgl, and notchless* only in PCL and PLA treatments Skeletogenic-related genes relative expression ↓ *nec* and *sm50* in PCL, PHB, and PLA treatments ↓ *p16* in PHB, *c-jun*, and *p19* in PHB and PLA,	

Groups	Species	Exposure Time [hours (days)]	MNPs				Main findings[d]	Ref.
			Bioplastic Type (µm)[a]	Petroleum-based Plastic Type(µm)[a]	Tested Concentration	Characterization Techniques[c]		
							bmp5-7 in PLA and *sm30* only in PCL treatments ↑ *p16* in PLA treatment <u>Detoxification-related genes relative expression</u> ↓ *mdr1*, *mt5*, *mt6*, and *mt8* in PCL, PHB, and PLA treatments ↓ *mt4* and *mt5* in PHB and PCL treatments ↓ *mt* in PHB and PLA treatments	
Insecta	*Chironomous kiinensis*	600 (25)	Spherical PLA (0.25)	none	0.5 g/kg	TEM, DLS and ELS, FTIR, X-ray photoelectron spectroscopy	↓ Gut epithelia histological alterations in PLA treatments ↓ GSH/GSSG balance Negative correlation between DAO and D-lac	[63]

(*Continued*)

Table 10.1 (*Continued*)

Groups	Species	Exposure Time [hours (days)]	MNPs				Main findings[d]	Ref.
			Bioplastic Type (μm)[a]	Petroleum-based Plastic Type(μm)[a]	Tested Concentration	Characterization Techniques[c]		
							(gut flora biomarkers) and particle size Positive correlation between DAO and D-lac and oxidative potential in PLA treatment ↓ In microbiome keystone phyla (e.g., Actinobacteriota) ↓ In genus diversity across PLA treatments	
Chordata	*Lates calcifer*	48 (2)	Spherical PHB (1)	Spherical PS (1)	0.5 and 1 mg/L	HR-SEM, FS, FTIR, DLS, and ELS	Both MPs were assimilated into various tissues and caused histological damage ↑ Muscle tissues alterations (e.g., dystrophic changes,	[64]

Groups	Species	Exposure Time [hours (days)]	MNPs				Main findings[d]	Ref.
			Bioplastic Type (μm)[a]	Petroleum-based Plastic Type(μm)[a]	Tested Concentration	Characterization Techniques[c]		
							muscle bundle degeneration among others) were observed in both MPs treatment ↑ Degenerative changes in endocardium and necrosis were observed in PS and PHB treatment	
	Oreochromis mossambicus	336 (14)	Spherical PLA (2.52)	Spherical PVC (1.58)	0.1 mg/L	SEM, FTIR	↑ Bioaccumulation in both MP treatments and PVC had the higher nº of particles per mg of tissue. ↑ MDA content at 7-day sampling in PVC treatment and - MDA content had no differences at 14-day sampling.	[65]

(*Continued*)

Table 10.1 (*Continued*)

Groups	Species	Exposure Time [hours (days)]	MNPs				Main findings[d]	Ref.
			Bioplastic Type (µm)[a]	Petroleum-based Plastic Type(µm)[a]	Tested Concentration	Characterization Techniques[c]		
							- SOD activity in both MPs at both sampling timepoints ↓ Cell integrity, mitochondrial swelling, and impairment of the gut barrier in PLA were severe compared to PVC treatment. ↓ Altered gut metabolic profile in both PLA and PVC treatment	
	Danio rerio	260 (15)	Irregular PLA (135.12)	Irregular PET (111.12)	10 mg/L PET 17.5 mg/L PLA	SEM	5-day ingestion and excretion assay ↑ Ingestion of higher PLA MPs than PET, where after excretion both MPs were easily eliminated.	[66]

Groups	Species	Exposure Time [hours (days)]	MNPs				Main findings[d]	Ref.
			Bioplastic Type (μm)[a]	Petroleum-based Plastic Type(μm)[a]	Tested Concentration	Characterization Techniques[c]		
							15-day exposure ↑ Intestinal weight in PLA treatment compared to PET ↓ Length of intestinal villi in PLA compared to PET treatment ↓ Goblet-like cells in PLA compared to PET treatment ↑ Vacuolization and cilia defects in PLA treatment ↑ PLA induced greater changes in the gut microbiome, altering the dominant phyla to Proteobacteria, while Firmicutes comminated in PET and control group	

(Continued)

Table 10.1 (*Continued*)

Groups	Species	Exposure Time [hours (days)]	MNPs				Main findings[d]	Ref.
			Bioplastic Type (μm)[a]	Petroleum-based Plastic Type(μm)[a]	Tested Concentration	Characterization Techniques[c]		
		849 (35)	Almost spherical PLA (100)	None	5 mg/L	SEM, IR-spectroscopy, ELS	↓ In testosterone and oestradiol levels in PLA treatment[º] compared to control ↑ Metabolomic alterations affecting lipid, amino acid, and purine metabolism in PLA treatment Offsprings exhibited ↑ heart rate and ↑ mortality in PLA treatment	[67]

↓ Decrease; ↑ Increase; - No effect. [a] Polystyrene (PS); polypropylene (PP); polyethylene (PE); low-density polyethylene (LDPE); high-density polyethylene (HDPE); polyvinyl chloride (PVC); polyterephthalate (PET); polyvinylidene difluoride (PVDF). [b] Polylactic acid (PLA); polyhydroxybutyrate (PHB); polycaprolactone (PCL); polybutylene succinate (PBS); poly(butylene succinate-co-adipate) (PBSA); polybutylene adipate terephthalate (PBAT). [c] Nanoparticle tracking analysis (NTA); scanning electron microscopy (SEM); field emission scanning electron microscopy (FE-SEM); high resolution scanning electron microscopy (HR-SEM); fluorescence spectrophotometer (FS); infrared (IR); transmission electron microscopy (TEM); laser particle size distribution analysis (LPSA); laser diffraction analysis (LDA); dynamic light scattering (DLS); Fourier-transform infrared spectroscopy (FTIR); attenuated-total reflection Fourier-transform infrared spectroscopy (ATR-FTIR); matrix-assisted laser desorption and ionization (time-of-flight)^2 (MALDI-TOF/TOF); electrophoretic light scattering (ELS) [d] Cholinesterase (ChE); acetylcholinesterase (AChE); glutathione (GSH); glutathione disulfide (GSSG); glutathione-S-transferase (GST); glutathione peroxidase (GPx); catalase (CAT); superoxide dismutase (SOD); malondialdehyde content (MDA); vitellogenin (vtg) reactive oxygen species (ROS); lipid peroxidation (LPO); lactate dehydrogenase (LDH); dissolved inorganic carbon (DIC); chlorophyl a (Chl a); phycobiliproteins (PBPs); microcystin-LR (MC-LR); inhibition response (IR); pyruvate kinase (PK); succinate dehydrogenase (SDH); adenylate kinase (AK); kaolin particles (KAO); P-glycoprotein (P-gp); multidrug resistance protein (MRP) availability of carbon sources are pointed out as the main causes of increased cell proliferation and population growth in the presence of PLA. Exposure to PLA has also been shown to induce alterations in community composition, decreasing the population growth of *Microcystis aeruginosa* in co-culture with *Scenedesmus obliquus*, while decreasing the inhibitory effect of the latter species' populational growth, leading to competitive advantages [45].

PCL NPs, with a mean diameter inferior to 0.1 μm, significantly induce oxidative stress in cyanobacteria during a 72 h exposure period, particularly in *Synechococcus* sp., leading to increased reactive oxygen species (ROS) production, membrane potential disruption, and increased hydrogen peroxide activity and decreases in the Chl a content [46].

These reported effects, adding to the reported increases in microcystin production at least in one study [42], build interesting evidence on the biological effects of these biodegradable plastics, indicating an increasing potential risk of harmful algal bloom formation associated with many environmental and human health, as well as socioeconomic impacts. For example, altering aquatic ecosystems, either through biotoxin release or by altering environmental conditions (e.g., oxygen saturation, light conditions, among others), which may have detrimental effects on keystone species of fishes and other organisms, as well as impacts on wild and aquacultured economically relevant stock organisms used for human consumption (e.g., fish, mussels, clams) [47].

10.4.2 Cnidarians

Both bio- and petroleum-based plastics have been reported to impact cnidarians such as *Hydra viridissima* and *Aurelia* sp. Thus, significant impacts on mortality, feeding rates, malformations, and altered regenerative capacity have been reported, with effects depending on the type and concentration of MPs involved.

In a study with *H. viridissima*, PHB MPs, with a mean size of 0.2 μm, at concentrations ranging from 0.001 to 10 mg/L showed no significant effects on mortality, feeding rate, or malformations after 96 h exposure. However, after an 8-day exposure period, although mortality and malformations still did not differ from the controls, an increase in feeding rate was observed across all treatments (except for the 1 mg/L treatment). After 12 days, survival remained consistent, but there was a notable increase in malformations and feeding rates [48].

Tamayo-Belda et al. [49] exposed *H. veridissima* to PLA (0.078 μm), PP (0.142 μm), PS (0.122 μm), and LDPE (0.091 μm) at concentrations between 0.001 and 100 mg/L and reported different effects after 96 h exposure. PP and LDPE at 100 mg/L induced increased mortality, and PP, LDPE, and PS treatments, at 10 and 100 mg/L, induced malformations. PLA exposure led to mortality and malformations but also showed, compared to other treatments, an ability to reduce feeding rates. Furthermore, LDPE (at 10 mg/L) showed the ability to impair the regenerative capacity of *H. viridissima*.

For *Aurelia* sp., a 24 h exposure to PLA and polyvinylidene difluoride (PVDF) MPs, ranging from 50 to 500 μm, at concentrations from 1 to 100 mg/L, led to observable particle uptake at 100 mg/L. PLA and PVDF treatments increased mortality, and the pulsation frequency of *Aurelia* sp. ephyra was reduced in PLA and PVDF treatments, indicating a negative impact on their locomotive and possibly their physiological function [50].

Overall, these studies suggest that bioplastics such as PHB and PLA, as well as petroleum-based plastics such as PP, PS, and LDPE, can significantly affect cnidarians, with varying degrees of severity depending on the type, concentration, and duration of exposure.

10.4.3 Molluscs

Over 31 days of exposure to PLA MPs and a mixture of petroleum-based plastics (PVC, PET, PS, and PP MPs), all approximately 50 μm in size, at concentrations ranging from 75 to 75,000 particles per kilogram, led to significant morphological alterations in the tissues of the marine gastropod *Ilyanassa obsoleta*. Both PLA and the petroleum-based plastic mixture caused an increase in the number of intercellular gaps and a higher distribution of granules at extreme concentrations. Additionally, both plastic types resulted in an increased number of mature hemocytes at high concentrations. However, PLA uniquely led to a decrease in the number of embryos per egg capsule, indicating a specific reproductive impact not observed with petroleum-based plastics [51].

Moreover, Zhong et al. [52], who exposed the mussel *Mytilus coruscus* to PLA (5 μm) and PS MPs (5 μm) at concentrations between 0.31 and 5.51 mg/L, over 14 days, reported different effects. Both PLA and PS reduced absorption efficiency, with PS having a more pronounced effect at medium and high concentrations. Clearance rates were also reduced at the highest concentrations of both MPs. In the gills, PLA and PS exposure led to decreased lactate dehydrogenase (LDH) and pyruvate kinase (PK) activities, with PLA showing a stronger effect. Succinate dehydrogenase (SDH) activity remained unchanged, while catalase (CAT) activity and glutathione (GSH) content increased in response to PS at medium and high concentrations. In the hemolymph, low concentrations of both plastics decreased LDH activity, while PK activity increased with medium and high concentrations of PS and medium concentrations of PLA. Glutathione peroxidase (GPx) activity was elevated at low concentrations of PLA and high concentrations of PS, while GST activity decreased across all plastic treatments.

These findings shed light on the potentially pernicious effects of bioplastics on molluscs, where PLA showed effects similar to other petroleum-based MPs but also induced distinct reproductive effects on the marine snail *I. obsoleta* that were not observed in MPs mixture treatment. Moreover, PLA and PS similarly affected physiological functions and enzyme activities related to energy metabolism in the mussel *M. coruscus*. Both types of MPs reduced absorption efficiency and clearance rate and LDH, PK, and SDH enzyme activity. These enzymes are of utmost importance in energy metabolism, where LDH catalyses the conversion of pyruvate to lactate in the glycolytic pathway under anaerobic conditions, while PK is a key element in maintaining glucose levels in hemolymph and SDH is crucial for the electron transport chain in aerobic respiration metabolism. The reported subcellular effects align with the physiological response of mussels, showing signs of physiological stress and energy intake. This review comprised only two articles reporting the effects of bioplastics in molluscs, highlighting the need for more tailored assessments to better understand their ecological impacts.

10.4.4 Crustaceans

The effects of petroleum-based bioplastics on various crustacean species were reviewed, which revealed distinct responses in terms of reproductive effects, behavioural changes, and oxidative stress responses.

In a study with *Tigriopus japonicus*, exposure for 48 h to either irregular PBAT/PLA copolymer or PE MPs, both ranging between 10 to 40 μm, at concentrations ranging from 20 to 500 mg/L, showed no significant effects on the copepod survival. This result suggests that, at these levels, both bioplastics and fossil-based MPs have a similar and non-lethal impact on this species [53].

Eurytemora affinis experienced microbial community structure changes, when exposed to 300 μg/L PBAT or LDPE MPs, with sizes between 1 and 10 μm, for 14 days, where microbial community alterations were independent of MP type and more correlated with the organism's exposed generations [54]. Ali et al. [55] reported that exposure to PLA (between 2 and 10 μm) for 14 days, at slightly higher concentrations (400 μg/L), induced a substantial decline in total population, clutch size, and prosome dimensions across all generations, along with increased premature hatching and reduced lipid droplets, indicating a more severe impact on reproductive and developmental endpoints than LDPE.

In a study with *Daphnia magna*, individuals were exposed to PLA NPs, with an approximate size of 0.27 μm, originating from different commercially available items (i.e., soup cup lids, 3D printer filaments, and plastic cups), and to synthesized engineered PLA NPs (0.17 μm). The authors observed that only PLA NPs originating from the plastic cup increased the survival of daphnids compared to control [56]. In another study, *D. magna* exposed for 21 days to higher concentrations of PLA MPs (160 μm) led to time-dependent decreases in superoxide dismutase (SOD) activity, increased CAT and GPx activity, and altered protein, lipid, and caloric content, indicating significant metabolic and oxidative stress responses [57]. PLA exposure also proved to alter *D. magna* behaviour, with reported increased movement and swimming speed [57]. In another study, *D. magna* exhibited slightly higher sensitivity to 50 μm PLA than to PE MPs during a 48 h exposure, with increased ROS production and lipid peroxidation (LPO) observed after exposure to both MP treatments. PLA exposure uniquely increased SOD activity compared to PE [58]. In a similar comparison between PET and PLA, sized between 1 and

80 μm, the latter had slightly lower, but not significantly different, EC_{50} values. However, during a chronic 21-day assay with 1 and 5 mg/L, PLA led to reduced survival, increased deformities, and fewer offspring per female, along with altered *vtg* and *ak* gene expression, indicating more severe long-term effects compared to PET [59]. The impact of 24 h exposure to 150 mg/L of standard and commercially available PHB and PLA MPs (<60 μm) on *D. magna* was also notable, with both MPs reducing basal locomotor activity, visual motor response, and maximal photomotor response compared to controls. Standard plastics, including HDPE, inhibited feeding activity, but PLA and PHB had more pronounced effects on these behavioural endpoints [60].

In a study with *Diaphanosoma celebensis*, reproductive endpoints were negatively affected by PLA, β-cyclodextrin-grafted PLA, PBAT, and LDPE MPs, ranging between 2 and 10 μm, across concentrations from 0.5 to 10 mg/L over a period of 24 h. PLA triggered increased oxidative stress, as evidenced by higher SOD and CAT activity and altered *cat1* gene expression, particularly in comparison to LDPE, which had milder effects on enzyme activity and gene expression, yet ROS production was mostly significant in LDPE. PLA and PBAT were associated with more significant alterations in oxidative stress markers and an increase in p-glycoprotein and multidrug resistance protein fluorescence biomarkers, indicating induction of multi-xenobiotic resistance genes for detoxification [61].

Overall, bioplastics such as PLA and PBAT tend to induce more severe oxidative stress, and reproductive and developmental effects in crustaceans compared to fossil-based plastics such as PE, PET, and LDPE. While the acute toxicity of bioplastics may be comparable to or slightly higher than fossil-based plastics, chronic exposure results in more profound physiological and behavioural disruptions, suggesting that bioplastics might pose greater long-term risks to aquatic invertebrates.

10.4.5 Echinoderms

In a study with *Paracentrotus lividus*, 48 h exposure to various bioplastic MPs, including PHB (0.64 μm), PBS (193 μm), PBSA (207.7 μm), PLA (335 μm), and PCL (164.9 μm), revealed several biological effects at concentrations ranging from 1 to 10 mg/L. Although egg fertilization was unaffected by any treatment, and PBS and PBSA did not induce any morphological changes, at a higher concentration (10 mg/L), PHB and PLA led to an increased time for the first cleavage and malformations, while PCL caused similar issues at both 5 and 10 mg/L. Stress-related genes such as *arf1*, DNA *MTase*, *hif1a*, and others showed decreased mRNA expression levels in PCL, PHB, and PLA treatments. The mRNA expression levels of specific stress-related genes such as *ercc3* and *hsp70* were downregulated only in PLA and PHB treatments. Developmental genes *cm-k* and *tak1* were upregulated, while others like *bp10* and *nodal* were downregulated by PCL and PLA. Skeletogenic genes *nec* and *sm50* were downregulated across PCL, PHB, and PLA treatments, with specific decreases observed in other genes depending on the plastic type. Lastly, detoxification-related genes showed reduced expression levels across various treatments, particularly in PCL, PHB, and PLA exposures [62]. In this literature review, a single study regarding the effects of MNBPs on echinoderms was found, highlighting the need for more research to fully understand the effects of bioplastics on this taxon.

10.4.6 Insects

Exposure to bioplastics such as PLA can have a significant impact on insect gut health and microbiome composition. Zhang et al. [63] exposed *Chironomus kiinensis* larvae to 0.5 g/kg PLA NPs (0.25 μm) over 25 days and reported histological alterations in the gut epithelia and a disruption in the GSH/GSSG balance, indicating oxidative stress. A negative correlation was observed between gut inflammation biomarkers and metabolic disorders of the gut flora biomarker (i.e., diamine oxidase and D-lactate) and particle size, while a positive correlation was found between these biomarkers and the oxidative potential of the PLA treatments. Additionally, a decline in the abundance of microbiome keystone phyla, such as Actinobacteriota, and a reduction in genus diversity were noted across PLA treatments. However, it should also be highlighted that a single study focusing on insects was found, emphasizing the need for more research to fully understand the effects of bioplastics on insect health and ecosystems.

10.4.7 Fish

Bioplastics and fossil-based plastics show significant differences in effects on tissue damage, bioaccumulation, metabolic changes, and impacts on offspring. In a 48 h study with *Lates calcarifer*, 1 mg/L PHB and PS MPs with 1 μm were assimilated into various tissues, leading to notable histological damage. Muscle tissue alterations, such as dystrophic changes and muscle bundle degeneration, were observed in PHB and PS treatments. Additionally, degenerative changes in the endocardium and necrosis were noted, indicating a strong pathological response to these MPs [64].

In a 14-day study with *Oreochromis mossambicus*, PLA (2.52 μm) and PVC (1.58 μm) MPs, at 2.52 mg/L and 1.58 mg/L, respectively, were bioaccumulated, with PVC showing a higher number of particles per mg of tissue. Interestingly, while malondialdehyde content increased after 7 days in the PVC treatment, no significant changes were observed after 14 days. SOD activity remained unchanged in all types of MPs. However, cell integrity, mitochondrial swelling, and gut barrier impairment were more severe in the PLA treatment than in PVC, with both treatments also causing alterations in the gut metabolic profile [65].

Duan et al. [66], who performed a 5-day ingestion and excretion assay with *Danio rerio*, revealed that PLA MPs (135.12 μm) were ingested at higher rates than PET (111 μm) MPs, though both were easily eliminated after excretion. In a separate 15-day exposure to 10 mg/L of PET or 17.5 mg/L of PLA, the latter caused a greater increase in intestinal weight compared to PET, along with shorter intestinal villi, fewer goblet-like cells, and more pronounced vacuolization and cilia defects [66]. Furthermore, PLA induced significant changes in the gut microbiome, altering the dominant phyla to Proteobacteria, whereas Firmicutes were dominant in both the PET and control groups [66]. In another study, exposure for 35 days to 5 mg/L PLA (100 μm) MPs led to a decrease in testosterone and estradiol levels and caused significant metabolomic alterations affecting lipid, amino acid, and purine metabolism. Offspring from PLA-exposed parents exhibited an increased heart rate and higher mortality rates, highlighting the potential long-term reproductive and developmental risks associated with PLA exposure [67].

Moreover, Tamayo-Belda et al. [49] exposed *D. rerio* embryos to PLA, PP, PS, and LDPE NPs for 96 h to concentrations ranging from 0.001 to 100 mg/L and reported a decrease in heart rate for organisms exposed to PLA and PP, while an increase in heart rate for PS NPs treatment. These findings correlate with behavioural changes in the locomotor activity assay, where PLA and PP induced more alterations compared to the other NPs tested, causing a decrease in all the parameters during the light period, even at low concentrations, while PS was associated with an overall increase in locomotor activity parameters and LDPE barely altered any parameter analysed for larvae behaviour.

10.5 An Overlooked Side of Bioplastics? Additives, Oligomers, and Monomers

Plastic production and processing rely on the polymerization of repeating monomer units (low molecular weight molecules) during which several chemical substances are introduced. Additives, such as plasticizers (to increase flexibility), antioxidants (to protect the polymer matrix from oxidation), and flame retardants (to impart fire resistance), are incorporated to enhance, maintain, or add new properties to everyday plastics. Additionally, processing aids, such as polymerization catalysts, solvents, or lubricants, may be required to facilitate production. These substances are classified as intentionally added substances (IAS) [68]. However, non-intentionally added substances (NIAS) can also arise through contamination, by-products, or breakdown products during manufacturing [69]. Over 10,000 substances, including monomers, additives, and processing aids, that are not chemically bound to the polymer matrix have been identified, many of which may have been previously overlooked [68, 70]. These substances may significantly affect the biological effects of micro- and nanoplastics and could be more harmful than the polymer itself. The growing use of additives to bridge the performance gap between bioplastics and conventional plastics could exacerbate their potential adverse effects. Although there is increasing concern about the impact of these chemicals, more research is needed to fully understand their ecological consequences (Fig. 10.4).

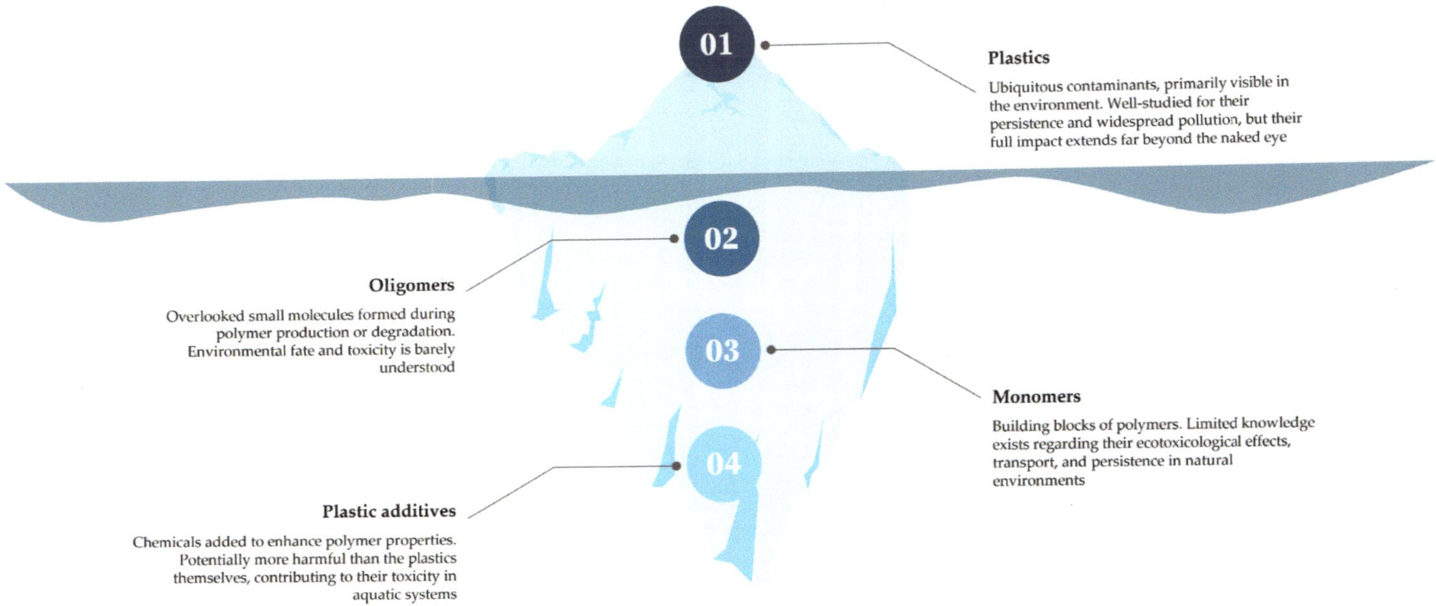

Figure 10.4 Layers of plastic contamination: from visible contaminants to less-studied xenobiotics.

Oligomers, defined as molecules with a small number of repeating monomer units (typically below 1000 Da), possess migration properties [71]. These polymer oligomers, typically containing 2 to 40 repeating units, are considered primary forms of NIAS and IAS in polymer production [72]. They can form during polymer production (primary oligomers) or through abiotic and biotic degradation (secondary oligomers) [73]. Oligomers can be found in significant amounts within the polymer matrix, with studies reporting that they constitute 0.5 to 1.3% of PET products [74]. Despite their presence, there is limited research on oligomers, with only 368 studies published between 2004 and 2023, highlighting a lack of scientific data on their potential environmental and toxicological effects [71].

The formation of secondary oligomers through plastic degradation and ageing is complex. Abiotic factors, such as UV radiation, temperature, and biodegradation, contribute to the breakdown of polymer chains into lower-molecular-weight oligomers [75, 76]. Environmental conditions such as polarity, pH, and salinity may further influence this process [77, 78]. It is already known that different PET oligomers, both circular and linear, can migrate in the presence of food simulants used in packaging studies. As plastics degrade, the presence of oligomers in the environment is expected to increase, with their distribution likely mirroring that of plastic pollution itself [79]. There is a limited understanding of oligomer toxicity. When assessing the ecotoxicological effects of micro- and nanoplastics, the presence of oligomers may be overlooked, especially in studies focused on secondary MPs [71]. It is essential to consider oligomers when evaluating the biological effects of plastic degradation. For example, styrene oligomers have been shown to cause inflammation and act as endocrine disruptors [80].

Bioplastics, while marketed as environmentally friendly due to their biodegradability, may be more prone to forming degradation by-products, such as oligomers, especially in environmental contexts where conditions are not optimal for full degradation. For instance, while bioplastics can break down in industrial composting facilities, in natural environments such as oceans or soils, they may degrade incompletely, leading to persistent microplastics and oligomers [22, 81]. For example, research has shown that oligomers from hydrolytic degradation of PCL can negatively impact nitrogen fixation in cyanobacteria [46]. The diversity of these oligomers, such as variations in the number of repeating units and functional groups, can influence their physicochemical properties and environmental behaviour. Additionally, bioplastics are often commercially available as blends, copolymers, or grafted polymers, further increasing the variety of oligomers that may enter the environment.

Monomers also have limited data regarding their ecotoxicological effects, transport, and fate. For example, styrene monomers have been classified as Group 2A (probably carcinogenic to humans) by the International Agency for Research on Cancer (IARC) [82]. This underscores the need for more research to understand the potential biological effects of these materials on non-target aquatic species. Furthermore, biobased plastics often contain monomers that are naturally occurring molecules (e.g., lactic acid), which could disrupt energy metabolism and cause metabolic disorders in certain organisms. Overall, it becomes clear that research on the effects of plastics should also consider the molecules to allow a better understanding of the mechanisms of toxicity.

10.6 Conclusion and Future Directions

The research on bioplastics, often considered eco-friendlier than the "traditional" fossil-based plastics, is rapidly growing and expected to continue expanding. However, the limited available studies on the ecotoxicity of these materials indicate that their use as sustainable alternatives should be considered with caution. The main finding of the recent studies mentioned previously that considered both types of plastics suggests that bioplastics may have similar, if not greater, effects on aquatic organisms than conventional plastics. In some cases, bioplastics have been found to exert unique effects that have not been associated with their fossil-based counterparts. Moreover, many of the existing studies lack environmental relevance, further highlighting the need for more comprehensive research.

Future research should focus on the following: (a) assessing the effects of MNBPs at environmentally relevant concentrations, (b) evaluating the persistent effects of both acute and prolonged exposures to MNBPs, (c) examining the impact of MNBPs on ecosystem communities, and (d) exploring how climate change scenarios and ageing, caused by biotic and abiotic factors, may influence the effects of MNBPs.

Additionally, new research directions should be considered when addressing the impacts of plastic pollution, with a particular emphasis on bioplastic contamination and its contribution to the release of other environmental contaminants, such as oligomers, monomers, and plastic additives. Polymer oligomers and monomers are underexplored areas, which may outnumber different types of polymers themselves. This research area urgently requires extensive screening, database development, and assessments of persistence, bioaccumulation, and toxicity to better understand their environmental and biological impacts.

CRediT authorship contribution statement

Márcio Barreto: Conceptualization, Investigation, Writing - original draft, Writing - review & editing. **Isabel Lopes**: Writing - review & editing. **Miguel Oliveira**: Conceptualization, Investigation, Writing - review & editing, Supervision.

Declaration of competing interest

The authors declare that they have no known competing financial interests or personal relationships that could have appeared to influence the work reported in this chapter.

Acknowledgements

Thanks are due to financial support through national funding by OE and FCT, Foundation for Science and Technology, and FEDER, within the PT2020 Partnership Agreement and Compete 2020, through the research project Nanoplanet (2022.02340.PTDC) and financial support to CESAM (UIDB/50017/2020+UIDP/50017/2020+LA/P/0094/2020). Support was also provided by the European Union under BioUptake project (GA 101057049).

References

1. Andrady, A.L., *Plastics and Environmental Sustainability*. 2015, Hoboken, NJ: John Wiley & Sons, Inc.
2. Oliveira, M., and M. Almeida, The why and how of micro(nano)plastic research. *TrAC Trends in Analytical Chemistry*, 2019. **114**: 196–201.

3. Hernandez, L.M., N. Yousefi, and N. Tufenkji, Are there nanoplastics in your personal care products? *Environmental Science & Technology Letters*, 2017. **4**(7): 280–285.
4. Carpenter, E.J., and K.L. Smith, Plastics on the Sargasso Sea surface. *Science*, 1972. **175**(4027).
5. Yin, K., et al., A comparative review of microplastics and nanoplastics: Toxicity hazards on digestive, reproductive and nervous system. *Science of The Total Environment*, 2021. **774**: 145758–145758.
6. Gigault, J., et al., Nanoplastics are neither microplastics nor engineered nanoparticles. *Nature Nanotechnology*, 2021. **16**(5): 501–507.
7. Mattsson, K., et al., Nanoplastics in the aquatic environment, in *Microplastic Contamination in Aquatic Environments*. 2018, Elsevier. pp. 379–399.
8. Barreto, M., I. Lopes, and M. Oliveira, Micro(nano)plastics: A review on their interactions with pharmaceuticals and pesticides. *TrAC Trends in Analytical Chemistry*, 2023. **169**: 117307–117307.
9. Petersen, E., et al., Evaluation of bioaccumulation of nanoplastics, carbon nanotubes, fullerenes, and graphene family materials. *Environment International*, 2023. **173**: 107650–107650.
10. European Commission, *The European Green Deal*. 2019.
11. European Commission, *Circular Economy Action Plan*. 2020.
12. Venâncio, C., I. Lopes, and M. Oliveira, Bioplastics: Known effects and potential consequences to marine and estuarine ecosystem services. *Chemosphere*, 2022. **309**: 136810–136810.
13. ETC/CE, Pathways towards circular plastics in Europe—good practice examples from countries, business and citizens. https://www.eea.europa.eu/publications/pathways-towards-circular-plastics-in
14. Savva, K., et al., Cytotoxicity assessment and suspected screening of PLASTIC ADDITIVES in bioplastics of single-use household items. *Chemosphere*, 2023. **313**: 137494–137494.
15. Zimmermann, L., et al., Are bioplastics and plant-based materials safer than conventional plastics? In vitro toxicity and chemical composition. *Environment International*, 2020. **145**: 106066–106066.
16. EN, *EN 13432, Certified Bioplastics Performance in Industrial Composting*, 2000.
17. Viera, J.S.C., et al., Are biodegradable plastics an environmental rip off? *Journal of Hazardous Materials*, 2021. **416**: 125957–125957.
18. Kokalj, A.J., et al., Quality of nanoplastics and microplastics ecotoxicity studies: Refining quality criteria for nanomaterial studies. *Journal of Hazardous Materials*, 2021. **415**: 125751–125751.

19. Page, M.J., et al., The PRISMA 2020 statement: an updated guideline for reporting systematic reviews. *BMJ*, 2021. **372**: n71.
20. European Bioplastics, *Bioplastics Market Development Update 2023*. 2024.
21. Ghasemlou, M., C.J. Barrow, and B. Adhikari, The future of bioplastics in food packaging: An industrial perspective. *Food Packaging and Shelf Life*, 2024. **43**: 101279–101279.
22. Kim, M.S., et al., A review of biodegradable plastics: Chemistry, applications, properties, and future research needs. *Chemical Reviews*, 2023. **123**(16): 9915–9939.
23. Taib, N.-A.A.B., et al., A review on poly lactic acid (PLA) as a biodegradable polymer. *Polymer Bulletin*, 2023. **80**(2): 1179–1213.
24. Yu, J., et al., PLA bioplastic production: From monomer to the polymer. *European Polymer Journal*, 2023. **193**: 112076–112076.
25. Raquez, J.-M., et al., Polylactide (PLA)-based nanocomposites. *Progress in Polymer Science*, 2013. **38**(10–11): 1504–1542.
26. Chamas, A., et al., Degradation rates of plastics in the environment. *ACS Sustainable Chemistry & Engineering*, 2020. **8**(9): 3494–3511.
27. Westlie, A.H., et al., Synthetic biodegradable polyhydroxyalkanoates (PHAs): Recent advances and future challenges. *Progress in Polymer Science*, 2022. **134**: 101608–101608.
28. Ladhari, S., et al., Recent development of polyhydroxyalkanoates (PHA)-based materials for antibacterial applications: A review. *ACS Applied Bio Materials*, 2023. **6**(4): 1398–1430.
29. Torabi, H., et al., Effects of rheological properties on 3D printing of poly(lactic acid) (PLA) and poly(hydroxy alkenoate) (PHA) hybrid materials. *ACS Applied Polymer Materials*, 2023. **5**(6): 4034–4044.
30. Soroudi, A., and I. Jakubowicz, Recycling of bioplastics, their blends and biocomposites: A review. *European Polymer Journal*, 2013. **49**(10).
31. Barletta, M., et al., Poly(butylene succinate) (PBS): Materials, processing, and industrial applications. *Progress in Polymer Science*, 2022. **132**: 101579–101579.
32. Jariyasakoolroj, P., P. Leelaphiwat, and N. Harnkarnsujarit, Advances in research and development of bioplastic for food packaging. *Journal of the Science of Food and Agriculture*, 2020. **100**(14): 5032–5045.
33. Puchalski, M., et al., Molecular and supramolecular changes in polybutylene succinate (PBS) and polybutylene succinate adipate (PBSA) copolymer during degradation in various environmental conditions. *Polymers*, 2018. **10**(3): 251–251.

34. Kato, S., et al., BioPBS™ (polybutylene succinate), in *Synthetic Biodegradable and Biobased Polymers*, K. Andreas, et al., Ed. 2023, Springer. pp. 269–304.

35. Zhang, J., et al., Ring-opening polymerization of ε-caprolactone with recyclable and reusable squaric acid organocatalyst. *European Polymer Journal*, 2021. **157**: 110643–110643.

36. Arrieta, M.P., et al., Degradation under composting conditions of lysine-modified polyurethane based on PCL obtained by lipase biocatalysis. *Polymer Degradation and Stability*, 2018. **152**: 139–146.

37. Thakur, M., et al., Poly(ε-caprolactone): A potential polymer for biodegradable food packaging applications. *Packaging Technology and Science*, 2021. **34**(8): 449–461.

38. Jian, J., Z. Xiangbin, and H. Xianbo, An overview on synthesis, properties and applications of poly(butylene-adipate-co-terephthalate)–PBAT. *Advanced Industrial and Engineering Polymer Research*, 2020. **3**(1): 19–26.

39. Rebelo, R.C., et al., Increased degradation of PLA/PBAT blends with organic acids and derivatives in outdoor weathering and marine environment. *Polymer*, 2022. **256**: 125223–125223.

40. Peng, B.-Y., et al., Biodegradation and carbon resource recovery of poly(butylene adipate-*co*-terephthalate) (PBAT) by mealworms: Removal efficiency, depolymerization pattern, and microplastic residue. *ACS Sustainable Chemistry & Engineering*, 2023. **11**(5): 1774–1784.

41. Qin, P., et al., Superior gas barrier properties of biodegradable PBST vs. PBAT copolyesters: A comparative study. *Polymers*, 2021. **13**(19): 3449–3449.

42. Kim, J.Y., et al., Effects of micro-sized biodegradable plastics on *Microcystis aeruginosa*. *Science of the Total Environment*, 2024. **912**.

43. Su, Y., et al., Biodegradable and conventional microplastics posed similar toxicity to marine algae *Chlorella vulgaris*. *Aquatic Toxicology*, 2022. **244**: 106097–106097.

44. Tang, B., et al., Revealing the environmental hazard posed by biodegradable microplastics in aquatic ecosystems: An investigation of polylactic acid's effects on *Microcystis aeruginosa*. *Environmental Pollution*, 2024. **344**.

45. Zhen, Z., et al., The competitive advantage of *Microcystis aeruginosa* over *Scenedesmus obliquus* weakened by exposure to polylactic acid microplastics. *Ecotoxicology and Environmental Safety*, 2023. **267**.

46. Tamayo-Belda, M., et al., Identification and toxicity towards aquatic primary producers of the smallest fractions released from hydrolytic degradation of polycaprolactone microplastics. *Chemosphere*, 2022. **303**: 134966–134966.

47. McGowan, S., Harmful algal blooms, in *Biological and Environmental Hazards, Risks, and Disasters*. 2023, Elsevier. pp. 9–53.

48. Santos, A., et al., Short- and long-term toxicity of nano-sized polyhydroxybutyrate to the freshwater cnidarian *Hydra viridissima*. *Science of the Total Environment*, 2024. **917**: 170282–170282.

49. Tamayo-Belda, M., et al., Effects of petroleum-based and biopolymer-based nanoplastics on aquatic organisms: A case study with mechanically degraded pristine polymers. *Science of the Total Environment*, 2023. **883**.

50. Di Giannantonio, M., et al., Ecotoxicity of polyvinylidene difluoride (PVDF) and polylactic acid (PLA) microplastics in marine zooplankton. *Toxics*, 2022. **10**(8): 479–479.

51. Hewins, B., and G. Gibson, Petroleum-based and biodegradable microplastics alter tissue structure and fecundity in the eastern mudsnail (*Ilyanassa obsoleta*). *Canadian Journal of Zoology*, 2022. **100**(12): 776–788.

52. Zhong, Z., et al., Bio-based microplastic polylactic acid exerts the similar toxic effects to traditional petroleum-based microplastic polystyrene in mussels. *Science of the Total Environment*, 2024. **946**: 174386–174386.

53. Tong, H., et al., Micro- and nanoplastics released from biodegradable and conventional plastics during degradation: Formation, aging factors, and toxicity. *Science of the Total Environment*, 2022. **833**: 155275–155275.

54. Thery, J., et al., Multigenerational exposure of microplastics on the microbiota of *E. affinis* (copepod): A comparative study between biodegradable and nonbiodegradable microplastics. *Frontiers in Ecology and Evolution*, 2023. **11**: 1231346–1231346.

55. Ali, W., et al., Acute and multigenerational toxicity of polylactic acid microplastics on a copepod bioindicator. *Environmental Chemistry Letters*, 2024.

56. Kelpsiene, E., et al., Prolonged survival time of *Daphnia magna* exposed to polylactic acid breakdown nanoplastics. *PLoS ONE*, 2023. **18**(9 September).

57. De Felice, B., et al., Multi-level toxicity assessment of polylactic acid (PLA) microplastics on the cladoceran *Daphnia magna. Aquatic Toxicology*, 2024. **272**: 106966–106966.

58. Luangrath, A., et al., Ecotoxicity of polylactic acid microplastic fragments to *Daphnia magna* and the effect of ultraviolet weathering. *Ecotoxicology and Environmental Safety*, 2024. **271**.

59. An, G., et al., Chronic toxicity of biodegradable microplastic (polylactic acid) to *Daphnia magna*: A comparison with polyethylene terephthalate. *Aquatic Toxicology*, 2024. **266**.

60. Savva, K., M. Farré, and C. Barata, Sublethal effects of bio-plastic microparticles and their components on the behaviour of *Daphnia magna. Environmental Research*, 2023. **236**.

61. Ali, W., et al., The comparative toxicity of biobased, modified biobased, biodegradable, and petrochemical-based microplastics on the brackish water flea *Diaphanosoma celebensis. Science of the Total Environment*, 2024. **944**.

62. Viel, T., et al., Effects of biodegradable-based microplastics in *Paracentrotus lividus* Lmk embryos: Morphological and gene expression analysis. *Environmental Pollution*, 2023. **334**: 122129–122129.

63. Zhang, J., et al., Photoaging of biodegradable nanoplastics regulates their toxicity to aquatic insects *(Chironomus kiinensis)* by impairing gut and disrupting intestinal microbiota. *Environment International*, 2024. **185**.

64. Sai, S., et al., Risk assessment of potential toxicity induced by bio and synthetic plastic microspheres in *Lates calcarifer. Chemosphere*, 2022. **298**.

65. Bao, R., et al., Effects of biodegradable and conventional microplastics on the intestine, intestinal community composition, and metabolic levels in tilapia (*Oreochromis mossambicus*). *Aquatic Toxicology*, 2023. **265**.

66. Duan, Z., et al., Diet preference of zebrafish (Danio rerio) for bio-based polylactic acid microplastics and induced intestinal damage and microbiota dysbiosis. *Journal of Hazardous Materials*, 2022. **429**: 128332–128332.

67. Zhang, L., et al., Enhanced reproductive toxicity of photodegraded polylactic acid microplastics in zebrafish. *Science of the Total Environment*, 2024. **912**.

68. Wiesinger, H., Z. Wang, and S. Hellweg, Deep dive into plastic monomers, additives, and processing aids. *Environmental Science & Technology*, 2021. **55**(13): 9339–9351.

69. Geueke, B., *Dossier-Non-Intentionally Added Substances (NIAS)*, 2nd ed. 2018.

70. Karapanagioti, H.K., and H. Takada, Hazardous chemicals associated with plastics in the marine environment, in *The Handbook of Environmental Chemistry*, H. Takada and H.K. Karapanagioti, Eds. 2019, Springer International Publishing: Cham.

71. Shi, C., et al., Oligomers from the synthetic polymers: Another potential iceberg of new pollutants. *Environment & Health*, 2023. **1**(4): 228–235.

72. Nerin, C., et al., The challenge of identifying non-intentionally added substances from food packaging materials: A review. *Analytica Chimica Acta*, 2013. **775**: 14–24.

73. Jehanno, C., et al., Critical advances and future opportunities in upcycling commodity polymers. *Nature*, 2022. **603**(7903): 803–814.

74. Hoppe, M., et al., Migration of oligomers from PET: Determination of diffusion coefficients and comparison of experimental versus modelled migration. *Food Additives & Contaminants: Part A*, 2017. **34**(7): 1251–1260.

75. Gewert, B., M.M. Plassmann, and M. MacLeod, Pathways for degradation of plastic polymers floating in the marine environment. *Environmental Science: Processes & Impacts*, 2015. **17**(9): 1513–1521.

76. Ivar do Sul, J.A., and M.F. Costa, The present and future of microplastic pollution in the marine environment. *Environmental Pollution*, 2014. **185**: 352–364.

77. Kwon, B.G., et al., Global styrene oligomers monitoring as new chemical contamination from polystyrene plastic marine pollution. *Journal of Hazardous Materials*, 2015. **300**: 359–367.

78. Ubeda, S., et al., Migration of oligomers from a food contact biopolymer based on polylactic acid (PLA) and polyester. *Analytical and Bioanalytical Chemistry*, 2019. **411**(16): 3521–3532.

79. Lim, X., Microplastics are everywhere: But are they harmful? *Nature*, 2021. **593**(7857): 22–25.

80. Ohyama, K.I., F. Nagai, and Y. Tsuchiya, Certain styrene oligomers have proliferative activity on MCF-7 human breast tumor cells and binding affinity for human estrogen receptor. *Environmental Health Perspectives*, 2001. **109**(7): 699–703.

81. Narancic, T., et al., Biodegradable plastic blends create new possibilities for end-of-life management of plastics but they are not a panacea for plastic pollution. *Environmental Science & Technology*, 2018. **52**(18): 10441–10452.

82. WHO (World Health Organization), 2024. *IARC Monographs on the Identification of Carcinogenic Hazards to Humans*. 2024.

Index